Structures Under Crash and Impact

Continuum Mechanics, Discretization and Experimental Characterization

Stefan Josef Hiermaier

Structures Under Crash and Impact

Continuum Mechanics, Discretization and Experimental Characterization

Springer

Stefan Josef Hiermaier
Fraunhofer Institut für Kurzzeitdynamik
Ernst-Mach-Institut
Eckerstr. 4
79104 Freiburg
Germany

Library of Congress Control Number: 2007935208

ISBN 978-0-387-73862-8 e-ISBN 978-0-387-73863-5

Printed on acid-free paper.

9 8 7 6 5 4 3 2 1

springer.com

To the Memory of

Larry D. Libersky

A Fine Scientist, True Friend and Upright Voice of the Lord.

Preface

Over the past eleven years I had the privilege to form a group of scientists and technicians at the Ernst-Mach-Institute (EMI) in Freiburg focused on the description of structures under dynamic loading conditions. The scientific background of the group covers both the field of experimental material characterization as well as the development and application of numerical methods. That diversity of competences engaged in working with highly sophisticated hardware and/or in the virtual world of mathematical modelling reflects the intended mutual benefit of experimental and theoretical investigations of dynamic deformation processes. Encouraged by the continuous support from the institute director Klaus Thoma and further enabled by positive feed-back from industry, the concept of a combined experimental-numerical approach could be followed and refined over the years. And the group could grow to a department of now more than fifty persons.

With this book I tried to collect the scientific corner stones for a combined approach to analyzing structures under crash and impact loads. The related theories on continuum mechanics, numerical discretization and experimental material characterization range over a wide spectrum. Consequently, each topic could only be covered selectively. At the same time, this book has been influenced by the research work of many colleagues which I hope to have reproduced in an adequate and correct way. Specifically the many achievements documented in doctoral thesis of Werner Riedel, Martin Sauer, Ingmar Rohr, Michael Junginger, John Corley, Jochen Peter, Markus Wicklein, Harald Schuler, Jan Jansen, Shannon Ryan and Thomas Meenken are indispensable sources for this book and would have each deserved deeper attention.

I greatly appreciated the many discussions of the various topics concerning numerical simulation in general with Arno Klomfass and Heinz Werner and the specific area of shock loaded composites with Rade Vignjevic and Richard Clegg. To Hartwig Nahme I owe a lot of his time that he spent for the manifold introductions and discussions on dynamic material testing.

Since the days of my first own programming of meshfree methods I gained understanding and attraction to the method from the intense exchange with Colin Hayhurst, Chuck Wingate and Larry Libersky. I won't forget the unbreakable enthusiasm of Larry towards SPH which finally lead to a stabilized and well established numerical methodology. It was Larry's endurance and his sharp mind that inspired and guided many other researchers around the world in the field of meshfree methods.

To my academic mentor, Diethard Könke, I owe the fascination in teaching and describing the theoretical basics of numerical methods. The idea to actually write the book emerged during a three months stay as visiting scientist at M.I.T.. Working in the Impact and Crashworthiness Laboratory of Tomasz Wierzbicki was an exceptional experience. During that time and motivated by Tom's fascinating verve in conceiving ever new scientific projects, the vague idea of a book became a concept. Elaine Tham of Springer finally set the spark to actually launch the book project.

Writing this book was a welcome opportunity for me to collect the basics in one of the fundamental research areas of EMI. I want to express my earnest gratitude to Klaus Thoma who not only sent me to M.I.T. but also gave me the necessary time and support to compile the book.

Last, not least, I want to thank my love Claudia and my daughter Paula Lou for their persistent understanding and support over the last eighteen months. I know that it was anything but amusing to live with somebody who was either writing or pondering how to write what he had in mind. Thank you!

Günterstal, July 2007 *Stefan Hiermaier*

Contents

1

Introduction

Crash and Impact - two keywords moving steadily into the foreground of our consciousness. Advertisement campaigns of automotive companies emphasize new technologies protecting occupants and pedestrians in the case of accidents. *Crashworthiness* has become a significant quality attribute of cars. And as a result the yearly casualties due to traffic accidents have been reduced drastically. Car engineering self-evidently comprises optimization of kinetic energy absorption via highly sophisticated combinations of materials and structural design. The exact same physical fundamentals are applied to improve the protective standards of trains, helicopters and airplanes. Common to all these applications is the challenge of building structures at the lowest possible weight and the highest achievable level of protection. Two conflicting and at the same time equally inevitable requirements which need to be met in ever shorter times of development.

Compared to the collision velocities of automotive crash which are typically between 10 and 50 $m\,s^{-1}$, impact problems deal with velocities of several hundred $m\,s^{-1}$. The related scenarios range from bird or other foreign object impact against helicopter rotors, wing structures, landing gears or empennage over military armour and anti-armour problems to the protection of space crafts against micro-meteoroids and space debris as well as planetary impact. With the latter processes typical impact velocities of several kilometers per second are observed. Public attention is called to the mentioned scenarios via science fiction films as well as popular science media illustrating the statistics and effects of possible meteoroid impacts. Actual planetary collisions like the Shoemaker-Levy 9 comet impacting Jupiter underline how real the threat is.

But not only vehicles of various kind are subject to crash and impact. At the latest since the Oklahoma City blast, the September 11 terrorist attacks against the United States or the bombings in Madrid and London the vulnerability of buildings and people in buildings became a topic of public interest and intensified research. In the United Kingdom protective precaution for public buildings became a major topic already since the early days of the devastating IRA bombings. Along with the protective design for buildings *blast loading* as a third term needs to be reflected in addition to crash and impact. Air blast is the propagated pressure wave initiated by detonation. Its interaction with structures can lead to their partial or total destruction and needs to be taken into account as an additional load case by civil engineers.

With respect to the *loading time duration* of the related physical processes, the basic problems of crash and impact can be separated into the milli-seconds and the micro-seconds regimes, respectively. The blast loading is located in both regimes since the interaction duration typically takes milliseconds, the load rise times, however, are in the micro-seconds regime or below.

Closely related to the time duration of their loading processes is the so called *strain rate*, besides pressure amplitudes and impulses another major attribute for the classification of dynamic processes. The strain rate measures the time rate of change of strain, usually taken in units of s^{-1}. Hence it describes the rapidness of deformation processes. Its significance turns out when materials are loaded at deformation speeds of varying strain rates. Most materials show fundamental sensitivity to the strain rate. A fact that needs to be well known in the course of the design of dynamically loaded structures. Whereas strain rates in automotive collisions range between 10^{-2} s^{-1} and 10^2 s^{-1}, *hypervelocity impact* processes, i.e. the space craft relevant kilometer per second impact velocity regime, exhibit strain rates of 10^6 s^{-1} or more. Note that if we consider the longitudinal stretching of a one-dimensional beam, a strain rate of 10^0 s^{-1} already means 100 % change in length per second.

Common to all of the mentioned processes is the time dependency of the solutions to their related mathematical descriptions. In the case of scenarios involving strain rates of 10^3 s^{-1} and more, the evolution and propagation of discontinuous compressive waves, called *shock waves*, becomes more and more important with respect to the type and speed of deformation.

Coming back to the task of designing structures for vehicles or buildings under dynamic loading conditions like crash, impact or blast, we realize that virtually all fields of application are nowadays supported if not driven by *numerical simulation*. Along with the rapid development of computer power, utilization of numerical methods as a tool to design structures for all kinds of loading conditions evolved. Simulation of the expected structural response to certain loadings is motivated by the wish

- to optimize the design
- and to better understand the physical processes.

For both intentions the *predictive capability* of the codes is an indispensable quality. In fact, the predictive capability separates numerical tools from graphical visualization. It means nothing less than the ability to calculate physical processes without experimental results at hand to a sufficient degree of precision.

Three fundamental aspects need to be stated hereto:

(A) The value of a numerical tool fully depends on the validity of its mathematical basics, i.e. the describing partial differential equations.
(B) By its nature and with very few unrepresentative exceptions, numerical simulation can only produce *approximations*. To estimate the quality of the approximate solution, the user needs to know the basic theory, its applicability and the limitations of the code used.
(C) No simulation of physical processes can work without input from experiments. It is the objective derivation and description of generalized physical relations, e.g. material behaviour, that provides the applicability of a code to varying initial and boundary conditions. This includes for example variations of geometries, loads and confinements without loss of predictive capabilities.

What does that mean for the simulation of structures under crash or impact loads? Requirement (A) demands that all relevant dependencies are covered by the describing equations. Trivially, a formulation without time-dependency cannot be used to describe wave propagation and interaction. In addition, the solution strategy needs to be adequately chosen. This will be topic of section 5.3.

The approximate nature stated in (B) requires a thorough approach to solutions including assessments of

- the used types of discretization,
- its resolution,
- the validity of assumed linearizations, e.g. in terms of formulating the kinematic relations between displacements and strains,
- and of the adequacy of the chosen constitutive laws describing relations between stresses and strains.

Finally, the objectivity statement (C) means that the needed input in terms of parameters for the constitutive relations as well as the constitutive relations themselves need to be of unique and general quality. A *unique relation* between stresses and strains, say at a given strain rate and temperature, must be derived by experiments. Only if these experiments are performed under *well defined states of stress and strain*, a predictive relation will arise that can be applied to arbitrary structures and to loading conditions covered by the conditions examined in the material tests. Otherwise, the relation will only be valid for one specific test. Application to a more complex structure or different loading conditions will fail in predicting the structural response.

In the context of material characterization, growing importance must be attributed to very recent methods of the so called *multi-scale material modelling*. Simulation of deformation processes on a micromechanical, molecular or even atomistic level allows for investigations providing macroscopic material data in a totally different way. Again, empirically derived or otherwise postulated relations for the material behaviour on the respective scale are needed. The impact on understanding the physics of materials from a new perspective, however, is tremendous.

In general, an ever closer combination of physical material testing and accompanying numerical simulation, on various scales, can be observed and is for sure of positive influence on the predictive capability of the resulting material models.

With this book a spectrum of the basic theories shall be collected that is needed to investigate structures under dynamic loading conditions. This covers the continuum mechanics, material modelling, numerical discretization and experimental material characterization.

Chapter 2 deals with material independent fundamental formulations. Starting with the kinematic descriptions of deformations as used in structural mechanics and in fluid dynamics, stress measures as well as descriptions of static equilibrium will be derived. The time dependency of the equations of motion will be invented with the section on the conservation equations. For both equilibrium and balance equations variational formulations with the intention of finding more generalized solutions will be introduced. Figure 1.1 illustrates the discussed alternatives to solve static or dynamic problems via variational formulations. The fact that fluid dynamical approaches are repeatedly used in this book is deduced from the origins of the so called *hydrocodes* which are the most common codes used for the simulation of dynamic processes.

Material specific constitutive equations are topic of chapter 3. Relations describing the elastic, hyperelastic and inelastic deformation of materials will be formulated. Specific attention will be paid to strain rate dependent models. With the example of polymer materials the elegance of physically based constitutive models will be found. In that context, the derivation of elastic potentials for polymers using statistical mechanics seems to be a perfect example. Furthermore, meso-mechanical modelling of an aluminum foam material illustrates the expanded spectrum of material characterization if a combined experimental-numerical path is followed.

Their significant influence on the material and structural response under extremely dynamic loading conditions requires the treatment of *shock waves* by a specifically dedicated chapter 4. Dispersion and the related non-linearity

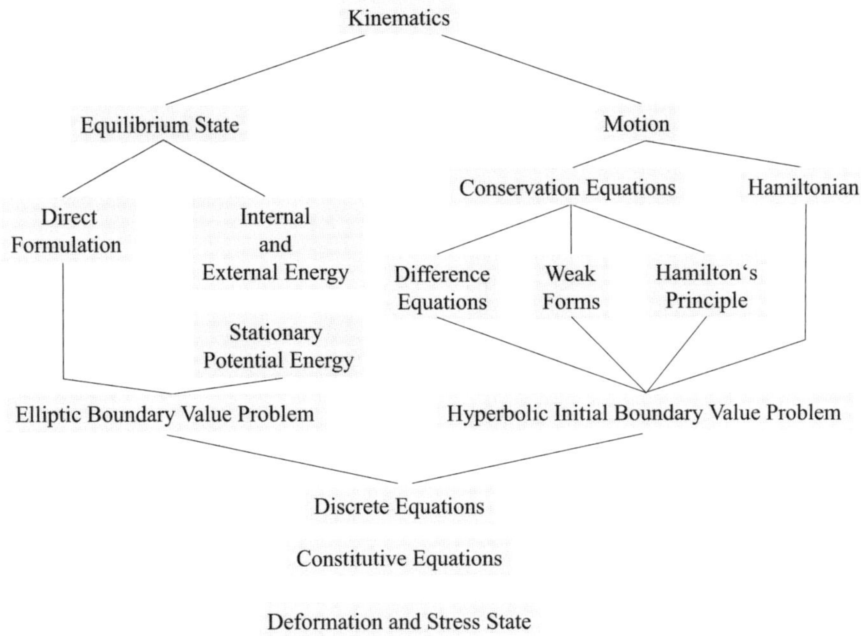

Fig. 1.1. Continuum Mechanical Solution Paths for Static and Dynamic Problems.

of equations of state as a precondition for shock waves to be initiated and to propagate as well as the thermo-mechanical basics of shocks in fluids and solids will be derived. A straight path to derive non-linear equations of state via the Rankine-Hugoniot equations completed by an additional relation derived from the solution of the Riemann problem will be explained. Various forms of nonlinear equations of state and their limitations are discussed.

Hydrocodes will be introduced in chapter 5 starting with a collection of the needed components and a cyclic procedure followed to achieve a marching solution in time. Basic methods of discretization in time and space will be introduced. Besides the grid-based methods of finite differences, volumes and elements, meshfree methods will be discussed. The specific task of providing solutions that include massive deformation, multiple fragmentation and possibly phase changes illustrates the need for alternative discretization techniques provided by meshfree methods. As a particular emphasis, the usefulness of coupled and adaptive methods will be stressed. As shown by the example of Figure 1.2, simulation of dynamic deformation processes demands for a representation of multiple non-linearities originating from large deformations, non-linear material behaviour and contact forces. More complex deformation and failure mechanisms will be illustrated by several examples in the course of chapter 5.

Fig. 1.2. Dynamic axial crushing of a cylindrical tube at a deformation speed of 10 $\mathrm{m\,s^{-1}}$ simulated with AUTODYN. The numerical simulation of the structure requires the treatment of strain rate dependent non-linear material behavior, large deformations and multiple contact.

Intentionally decoupled from the constitutive equations introduced in chapter 3, *failure modelling* will be the topic of chapter 6. The separate introduction after the chapter on hydrocodes is due to the intended emphasis on the need for a closer coupling of material models with discretization techniques. Consequently, the chapter on failure models will only cover a very selected number of models. The need and specific advantage of a material state related adaptive discretization will be illustrated by the example of composite failure.

With the final chapter several aspects of advanced dynamic material characterization will be outlined. The indispensable need for objectivity in the material characterization will be stressed. Experimental set-ups for low dynamic, moderate as well as extreme dynamic strain rates will be introduced along with advanced instrumentation techniques which allow for improved or enhanced measuring capabilities.

Having in mind that the nowadays standard university education provides the students with basic knowledge concerning formulation of static equilibrium and numerical solution techniques for that regime, both the static and the dynamic formulations will be covered in this book. The emphasis will of course be on the dynamic side. But the static formulations given in parallel are intended to provide with both contrast and amendment. In that context, the growing need for coupled static-dynamic solvers should be mentioned. Quasi-statically pre-stressed structures which are afterwards loaded dynamically are rather standard than exception. On the other hand, structures with intense dynamic loading history are for example found in a blast loaded building. An investigation of its residual load carrying capability can be the task of a quasi-static solution based on the result of a dynamic one.

This book is written to support two overall ideas: one is the combination of experimental and numerical methods not only to solve a specific problem in structural mechanics, but also in the course of material characterization. The second one is the use of discretization techniques which are adequately chosen for the actual material state. In other words, coupled and adaptive use of discretizations.

2

Thermo-Mechanical Basics

The thermo-mechanical behaviour of materials under dynamic loads ranges from (visco-)elastic-plastic deformation over micro-mechanical damage and macroscopic failure to phase changes under high pressure and temperature. In order to describe processes involving some or all of these effects, a set of functional relations is needed providing a mathematical basis. To give an overview on the common basics as well as the differences between static and dynamic analyses, the following topics will be covered in this chapter:

- *Kinematic equations* as an objective, material independent description of deformations, strains and strain rates,
- objective *stress measures* and their rate dependent formulations,
- static equilibrium described via direct formulations and variational approaches,
- still material independent *first principles*, e.g. conservation equations, principles of mechanics and thermodynamics, defining admissible conditions of the continuum with regard to mass, momentum, energy and entropy,
- variational formulations for the balance equations needed later for finite element solutions and
- thermodynamic laws and potentials along with the conept of internal state variables and consequences for constitutive laws.

Material dependent constitutive laws will then be discussed in the next chapter 3.

2.1 Kinematic Equations

It is the basic aim of continuum mechanics to describe the motion and deformation of structures as thermo-mechanical systems. To achieve this, the investigated media are regarded as continuous accumulations of material points for which Truesdell demands *smoothness*[1]. Accordingly, Macvean [260] defines a continuum as body at time t consisting of material points to which at any other time well defined corresponding points exist. The related *axiom of continuity* means that two neighbouring points will be at neighbouring positions at any time, further that one point cannot occupy multiple positions and finally that one position can never be occupied by multiple points.

In the course of static analyses, the interest is generally focused on a final deformation state of the investigated structure under a given loading condition. That deformation state is characterized by equilibrium of the forces. It may consist of small or large deformations and lead to a pure elastic state or include more or less expanded plastic deformation zones. Also, the history of the deformation process may have strong influence on the material state and, thus, on the equilibrium. However, the precise physical manner of how the information about the loading condition is communicated through the structure, i.e. the propagation of pressure and release waves, is not influencing the static equilibrium and therefore not part of the solution.

Completely different is the situation for dynamic deformation processes. Here, the final deformation state in most cases depends on the deformation velocity and, specifically under impact loads, on the propagation of pressure or shock waves. As for quasi-static loadings, the final deformation can be a result of repeated loading and unloading processes during which the material undergoes various strain and stress states including plastic deformation and the development of damage and failure mechanisms. The loading process itself is often a sequence and superposition of wave propagations and their reflections. Equilibrium conditions used for static analyses are amended by an acceleration term leading to the time dependent equation of motion.

Therefore, the kinematic description of a dynamic process necessarily must include the time resolved motion of the structure and the related material states. The time resolution must be sufficiently high to cover the wave speeds if wave propagation is relevant. The spatial description of the whole deformation process by kinematic equations can be extremely complex. Its discretization in space and time must reflect large deformations possibly including multiple fragmentation. The discretized equations are solved at a finite number of points and elements in the structure at discrete instants of time.

2.1.1 Coordinates and Displacements in Reference Systems

Positions of points in space are identified by position vectors with respect to a coordinate system. Though the choice of coordinate systems is arbitrary

[1] Truesdell [397]: "The two fundamental properties of a continuum B are then: 1) B consists of a finite number of parts which can mapped smoothly onto cubes in Euclidean space. 2) B is a measure space."

for vector and tensor equations, the most commonly used ones in continuum mechanics are three-dimensional rectangular Cartesian coordinates or the curvilinear systems of cylindrical and spherical type. In this book, components of vectors and tensors are basically formulated in the orthonormal system of Cartesian basis vectors $\mathbf{e}_1$, $\mathbf{e}_2$, $\mathbf{e}_3$, . For that system we demand:

$$\mathbf{e}_i \cdot \mathbf{e}_j = \delta_{ij} = \mathbf{I} = \begin{cases} 0 & \text{if} \quad i \neq j \\ 1 & \text{if} \quad i = j \end{cases} \tag{2.1}$$

using the *Kronecker symbol* δ_{ij}, and:

$$\mathbf{e}_i \times \mathbf{e}_j = \epsilon_{ijk}\, \mathbf{e}_k \tag{2.2}$$

applying the *Levi-Civita symbol* to represent the permutation:

$$\epsilon_{ijk} = \begin{cases} 1\,, & \text{for even permutations of (i, j, k), (i.e. 123, 231, 312)} \\ -1\,, & \text{for odd permutations of (i, j, k), (i.e. 132, 321, 213)} \\ 0\,, & \text{for repeated indices} \end{cases} \tag{2.3}$$

Theory of Dual Spaces

Physical laws described by continuum mechanics involve tensors which cannot be limited to orthonormal coordinate systems. Instead, arbitrary other sets of independent base vectors $\mathbf{g}_i$ are used. In order to be able to apply within these systems the same calculus as in the Euclidean base $\mathbf{e}_i$, the theory of a *dual vector space* is usually applied. The two reciprocal bases are called *covariant* $\mathbf{g}_i$ and *contravariant* base $\mathbf{g}^i$, respectively. That theory allows for a unique representation of a vector $\mathbf{u}$ through a linear combination in either one of the bases as:

$$\mathbf{u} = u_i\, \mathbf{g}^i = u^i\, \mathbf{g}_i\,, \tag{2.4}$$

where u_i are called the covariant components of the vector $\mathbf{u}$, and u^i its contravariant ones. Since the two bases fulfill the requirement

$$\mathbf{g}^i \cdot \mathbf{g}_j = \delta_i^j \tag{2.5}$$

a orthogonality between vectors of respective bases is guaranteed. The product of the length of two basis vectors is also referred to through the so called *metric coefficients* g_{ij} and g^{ij}:

$$\mathbf{g}_i \cdot \mathbf{g}_j = g_{ij}\,, \quad \mathbf{g}^i \cdot \mathbf{g}^j = g^{ij} \tag{2.6}$$

In the course of this text, there will generally be no need to distinguish between covariant and contravariant bases and it will otherwise be indicated specifically. Further introductions to the tensor calculus and the use of the dual space theory can be found in Malvern [263] and Holzapfel [194].

Material and Spatial Configurations

The position of a body shall be defined by the position of a finite number of material points in a coordinate system and the body's motion is traced as sequence of such positions along a pathline. For the observation of motions it is useful to note the material coordinates, that means the coordinates of a particle's position in a reference configuration, X_1, X_2, X_3 by upper case letters and the spatial coordinates denoting the position of a point in space at time t by lower case letters x_1, x_2, x_3. In this context, the term material point or particle will be used for an infinitesimal portion of matter in a continuous body to distinguish it from a spatial position.
The motion of a continuous body in non-relativistic kinematics may be analyzed by various distinguished ways of description:

- A *Referential description* with the position $\mathbf{X}$ of a particle and the time t as independent variables. This description is called *Lagrangean* when the initial configuration is chosen to be the reference configuration for positions $\mathbf{X}$. The focus is on a particle travelling through space. Lagrangean descriptions are preferably chosen in structural mechanics. The referential description is often also called *material description.* Although this is not fully precise[2], we will assume the terms *material* and *referential* being equivalent in this book.
- The *Spatial description* that observes a spatial position $\mathbf{x}$ which is at time t occupied by a particle. Here, the focus is on a position or volume in space that may over time be occupied by various particles. This predominantly in fluid mechanics applied description is also called *Eulerian.*

There are various attempts and ways to be found in literature illustrating the differences between the Lagrangean and the Eulerian description. Figure 2.1 may help to understand the fundamental difference and clarify the sometimes arising confusions. First of all, we have to recall the before mentioned difference between positions in space $\mathbf{x}$ and locations of material particles $\mathbf{X}$. With that discrimination in mind we can denote a spatial description by lower case letters and a material description by upper case notations.
What makes the difference, is that Eulerian descriptions, as shown in the upper half of Figure 2.1, observe spatial positions ${}^t\mathbf{x}$ with respect to the change

[2] According to Truesdell [397] there are two further ways which however are of less importance and use compared to the above mentioned Lagrangean and Eulerian description:

- A *Material description* of a point X at time t where X is the label not the position $\mathbf{X}$ of the point.
- And, finally, a *Relative description* of a point at position $\mathbf{x}$. The variable time τ references to a time when the point occupied a former position ξ. It is distinguished from the Lagrangean description since it references to $\mathbf{x}$ at time t instead of $\mathbf{X}$ at time $t = 0$.

of field variables at that position. The identity of a particle that currently occupies that position ${}^t\mathbf{x}$ is found through its position in the referential configuration ${}^0\mathbf{X}$. In this context, one could also call ${}^0\mathbf{X}$ the particle's "name" which was attributed at time $t = 0$ through its then spatial position ${}^0\mathbf{x}$. Therefore, the independent variables in Eulerian descriptions are the coordinates $\mathbf{x}$ and the time t.
Lagrangean formulations on the other hand observe the material particle in the initial configuration and trace its spatial coordinates ${}^t\mathbf{x}({}^0\mathbf{X})$ over time. Thus, in contrast to the Eulerian description, the coordinates $\mathbf{x_i}$ are now mapped to the particles' positions. The independent variables are now the material particle positions $\mathbf{X}$ and of course the time t. What we get, are body fixed coordinates which move and deform with the body.
In Figure 2.1, material particles are denoted by black dots and spatial positions by position vectors. Of course, the same particles shown in the deformed Eulerian part of Figure 2.1 are in the same way existing in the deformed Lagrangean. As well as in the deformed Eulerian regime also position vectors have their normal meaning. Just in order to emphasize the different approaches, only the dominant notation is drawn in the respective parts. In fact, we will need both notations later when current configurations need to be compared to initial ones in order to describe deformation and strain measures. The basic kinematic action of mapping back and forth between configurations will then utilize the different notations.

In the Eulerian description a particle $\mathbf{{}^tX({}^0x}, t)$ inside the body moves through the space fixed coordinate system. Consequently, it does not necessarily occupy an intersection of the coordinate grid at time t. In the Lagrangean description however, a position $\mathbf{{}^0x}$ originally located on an intersection will always be mapped on the same grid point in the mapped coordinate system. That is, why in Eulerian descriptions usually applied in fluid dynamic analyses the non deforming coordinate system is made up by a finite number of so called *cells* or *volumes* through which the material flux is transported. Whereas in Lagrangean structural dynamics the so called *nodes* carry a specific portion of the body's mass.

Thus, with the referential description it is straight-forward to observe how a structural body is moving and deforming in space, whereas the spatial description shows what occurs in terms of material fluxes to a specific region in space over time. Therefore, fluid dynamics preferably utilize the spatial description since the flow of matter in space is of primary interest. In the context of this book the main focus is concerned with structural dynamics and correspondingly the standard description will be Lagrangean to observe the deformation of bodies by referring the current location of a material point $\mathbf{{}^tx}$ to its initial position $\mathbf{{}^0X}$:

$$ {}^t x_i = {}^t x_i \left({}^0 X_i \, , \, t \right) \qquad \text{or} \qquad {}^t\mathbf{x} = {}^t\mathbf{x} \left({}^0\mathbf{X} \, , \, t \right) \tag{2.7} $$

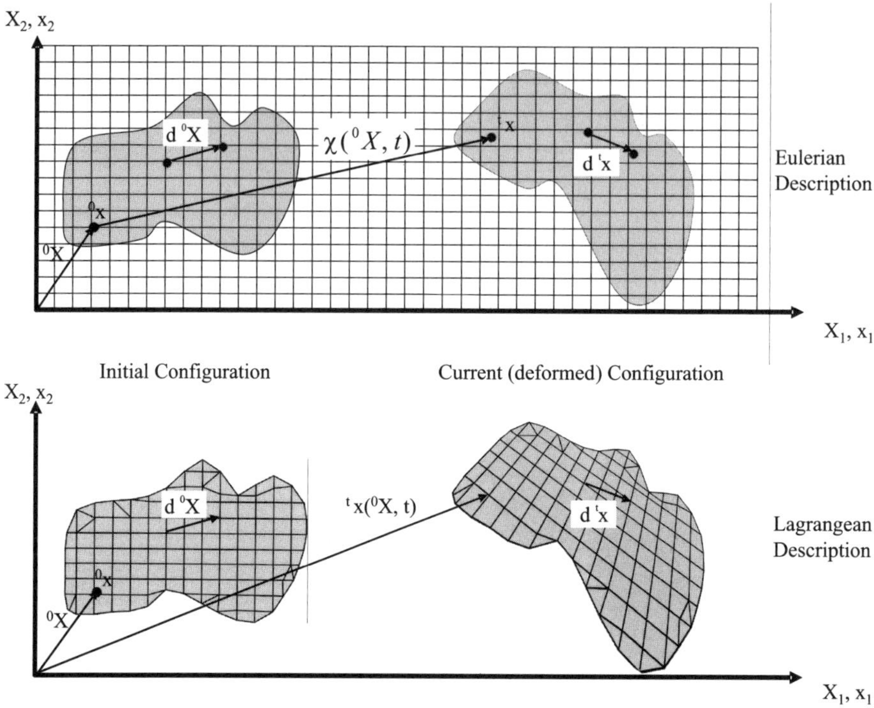

Fig. 2.1. Initial and current configurations of a body mapped by spatial and referential descriptions.

Nevertheless, alternative formulations with spatial descriptions will prove to be useful. The corresponding Eulerian description is:

$$ {}^{0}X_i = {}^{0}X_i \left({}^{t}x_i \, , \, t \right) \qquad \text{or} \qquad {}^{0}\mathbf{X} = {}^{0}\mathbf{X} \left({}^{t}\mathbf{x} \, , \, t \right) \tag{2.8} $$

The *consistency* between the two formulations is guaranteed through the condition:

$$ {}^{0}\mathbf{X} = {}^{0}\mathbf{x} \tag{2.9} $$

Since we regard the continuum as an accumulation of points in a coordinate system, the motion of an individual point over a certain time interval is defined by its displacement. In a Lagrangean description, the displacement vector U_i of a particle i at time t is defined as

$$ \mathbf{U}(\mathbf{X}, t) = \mathbf{x}(\mathbf{X}, t) - \mathbf{X} \tag{2.10} $$

whereas in the corresponding Eulerian form the same displacement is described through:

$$ \mathbf{u}(\mathbf{x}, t) = \mathbf{x} - \mathbf{X}(\mathbf{x}, t) \tag{2.11} $$

The values of both displacement measures are identical and the relation between (2.10) and (2.11) can be described using the motion $\chi: \ \mathbf{x} = \chi(\mathbf{X},\ t)$:

$$\mathbf{U}(\mathbf{X},t) = \mathbf{U}\left\{\chi^{-1}(\mathbf{x},t)\right\} = \mathbf{u}(\mathbf{x},t) \tag{2.12}$$

2.1.2 Deformation Gradients and Displacement Gradients

In general, the motion of a body can be seen as a combination of both rigid and deforming components. The rigid parts consist of translation and rotation whereas deformation can be split into shape changing distortion and shape keeping dilatation. Each of these components can be covered by kinematic equations and related measures but only non-rigid motion causes changes in the strain state and the related stress inside the body. Therefore, rigid body motions are of secondary interest for continuum mechanics. But in order to derive objective stress measures from deformation states, that is the stress resulting exclusively from non rigid motion, the motion components need to be separated adequately.

Material and Spatial Deformation Gradient

As a first step towards a description of body deformations, we observe the relative motion of two points inside a body. With the paths of these two moving points $\mathbf{x_i}$ and $\mathbf{x_j}$ over the time interval t we can formulate line vectors $\mathbf{d^0x}$ and $\mathbf{d^tx}$ connecting both points at the beginning and at the end of the motion, respectively. The relation between the line vectors can be expressed by:

$$\mathrm{d}\,{}^t\mathbf{x} = \mathbf{F}\,\mathrm{d}\,{}^0\mathbf{X} \quad \mathit{or} \quad \mathrm{d}\,{}^t x_i = F_{ij}\,\mathrm{d}\,{}^0 X_j \tag{2.13}$$

where the second-order tensor $\mathbf{F}$ with reference to the undeformed configuration is called *material deformation gradient*:

$$F_{ij} = \frac{\mathrm{d}\,{}^t x_i}{\mathrm{d}\,{}^0 X_j} \tag{2.14}$$

That means, the material deformation gradient is a tensor that transforms a line element from its reference configuration $\mathbf{d}\,{}^0\mathbf{X}$ to a current configuration $\mathbf{d}\,{}^t\mathbf{x}$. The inverse mapping of the line element in the current configuration to the line element in the reference configuration is represented by the *spatial deformation gradient*:

$$F_{ij}^{-1} = \frac{\partial\,{}^0 X_i}{\partial\,{}^t x_j} \tag{2.15}$$

The determinant of the form

$$J = \det\left(\frac{\partial^t x_i}{\partial^0 X_j}\right) = \det\left(F_{ij}\right) \tag{2.16}$$

is called *Jacobian determinant*. In order to map back a current configuration to the initial one, the Jacobian must not vanish.

The transformation of scalars, vectors or tensors between a referential and the current configuration by the deformation gradients can be formulated as instruments in both directions. These operations are called *push forward* Φ_* and *pull back* Φ_*^{-1}. Expressed by the deformation gradients, their application to a second order tensor $\mathbf{A}$ is defined by:

$$\Phi_*[\mathbf{A}] = \mathbf{F}^{-T}\mathbf{A}\,\mathbf{F}^{-1} \qquad \textit{Push Forward} \tag{2.17}$$

$$\Phi^*[\mathbf{A}] = \mathbf{F}^{-1}\mathbf{A}\,\mathbf{F}^{-T} \qquad \textit{Pull Back} \tag{2.18}$$

Displacement Gradients

In the same manner as for the deformation gradients, the gradients of the displacement vector (2.10) with respect to the initial and current configuration lead to the *material displacement gradient*

$$H_{ij} = \frac{\partial\, {}^t u_i}{\partial\, {}^0 X_j} \tag{2.19}$$

and to the *spatial displacement gradient*

$$h_{ij} = \frac{\partial\, {}^t u_i}{\partial\, {}^t x_j}\,, \tag{2.20}$$

respectively. With the deformation ${}^t u_i$ (2.10) we can express the current position of point i by

$${}^t x_i = {}^t x_i\left({}^0 X_i\,,\, t\right) = {}^0 X_i + {}^t u_i \tag{2.21}$$

and thus write the material deformation gradient as

$$F_{ij} = \frac{\partial\, {}^t x_i}{\partial\, {}^0 X_j} = \frac{\partial}{\partial\, {}^0 X_j}\left({}^0 X_i + {}^t u_i\right) = \delta_{ij} + \frac{\partial\, {}^t u_i}{\partial\, {}^0 X_j} \tag{2.22}$$

which explains the relation between the material deformation gradient and the material displacement gradient as follows:

$$F_{ij} = \delta_{ij} + H_{ij} \quad \textit{or} \quad \mathbf{F} = \mathbf{1} + \mathbf{H} \tag{2.23}$$

The above shown meaning of the deformation gradient as a mapping function is not restricted to material line elements $\mathrm{d}\,x_i$. For surface and volume elements corresponding procedures are applicable to map material surface $\mathrm{d}\,a_i$ or volume elements $\mathrm{d}\,v_i$ between current and initial configuration[3] :

[3] Compare Truesdell and Toupin [399] for reference and also Haupt [165] for proof.

$$\mathrm{d}\,{}^{t}\mathbf{a} = (\det \mathbf{F})\,\mathbf{F}^{-T}\,\mathrm{d}\,{}^{0}\mathbf{A} \tag{2.24}$$

$$\mathrm{d}\,{}^{t}v = (\det \mathbf{F})\;\mathrm{d}\,{}^{0}V = J\,\mathrm{d}\,{}^{0}V \tag{2.25}$$

where equation (2.25) also means that

$$J = \frac{d\,{}^{t}v}{d\,{}^{0}V} \tag{2.26}$$

Polar Decomposition of the Deformation Gradient

Still, the components of the deformation gradient (2.14) reflect contributions from two types of deformation. One is resulting from rigid body rotation, the other component leads to stretching the body. A separation of the deformation gradient into both of these contributions would be useful in order to achieve a measure for the strain tensor, which is by definition free of rigid body motion. A multiplicative *polar decomposition* of the deformation gradient is the mathematical tool to achieve that separation. The decomposition into a symmetric *right stretch tensor* **U** (referring to the initial configuration) or *left stretch tensor* **V** (referring to the current configuration), respectively, and an orthogonal[4] *rotation tensor* **R** is possible since the deformation gradient itself is a non-singular second-order tensor:

$$F_{ij} = R_{ik}U_{kj} = V_{ik}R_{kj} \tag{2.27}$$

The polar decomposition applied to the material deformation gradient provides information on the components of a deformed configuration compared to its initial state. Applied to the spatial deformation gradient the related information would be provided with respect to the current configuration. The separation of deformation into rotation and stretch and the two combinations of the related tensors shown in (2.27) are demonstrated schematically in Figure 2.2.

Right and Left Cauchy-Green Tensors

Both ways of performing rotation first to be followed by dilatation and vice versa lead to the identical final deformation and strain state. With the two tensors **U** and **V**, the so called *right Cauchy-Green tensor* **C** and *left Cauchy-Green tensor* **B** are defined as

$$\mathbf{C} = \mathbf{F}^{T}\mathbf{F} = \mathbf{U}^{2} \tag{2.28}$$

and

$$\mathbf{B} = \mathbf{F}\mathbf{F}^{T} = \mathbf{V}^{2} \tag{2.29}$$

[4] An orthogonal tensor **A** is characterized through the property $\mathbf{A}^{T}\,\mathbf{A} = \mathbf{A}\,\mathbf{A}^{T} = \mathbf{I}$ which implies that $\mathbf{A}^{-1} = \mathbf{A}^{T}$.

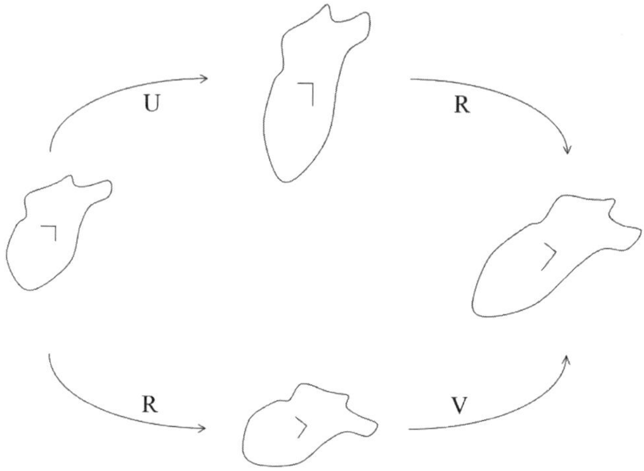

Fig. 2.2. Body deformation separated into rotation and dilatation

The right Cauchy-Green tensor can be understood as operator on the scalar product of two line elements $d\mathbf{X}\, d\mathbf{Y}$ transforming it to the scalar product of the corresponding line elements in the current configuration $d\mathbf{x}\, d\mathbf{y}$:

$$d\mathbf{x}\, d\mathbf{y} = d\mathbf{X}\, \mathbf{C}\, d\mathbf{Y} \tag{2.30}$$

whereas the left Cauchy-Green tensor represents the opposite transformation:

$$d\mathbf{X}\, d\mathbf{Y} = d\mathbf{x}\, \mathbf{B}\, d\mathbf{y} \tag{2.31}$$

2.1.3 Strain Measures

Now, what is strain and how should a strain tensor be defined? Compared to the above discussed displacements which are an absolute measure of length in material or spatial coordinates, strain is a relative measure comparing displacement or angular distortion with a reference length or a reference angle, respectively. With the development of numerical methods as tools for structural mechanics, the understanding of strain changed towards a linear displacement gradient inside finite elements for small deformations and a corresponding nonlinear gradient as finite strain measure. Thus, the question is which reference to choose and whether small or finite deformations are to be covered.

Engineering Strain and True Strain

In a one-dimensional situation as illustrated in Fig. 2.3 , a classical engineering approach to measure strain formulates a strain increment as change in length $\mathrm{d}\,x$ compared to the length in the initial configuration X:

$$\mathrm{d}\varepsilon = \frac{\mathrm{d}x}{X} \tag{2.32}$$

and thus, the corresponding one dimensional *engineering strain* is:

$$\varepsilon_e = \int_X^x \frac{1}{X}\,\mathrm{d}x = \frac{x - X}{X} = \frac{x}{X} - 1 \tag{2.33}$$

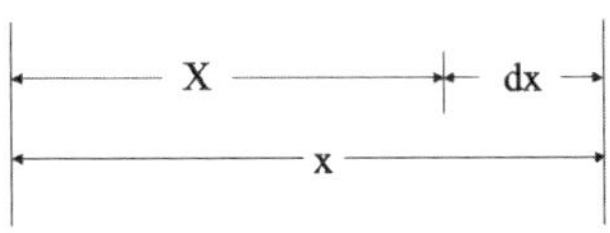

Fig. 2.3. One dimensional length components to define strain.

If we instead reference the change in length $\mathrm{d}\,x$ to the current length at the end of the motion, an alternative incremental strain develops:

$$\mathrm{d}\varepsilon_t = \frac{\mathrm{d}\,x}{x} \tag{2.34}$$

and consequently a logarithmic strain measure, called *natural* or *true strain*, originally introduced by Ludwik [256] in 1909:

$$\varepsilon_t(\mathbf{x}, t) = \int_X^x \frac{1}{x}\,\mathrm{d}x = \ln(x) - \ln(X) = \ln\left(\frac{x}{X}\right) = \ln\left(\varepsilon_e + 1\right) \tag{2.35}$$

The two measures for engineering (2.33) and true strain (2.35) result in approximately the same values for very small strains and significant differences for larger ones. A very common understanding of uni-axial strain would describe a deformation state that leads to double the original length as 100 percent strain, which is equal to the engineering strain. The corresponding true strain value for the same deformation yields approximately 69 percent.

Basically, there is no advantage of preferring the initial over the current dimension as reference or vice versa since both are known at any time. In contrast to the engineering strain however, the true strain measure is symmetric regarding tensile and compressive relative deformations. Whereas a compressive strain of 100 percent would mean infinite compression to zero length in terms of engineering strain, the same values with opposite sign represent true strain for double or half the original lengths (see Table 2.1).

Table 2.1. Comparison of engineering and true strains.

L_0	1	1	1	1	1	1	1	1	1	1	1
l	0.5	0.8	0.95	0.97	0.99	1	1.01	1.03	1.05	1.2	2.0
ε_e [%]	-50	-20	-5.0	-3.0	-1.0	0.0	1.0	3.0	5.0	20.0	100.0
ε_t [%]	-69.3	-22.3	-5.13	-3.05	-1.01	0.0	0.995	2.95	4.88	18.2	69.3

Infinitesimal Strain Tensors

In order to describe strain conditions in the two or three dimensional case, the above derived strain measures need to be extended by shear components. The two separate deformation modes demonstrated in Fig. 2.4 represent a longitudinal extension in one direction and a shear deformation. For the longitudinal strains we find:

$$\varepsilon_{xx} = \frac{\partial u_x}{\partial X} \quad \text{and} \quad \varepsilon_{yy} = \frac{\partial u_y}{\partial Y} \tag{2.36}$$

The shear deformation is composed of the angular distortions by the angles α_1 and α_2 and it is called *pure shear* if the angles α_1 and α_2 are identical.

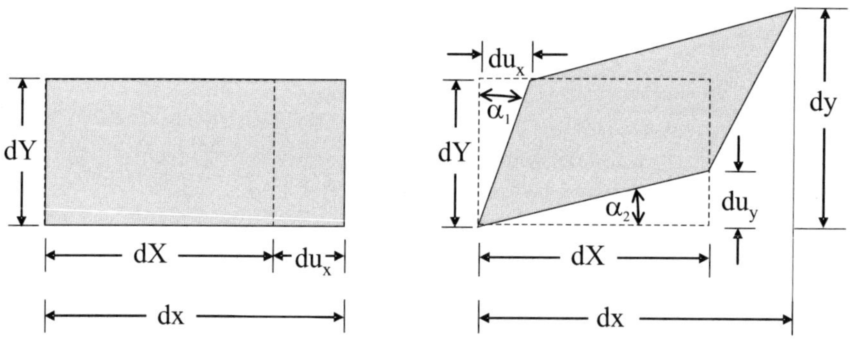

Fig. 2.4. Longitudinal and Shear Strain in Two Dimensions.

The total change in the originally right angle can be described for small angles by the linear approximation

$$\gamma_{xy} = \alpha_1 + \alpha_2 \approx \left(\frac{\partial u_x}{\partial Y} + \frac{\partial u_y}{\partial X} \right) \tag{2.37}$$

If we replace γ_{xy} by

$$\varepsilon_{xy} = \frac{1}{2}\,\gamma_{xy} = \frac{1}{2}\left(\frac{\partial u_x}{\partial Y} + \frac{\partial u_y}{\partial X} \right) \tag{2.38}$$

we can describe the two dimensional strain more elegantly as

$$\begin{bmatrix} \varepsilon_x \\ \varepsilon_y \\ \varepsilon_{xy} \end{bmatrix} = \begin{bmatrix} \frac{\partial u_x}{\partial X} \\ \frac{\partial u_y}{\partial Y} \\ \frac{1}{2}\left(\frac{\partial u_x}{\partial Y} + \frac{\partial u_y}{\partial X}\right) \end{bmatrix} \tag{2.39}$$

Thus, with the material displacement gradient (2.19) we find for an *infinitesimal Lagrangean strain tensor* the following expression:

$$\varepsilon_{ij} = \frac{1}{2}\left(\frac{\partial u_i}{\partial X_j} + \frac{\partial u_j}{\partial X_i}\right) = \frac{1}{2}\left(H_{ij} + H_{ij}^T\right) \tag{2.40}$$

Accordingly, with respect to the current configuration the *infinitesimal Eulerian strain tensor* is denoted by:

$$e_{ij} = \frac{1}{2}\left(\frac{\partial u_i}{\partial x_j} + \frac{\partial u_j}{\partial x_i}\right) = \frac{1}{2}\left(h_{ij} + h_{ij}^T\right) \tag{2.41}$$

Both the formulations for a strain tensor in (2.40) and (2.41) are at the same time *strain-displacement relations* and linear in the current form.

Finite Strain Formulations

Another access to strain measures that, in addition, is also capable of finite strains is found via the right and left Cauchy-Green tensors (2.28) and (2.29). In (2.30) we observed that the right Cauchy-Green tensor transforms the scalar product of two line elements $\mathrm{d}\mathbf{X}\,\mathrm{d}\mathbf{Y}$ in an initial configuration into the scalar product of the same line elements in the current configuration $\mathrm{d}\mathbf{x}\,\mathrm{d}\mathbf{y}$. What we expect from a strain tensor is the representation of the *change* in the scalar product of the two line elements. Following the referential description of the right Cauchy-Green tensor the following *Green-Lagrange strain tensor* $\mathbf{E}$ contains these changes in the difference of the squared current elements and compares them to the product of the referenced line elements:

$$\mathrm{d}\mathbf{x}\,\mathrm{d}\mathbf{x} - \mathrm{d}\mathbf{X}\,\mathrm{d}\mathbf{X} = 2\,\mathrm{d}\mathbf{X}\,\mathbf{E}\,\mathrm{d}\mathbf{X} \tag{2.42}$$

With the material deformation gradient (2.14) the Green-Lagrange strains can be written as:

$$\mathbf{E}(\mathbf{X},t) = \frac{1}{2}\left(\mathbf{F}^T\,\mathbf{F} - \mathbf{1}\right) = \frac{1}{2}\left(\mathbf{C} - \mathbf{1}\right) \tag{2.43}$$

which, in index writing, shows the difference between the finite strain formulation in the Green-Lagrange form and the linear strain that we have seen above in (2.40):

$$\mathbf{E}_{ij} = \frac{1}{2}\left(\frac{\partial u_i}{\partial X_j} + \frac{\partial u_j}{\partial X_i} + \frac{\partial u_k}{\partial X_i}\frac{\partial u_k}{\partial X_j}\right) \tag{2.44}$$

Again, if the reference frame is the current configuration we find an Eulerian form of the finite strain tensor in the *Euler-Almansi strain tensor* $\mathbf{A}$:

$$\mathrm{d}\mathbf{x}\,\mathrm{d}\mathbf{x} \;-\; \mathrm{d}\mathbf{X}\,\mathrm{d}\mathbf{X} = 2\,\mathrm{d}\mathbf{x}\,\mathbf{A}\,\mathrm{d}\mathbf{x} \tag{2.45}$$

and thus:

$$\mathbf{A}_{ij} = \frac{1}{2}\left(\frac{\partial u_i}{\partial x_j} + \frac{\partial u_j}{\partial x_i} - \frac{\partial u_k}{\partial x_i}\frac{\partial u_k}{\partial x_j}\right) \tag{2.46}$$

or symbolically:

$$\mathbf{A} = \frac{1}{2}\left(\mathbf{1} - \mathbf{F}^{\mathbf{T}^{-1}}\mathbf{F}^{-1}\right) = \frac{1}{2}\left(\mathbf{1} - \mathbf{B}^{-1}\right). \tag{2.47}$$

Up to here we derived strain tensors by the observation of line elements and their transformation from initial to current configurations. An alternative approach is described in detail by Haupt [165] and utilizes material surface elements for the derivation of a strain tensor. The change of both distance and orientation of material surface elements to each other is reflected with respect to the initial configuration by the so called *Piola strain tensor* $\mathbf{p}$:

$$\mathbf{p} = \frac{1}{2}\left(\mathbf{F}^{-1}\mathbf{F}^{\mathbf{T}^{-1}} - \mathbf{1}\right) = \frac{1}{2}\left(\mathbf{C}^{-1} - \mathbf{1}\right) \tag{2.48}$$

and to the current configuration by the so called *Finger strain tensor* $\mathbf{f}$:

$$\mathbf{f} = \frac{1}{2}\left(\mathbf{1} - \mathbf{F}\mathbf{F}^{\mathbf{T}}\right) = \frac{1}{2}\left(\mathbf{1} - \mathbf{B}\right). \tag{2.49}$$

A generalized formulation for finite strain tensors was introduced by Doyle and Ericksen [113][5]. The so called *Doyle-Ericksen generalized strain tensor* describes a whole family of strain measures through the parameter m as:

$$\begin{array}{lll} \frac{1}{m}\left(\mathbf{U}^m - \mathbf{1}\right), & \frac{1}{m}\left(\mathbf{V}^m - \mathbf{1}\right) & \text{for} \quad m \neq 0 \\ \ln \mathbf{U}, & \ln \mathbf{V} & \text{for} \quad m = 0 \end{array} \tag{2.50}$$

The generalization (2.50) leads, for example, to

- the Green-Lagrange strain tensor (2.43) for $m = 2$,
- the *Biot strain tensor* for $m = 1$
- the logarithmic *Hencky strains* for $m = 0$

With the above collected derivations we have a set of tensors describing the kinematics of continuous media. Table (2.2) provides an overview on these measures.

Figures 2.5 and 2.6 illustrate how much infinitesimal strain measures and finite strain measures correlate at small strain and how they diverge for growing deformation values. The comparison is taken for an element of 10 [mm] original length that is unidirectionally stretched.

[5] See also Seth [353], Hill [185] and Bažant and Cedolin [40]

Table 2.2. Material and spatial measures for deformation and strain.

Deformation	
Material Deformation Gradient	$F_{ij} = \frac{\partial\, {}^t x_i}{\partial\, {}^0 X_j}$
Spatial Deformation Gradient	$F^{-1}_{ij} = \frac{\partial\, {}^t X_i}{\partial\, {}^0 x_j}$
Material Displacement Gradient	$H_{ij} = \frac{\partial\, {}^t u_i}{\partial\, {}^0 X_j}$
Spatial Displacement Gradient	$h_{ij} = \frac{\partial\, {}^t u_i}{\partial\, {}^0 x_j}$
Right Cauchy-Green Tensor	$\mathbf{C} = \mathbf{F}^T\mathbf{F} = \mathbf{U}^2$
Left Cauchy-Green Tensor	$\mathbf{B} = \mathbf{F}\mathbf{F}^T = \mathbf{V}^2$
1D Strain	
Engineering Strain	$\varepsilon_e = \frac{1}{L_0}\int_{L_0}^{l} dl = \frac{\Delta l}{L_0}$
True Strain	$\varepsilon_t(\mathbf{x},t) = ln\left(\frac{X+dx}{X}\right)$ $= ln\,(1+\varepsilon_e)$
Infinitesimal Strain	
Lagrangean Strain Tensor	$\varepsilon_{ij} = \frac{1}{2}\left(\frac{\partial u_i}{\partial X_j} + \frac{\partial u_j}{\partial X_i}\right)$ $\varepsilon = \frac{1}{2}\left(\mathbf{H} + \mathbf{H}^T\right)$
Eulerian Strain Tensor	$e_{ij} = \frac{1}{2}\left(\frac{\partial u_i}{\partial x_j} + \frac{\partial u_j}{\partial x_i}\right)$ $\mathbf{e} = \frac{1}{2}\left(\mathbf{h} + \mathbf{h}^T\right)$
Finite Strain	
Green-Lagrange Strain Tensor	$E_{ij} = \frac{1}{2}\left(\frac{\partial u_i}{\partial X_j} + \frac{\partial u_j}{\partial X_i} + \frac{\partial u_k}{\partial X_i}\frac{\partial u_k}{\partial X_j}\right)$ $\mathbf{E} = \frac{1}{2}\left(\mathbf{F}^T\,\mathbf{F} - \mathbf{1}\right)$ $= \frac{1}{2}\left(\mathbf{C} - \mathbf{1}\right)$
Euler-Almansi Strain Tensor	$A_{ij} = \frac{1}{2}\left(\frac{\partial u_i}{\partial x_j} + \frac{\partial u_j}{\partial x_i} - \frac{\partial u_k}{\partial x_i}\frac{\partial u_k}{\partial x_j}\right)$ $\mathbf{A} = \frac{1}{2}\left(\mathbf{1} - \mathbf{F^T}^{-1}\mathbf{F}^{-1}\right)$ $= \frac{1}{2}\left(\mathbf{1} - \mathbf{B}^{-1}\right)$
Hencky Strain Tensor	$\mathbf{E}^H = \frac{1}{2}\,\ln\left(\mathbf{1} + 2\,\mathbf{E}\right)$ $= \frac{1}{2}\,\ln\mathbf{U}^2 = \ln\mathbf{U}$
Biot Strain Tensor	$\mathbf{E}^B = \left(\mathbf{U} - \mathbf{1}\right)$
Piola Strain Tensor	$\mathbf{p} = \frac{1}{2}\left(\mathbf{F}^{-1}\mathbf{F^T}^{-1} - \mathbf{1}\right) = \frac{1}{2}\left(\mathbf{C}^{-1} - \mathbf{1}\right)$
Finger Strain Tensor	$\mathbf{f} = \frac{1}{2}\left(\mathbf{1} - \mathbf{F}\mathbf{F^T}\right) = \frac{1}{2}\left(\mathbf{1} - \mathbf{B}\right)$

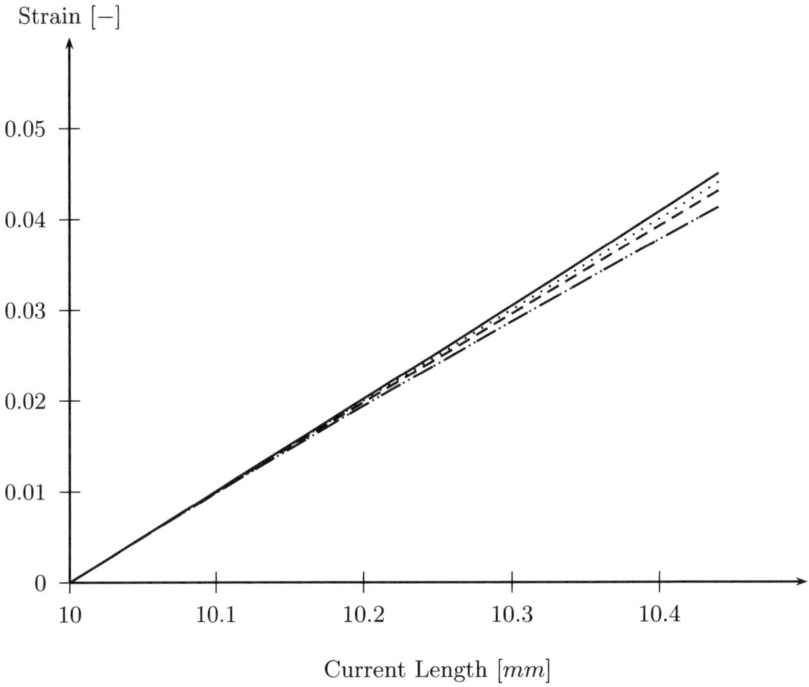

Fig. 2.5. Comparison of Engineering Strains (dotted), True Strains (dashed), Green-Lagrange Strains (solid) and Almansi Strains (dashed and dotted) for a uniaxial strain state in an element of 10 [mm] reference length at elongations up to 0.5 [mm].

Decomposition of Strain Tensors

Elastic-Plastic Separation

In the case of infinitesimal strains, the strain tensors can be separated into an elastic and a plastic part by additive terms according to

$$\varepsilon_{ij} = \varepsilon_{ij}^{el} + \varepsilon_{ij}^{pl} \tag{2.51}$$

Obviously, from the non-linear contributions in the finite strain formulations derived above it becomes evident that an additive decomposition is no longer possible than. In stead a multiplicative separation of the deformation gradient **F** is used, called *Kröner-Lee decomposition*

$$F_{ij} = F_{ij}^{el} F_{ij}^{pl} \tag{2.52}$$

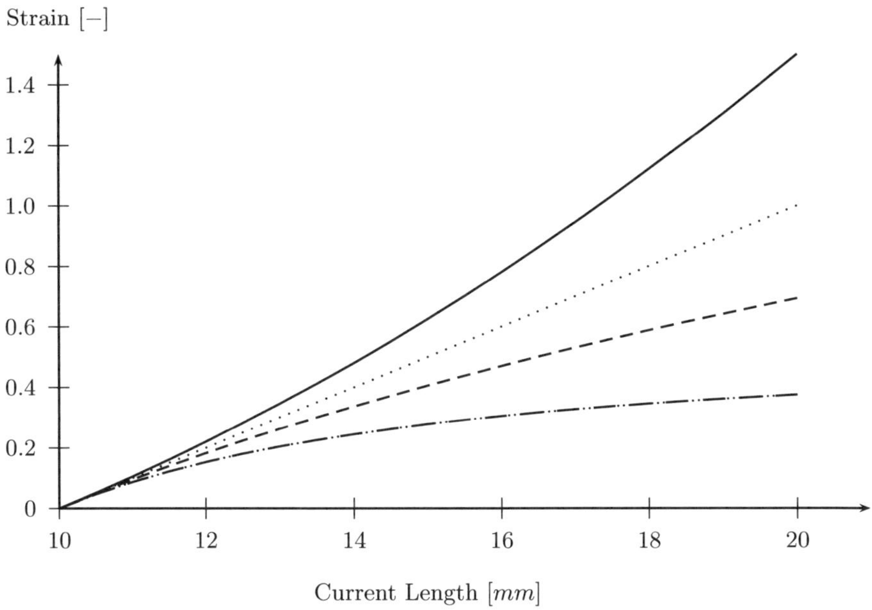

Fig. 2.6. Comparison of Engineering Strains (dotted), True Strains (dashed), Green-Lagrange Strains (solid) and Almansi Strains (dashed and dotted) for a uniaxial strain state in an element of 10 [mm] reference length at elongations up to 10 [mm].

according to which the elastic and plastic finite Lagrangean strains $\mathbf{E^{el}}$ and $\mathbf{E^{pl}}$, respectively, are calculated as:

$$E_{ij}^{el} = \frac{1}{2}\left(F_{ij}^{el^T} F_{ij}^{el} - \delta_{ij}\right) \tag{2.53}$$

and

$$E_{ij}^{pl} = \frac{1}{2}\left(F_{ij}^{pl^T} F_{ij}^{pl} - \delta_{ij}\right) \tag{2.54}$$

Polar decomposition of the separated elastic and plastic deformation gradients also allows to define, e.g., a plastic stretch tensor $\mathbf{V}^p$ through

$$\mathbf{F}^p = \mathbf{V}^p \, \mathbf{R} \tag{2.55}$$

Volumetric-Deviatoric Separation

As any other tensor, strain tensors can additively be decomposed into a spherical and a deviatoric tensor. Total deformation of a body in terms of finite strains E_{ij} can, thus, be described as sum of deviatoric and volumetric strains:

$$E_{ij} = E_{ij}^{dev} + \frac{1}{3}\left(\frac{V - V_0}{V_0}\right)\delta_{ij} = E_{ij}^{dev} + \frac{1}{3}\left(J - 1\right)\delta_{ij} \tag{2.56}$$

A decomposition of the right stretch tensor **U** is often used in hyperelastic models for rubbery materials[6] denotes the distortional component of **U** by

$$\mathbf{U}^* = J^{-\frac{1}{3}}\mathbf{U} \tag{2.57}$$

describing the dilatational part by the Jacobian:

$$J = \det \mathbf{U} \tag{2.58}$$

For infinitesimal strains the decomposition (2.56) can be approximated by

$$\varepsilon_{ij} \approx \varepsilon_{ij}^{dev} + \frac{1}{3}\,\varepsilon_{ij}\,\delta_{ij} \tag{2.59}$$

2.1.4 Material and Spatial Time Derivatives of Deformations

In chapter 2.1 we found that the condition for non rigid translation of a body was a non zero deformation gradient. The motion itself was characterized by an initial and a current configuration without regard of the exact path and the time the motion would take. If we are interested in history and rate effects towards the description of time resolved rate dependent material behavior, we need adequate measures for that time dependency. In other words, time derivatives must be formulated and applied to the deformation and strain measures found before.

It is the aim of this section to precisely define time derivatives for field variables both in material and spatial configurations. For that reason, a smooth field in a material frame shall be denoted by upper case letters $\mathbf{U}({}^0\mathbf{X},\ t)$, whereas a spatial field variable is expressed by $\mathbf{u}({}^t\mathbf{x},\ t)$.

Spatial Derivative of Spatial Field Variables

In order to find more general characteristics of time derivatives, the following discriminations shall provide the definitions needed for later application.

To start with, we find the *spatial time derivative* of a spatial field variable $\mathbf{u}({}^t\mathbf{x},\ t)$ representing the partial derivative with respect to time t at a fixed location ${}^t\mathbf{x}$:

$$\left.\frac{\partial\,\mathbf{u}(\mathbf{x},\ t)}{\partial\, t}\right|_{\mathbf{x}} \tag{2.60}$$

In other words, the *spatial derivative of a spatial field* is a measure for the time rate of change in $\mathbf{u}(\mathbf{x},\ t)$ at a specific fixed spatial position $\mathbf{x} = const.$ as a local observer would see it. Therefore, (2.60) is also called *local derivative of spatial fields.*

[6] Applied to rubber materials, this decomposition was used for example by Flory [127], Ogden [300] and Anand [4]

Material Derivative of Material Field Variables

Changing the type of observer from a fixed position in space to one that is traveling with an infinitesimal volume, we take the *substantial* or *material time derivative* of the field variable in the material frame $\mathbf{U}({}^0\mathbf{X},\ t)$ as:

$$\dot{\mathbf{U}}({}^0\mathbf{X},\ t) = \frac{\mathrm{D}\,\mathbf{U}({}^0\mathbf{X},\ t)}{\mathrm{D}\,t} = \left.\frac{\partial\,\mathbf{U}({}^0\mathbf{X},\ t)}{\partial\,t}\right|_{\mathbf{X}} \tag{2.61}$$

Since in a material frame the independent space variable $\mathbf{X}$ is constant, the substantial derivative of a material field variable is similar to the spatial derivative of a spatial field concerning the fact that there is no contribution from a derivative with respect to the space coordinates. Thus, for a material field variable the *total derivative* is identical to the partial time derivative:

$$\frac{\mathrm{D}\mathbf{U}({}^0\mathbf{X},\ t)}{\mathrm{D}\,t} = \frac{\partial\,\mathbf{U}({}^0\mathbf{X},\ t)}{\partial\,t} \tag{2.62}$$

Substantial Derivative of Spatial Field Variables

A different result is, however, achieved when the substantial derivative is applied to a spatial field $\mathbf{u}({}^t\mathbf{x},\ t)$. Now, the time derivative of the independent spatial variable ${}^t\mathbf{x}$ also contributes to the total derivative as convection term $\frac{\partial}{\partial\mathbf{x}}\frac{\partial\mathbf{x}}{\partial t}$. Therefore, the total time derivative is expressed by:

$$\dot{\mathbf{u}}({}^t\mathbf{x},\ t) = \frac{\mathrm{D}\,\mathbf{u}({}^t\mathbf{x},\ t)}{\mathrm{D}\,t} = \frac{\partial\,\mathbf{u}}{\partial t} + \frac{\partial\,\mathbf{u}}{\partial\,\mathbf{x}}\frac{\partial\,\mathbf{x}}{\partial\,t} = \frac{\partial\,\mathbf{u}}{\partial\,t} + \nabla\mathbf{u}\ \mathbf{v} \tag{2.63}$$

Thus, the substantial derivative of a spatial field consists of a *local derivative*, i.e. the first component in equation (2.63), and a *convective derivative* $\nabla\mathbf{u}\ \mathbf{v}$.

Lie-Oldroyd Derivative for Spatial fields

In addition, there is an alternative material time derivative for spatial fields $\mathbf{u}$, the so called *Lie time derivative*. In case of this specific operation, the field variable is first transformed to the initial configuration by a pull back operation (2.18). In the reference frame, the material time derivative is performed and the resulting term in turn pushed forward to the current configuration by (2.17):

$$\frac{\mathrm{d}_{Lie}\,[\mathbf{u}]}{\mathrm{d}\,t} = \Phi_*\left[\frac{\mathrm{d}}{\mathrm{d}\,t}\left(\Phi^*[\mathbf{u}]\right)\right] \tag{2.64}$$

With the spatial velocity gradient $\mathbf{L}$, the Lie rate (2.64) can alternatively be formulated as:

$$\frac{\mathrm{d}_{Lie}\,[\mathbf{u}]}{\mathrm{d}\,t} = \frac{\partial\,\mathbf{u}}{\partial\,t} + \mathbf{u}\mathbf{L} + \mathbf{L}^T\mathbf{u} \tag{2.65}$$

When applied to a strain field, the Lie derivative is called *Oldroyd rate*, as we will see in section 2.1.5.

Velocities and Velocity Gradients

Material and Spatial Velocities

Regarding the velocity field, we can formulate the *material velocity* of a particle $\mathbf{V}$ as time derivative of its motion χ with the material coordinates $\mathbf{X}$ as independent variables via the displacements $\mathbf{u} = \mathbf{x} - \mathbf{X}$:

$$\mathbf{V}(^{\mathbf{0}}\mathbf{X},\ t) = \frac{\partial\,\chi(\mathbf{X}, t)}{\partial\, t} = \frac{\partial\,\mathbf{x}}{\partial\, t} = \frac{\partial\,(\mathbf{u} + \mathbf{X})}{\partial\, t} = \frac{\partial\,\mathbf{u}}{\partial\, t} \tag{2.66}$$

noting that the initial configuration $\mathbf{X}$ is independent of time. Equally, we can describe the Eulerian velocity field $\mathbf{v}$ with the substantial derivative (2.63) to be:

$$\begin{aligned}\mathbf{v}(^{\mathbf{t}}\mathbf{x},\ t) &= \frac{\mathrm{D}\,\mathbf{u}(^{\mathbf{t}}\mathbf{x}, t)}{\mathrm{D}\, t} = \frac{\partial\,\mathbf{u}(^{\mathbf{t}}\mathbf{x}, t)}{\partial\, t} + \frac{\partial\,\mathbf{u}(^{\mathbf{t}}\mathbf{x}, t)}{\partial\, ^{\mathbf{t}}\mathbf{x}}\frac{\partial\,\mathbf{x}}{\partial\, t} \\ &= \frac{\partial\,\mathbf{u}(^{\mathbf{t}}\mathbf{x}, t)}{\partial\, t} + \frac{\partial\,\mathbf{u}(^{\mathbf{t}}\mathbf{x}, t)}{\partial\, ^{\mathbf{t}}\mathbf{x}}\,\mathbf{v}(^{\mathbf{t}}\mathbf{x},\ t)\end{aligned} \tag{2.67}$$

where the in (2.67) implicitly given Eulerian velocity $\mathbf{v}$ is also called the *instantaneous velocity field.*

The used convention of denoting material field variables with upper case letters and the corresponding spatial fields by lower case letters cannot be kept in full consequence. Therefore, whenever this convention is dropped the type of the used configuration will be noted. Specifically the velocity field will be described by the lower case letter $\mathbf{v}$ and occasional material configuration will be noted.

Material and Spatial Velocity Gradients

Differentiation of the found velocity fields with respect to space coordinates leads to a *material velocity gradient* $\dot{\mathbf{F}}$ defined as

$$\frac{\partial\,\mathbf{V}(\mathbf{X},\ t)}{\partial \mathbf{X}} = \nabla\ \mathbf{V}(\mathbf{X},\ t) = \dot{\mathbf{F}}(\mathbf{X},\ t) \tag{2.68}$$

Application of the inverse deformation gradient $\mathbf{F}^{-1}$ leads to the *spatial velocity gradient* $\mathbf{L}$ and illustrates its relation to the material gradient:

$$\mathbf{L} = \dot{\mathbf{F}}\,\mathbf{F}^{-1} = \nabla\ \mathbf{v}(\mathbf{x},\ t) \tag{2.69}$$

Rate of Deformation and Spin Tensors

It will turn out later to be useful decomposing the spatial velocity gradient $\nabla\,\mathbf{v}$ additively into a symmetric tensor $\mathbf{D}$ and a skew-symmetric tensor $\mathbf{W}$.

The *rate of deformation tensor* **D** describes the time rate of change in the deformation while the *spin tensor* **W** represents the vorticity or rotation rate:

$$\mathbf{L} = \nabla \mathbf{v} = \mathbf{D} + \mathbf{W} \tag{2.70}$$

where

$$D_{ij} = \frac{1}{2}\left(\frac{\partial v_i}{\partial x_j} + \frac{\partial v_j}{\partial x_i}\right)$$
$$\mathbf{D} = \frac{1}{2}\left(\mathbf{L} + \mathbf{L}^T\right) \tag{2.71}$$

and

$$W_{ij} = \frac{1}{2}\left(\frac{\partial v_i}{\partial x_j} - \frac{\partial v_j}{\partial x_i}\right)$$
$$\mathbf{W} = \frac{1}{2}\left(\mathbf{L} - \mathbf{L}^T\right) \tag{2.72}$$

2.1.5 Strain Rate Tensors

Strain rates, i.e. the time derivatives of strain measures, are the complementary variables to stress rates in the constitutive equations used for crash and impact simulations. Apparently, the various definitions of infinitesimal and finite strain measures in material and spatial configurations along with the corresponding formulations of time derivatives provide a variety of strain rate measures. To find the appropriate stain rate formulation for a chosen stress rate[7], common strain rates are discussed in the following.
In this context, it is worthwhile noticing how the above formulated rate of deformation tensor **D** compares to the material time derivatives of the Lagrangean strain tensor ε as well as with the rates of the linear and finite Eulerian strain tensors **e** and **A**. Fundamental derivations and transformations into each other can be found in Haupt [165].

Rates of Infinitesimal Strains

To begin with the infinitesimal Lagrangean strain tensor, we find the material time derivative which we call now the *Lagrangean strain rate*

$$\dot{\varepsilon}_{ij} = \frac{\mathrm{d}\,\varepsilon_{ij}}{\mathrm{d}t} = \frac{1}{2}\left\{\frac{\mathrm{d}}{\mathrm{d}\,t}\left(\frac{\partial u_i}{\partial X_j}\right) + \frac{\mathrm{d}}{\mathrm{d}\,t}\left(\frac{\partial u_j}{\partial X_i}\right)\right\} = \frac{1}{2}\left(\frac{\partial v_i}{\partial X_j} + \frac{\partial v_j}{\partial X_i}\right) \tag{2.73}$$

Thus, while the rate of deformation tensor **D** (2.71) operates as spatial derivative of the velocities with respect to the current configuration, the material

[7] The concept of *dual variables*, explicated in detail by Haupt [165], defines mutually applicable formulations.

time derivative of the infinitesimal Lagrangean strain tensor refers to the initial configuration.

$$D_{ij} = \frac{1}{2}\left(\frac{\partial v_i}{\partial x_j} + \frac{\partial v_j}{\partial x_i}\right) \neq \dot{\varepsilon}_{ij} \tag{2.74}$$

On the other hand, the *linear Eulerian strain rate tensor*

$$\dot{e}_{ij} = \frac{\mathrm{d}\, e_{ij}(\mathbf{x},t)}{\mathrm{d}t} = \frac{\mathrm{d}}{\mathrm{d}\,t}\left\{\frac{1}{2}\left(\frac{\partial u_i}{\partial x_j} + \frac{\partial u_j}{\partial x_i}\right)\right\} \tag{2.75}$$

is derived if we recall the convective terms for substantial time derivatives of spatial fields:
According to (2.63) the substantial derivative of $\mathbf{u}$ is

$$\frac{\mathrm{d}\,u_i}{\mathrm{d}\,t} = \frac{\partial u_i}{\partial t} + \frac{\partial u_i}{\partial x_p}\frac{\partial x_p}{\partial t} = \frac{\partial u_i}{\partial t} + \frac{\partial u_i}{\partial x_p} v_p \tag{2.76}$$

and thus

$$\begin{aligned}
\frac{\mathrm{d}\,v_i}{\mathrm{d}\,x_j} &= \frac{\partial}{\partial x_j}\left(\frac{\mathrm{d}\,u_i}{\mathrm{d}\,t}\right) = \frac{\partial}{\partial x_j}\left(\frac{\partial u_i}{\partial t} + \frac{\partial u_i}{\partial x_p} v_p\right) \\
&= \frac{\partial}{\partial t}\frac{\partial u_i}{\partial x_j} + \frac{\partial u_i}{\partial x_p}\frac{\partial v_p}{\partial x_j} + v_p \frac{\partial^2 u_i}{\partial x_p\,\partial x_j} \\
&= \underbrace{\left(\frac{\partial}{\partial t} + v_p \frac{\partial}{\partial x_p}\right)\frac{\partial u_i}{\partial x_j}}_{=\frac{\mathrm{d}}{\mathrm{d}\,t}\left(\frac{\partial u_i}{\partial x_j}\right)} + \frac{\partial u_i}{\partial x_p}\frac{\partial v_p}{\partial x_j}
\end{aligned} \tag{2.77}$$

gives for the components of the linear Eulerian strain rate tensor

$$\frac{\mathrm{d}}{\mathrm{d}\,t}\left(\frac{\partial u_i}{\partial x_j}\right) = \frac{\partial v_i}{d\,x_j} - \frac{\partial u_i}{\partial x_p}\frac{\partial v_p}{\partial x_j} \tag{2.78}$$

which shows implemented into (2.75)

$$\begin{aligned}
\dot{e}_{ij} &= \frac{1}{2}\left\{\frac{\mathrm{d}}{\mathrm{d}\,t}\left(\frac{\partial u_i}{\partial x_j}\right) + \frac{\mathrm{d}}{\mathrm{d}\,t}\left(\frac{\partial u_j}{\partial x_i}\right)\right\} \\
&= \frac{1}{2}\left\{\left(\frac{\partial v_i}{\partial x_j} - \frac{\partial u_i}{\partial x_p}\frac{\partial v_p}{\partial x_j}\right) + \left(\frac{\partial v_j}{\partial x_i} - \frac{\partial u_j}{\partial x_p}\frac{\partial v_p}{\partial x_i}\right)\right\} \\
&= \frac{1}{2}\left(\frac{\partial v_i}{\partial x_j} + \frac{\partial v_j}{\partial x_i}\right) - \frac{1}{2}\left(\frac{\partial u_i}{\partial x_p}\frac{\partial v_p}{\partial x_j} + \frac{\partial u_j}{\partial x_p}\frac{\partial v_p}{\partial x_i}\right) \\
&= D_{ij} - \frac{1}{2}\left(\frac{\partial u_i}{\partial x_p}\frac{\partial v_p}{\partial x_j} + \frac{\partial u_j}{\partial x_p}\frac{\partial v_p}{\partial x_i}\right)
\end{aligned} \tag{2.79}$$

the relation between $\mathbf{D}$ and $\dot{e}$. At this point it becomes obvious why $\mathbf{D}$ is often also called *strain rate tensor*. That term however should be specified since, as could be seen above, $\mathbf{D}$ is **not** equal to either one of both linear strain rates we introduced so far.

Finite Strain Rates

The Lagrangean and Eulerian strain rates in (2.73) and (2.75) are time derivatives of linearized strain measures. Of course, also for the finite strain measures related rate formulations exist. To establish the *Green strain rate tensor* $\dot{\mathbf{E}}$ we take the material time derivative of the Green-Lagrange strain tensor (2.42):

$$\mathrm{d}\mathbf{X}\,\dot{\mathbf{E}}\,\mathrm{d}\mathbf{X} = \frac{1}{2}\,\frac{\mathrm{d}}{\mathrm{d}t}\,(\mathrm{d}\mathbf{x}\,\mathrm{d}\mathbf{x} \;-\; \mathrm{d}\mathbf{X}\,\mathrm{d}\mathbf{X}) \tag{2.80}$$

and find

$$\dot{\mathbf{E}} = \frac{1}{2}\left(\dot{\mathbf{F}}^T\mathbf{F} + \mathbf{F}^T\dot{\mathbf{F}}\right) \tag{2.81}$$

To derive a *finite spatial strain rate tensor*, we take the time derivative of the Euler-Almansi strain tensor $\mathbf{A}$:

$$\begin{aligned}
\dot{\mathbf{A}} &= \frac{1}{2}\,\frac{\mathrm{d}}{\mathrm{d}t}\left(\mathbf{1} - \mathbf{F}^{-T}\,\mathbf{F}^{-1}\right) = -\frac{1}{2}\left(\dot{\mathbf{F}}^{-T}\,\mathbf{F}^{-1} + \mathbf{F}^{-T}\,\dot{\mathbf{F}}^{-1}\right)\\
&= \frac{1}{2}\left(\mathbf{F}^{-T}\,\dot{\mathbf{F}}^{T}\,\mathbf{F}^{-T}\,\mathbf{F}^{-1} + \mathbf{F}^{-T}\,\mathbf{F}^{-1}\,\dot{\mathbf{F}}\,\mathbf{F}^{-1}\right)\\
&= \frac{1}{2}\,\left(\mathbf{B}^{-1}\,\mathbf{L} + \mathbf{L}^T\,\mathbf{B}^{-1}\right)\frac{1}{2}\,\left(\mathbf{B}^{-1}\,\mathbf{L} + \mathbf{L}^T\,\mathbf{B}^{-1}\right)
\end{aligned} \tag{2.82}$$

The identity

$$\mathbf{A} = \frac{1}{2}\,\left(\mathbf{B}^{-1}\,\mathbf{L} + \mathbf{L}^T\,\mathbf{B}^{-1}\right) \tag{2.83}$$

and thus

$$\mathbf{B}^{-1} = \mathbf{1} - 2\,\mathbf{A} \tag{2.84}$$

finally yields the *Almansi strain rate tensor* in the form:

$$\dot{\mathbf{A}} = \frac{1}{2}\,\left(\mathbf{L} + \mathbf{L}^T - 2\,\mathbf{A}\,\mathbf{L} - 2\,\mathbf{L}^T\,\mathbf{A}\right) \tag{2.85}$$

Comparison of the Almansi strain rate tensor (2.85) with the spatial strain rates $\mathbf{D}$ shows that:

$$\dot{\mathbf{A}} = \mathbf{D} - \mathbf{A}\,\mathbf{L} - \mathbf{L}^T\,\mathbf{A} \tag{2.86}$$

As we experienced with the Lie time derivative (2.64), after application of a pull-back operation the material time derivative of Eulerian tensors can also be taken in the reference configuration. Applied to a strain tensor, the Lie derivative is called *Oldroyd rate*. Though its precise usefulness may at first not be very obvious, the Oldroyd rate of the Euler-Almansi strain tensor will proof material objectivity and applicability as conjugate strain rate to the Jaumann stress rate tensor. The Oldroyd rate of the Euler-Almansi strain tensor is defined as:

$$\breve{\mathbf{A}} = \mathbf{F}^{-T}\left\{\frac{\mathrm{d}}{\mathrm{d}t}\,\left(\mathbf{F}^{-1}\mathbf{A}\,\mathbf{F}^{-T}\right)\right\}\,\mathbf{F}^{-1} \tag{2.87}$$

Table 2.3. Material and spatial measures for velocities and strain rates.

Velocity	
Material Velocity Field	$\mathbf{V}(\mathbf{X},\,t) = \frac{\partial\,\chi(\mathbf{X},t)}{\partial\,t} = \frac{\partial\,\mathbf{x}}{\partial\,t}$
Spatial Velocity Field	$\mathbf{v}(\mathbf{x},\,t) = \frac{\partial\,\chi^{-1}(\mathbf{x},t)}{\partial\,t} = \frac{\partial\,\mathbf{X}}{\partial\,t} = \mathbf{V}(\mathbf{X},\,t)$
Material Velocity Gradient	$\dot{\mathbf{F}}(\mathbf{X},\,t) = \frac{\partial\,\mathbf{V}(\mathbf{X},\,t)}{\partial\,\mathbf{X}}$
Spatial Velocity Gradient	$\mathbf{L} = \frac{\partial\,\mathbf{v}(\mathbf{x},\,t)}{\partial\,\mathbf{x}} = \dot{\mathbf{F}}\,\mathbf{F}^{-1} = \mathbf{D} + \mathbf{W}$
Rate of Deformation Tensor	$D_{ij} = \frac{1}{2}\left(\frac{\partial\,v_i}{\partial\,x_j} + \frac{\partial\,v_j}{\partial\,x_i}\right)$
Spin Tensor	$W_{ij} = \frac{1}{2}\left(\frac{\partial\,v_i}{\partial\,x_j} - \frac{\partial\,v_j}{\partial\,x_i}\right)$
Infinitesimal Strain Rates	
Lagrangean Strain Rates	$\dot{\varepsilon}_{ij} = \frac{\mathrm{d}}{\mathrm{d}\,t}\left\{\frac{1}{2}\left(\frac{\partial u_i}{\partial X_j} + \frac{\partial u_j}{\partial X_i}\right)\right\}$ $= \frac{1}{2}\left(\frac{\partial v_i}{\partial X_j} + \frac{\partial v_j}{\partial X_i}\right)$
Eulerian Strain Rates	$\dot{e}_{ij} = \frac{\mathrm{d}}{\mathrm{d}\,t}\left\{\frac{1}{2}\left(\frac{\partial u_i}{\partial x_j} + \frac{\partial u_j}{\partial x_i}\right)\right\}$ $= \frac{1}{2}\left(\frac{\partial\,v_i}{\partial\,x_j} + \frac{\partial\,v_j}{\partial\,x_i}\right) - \frac{1}{2}\left(\frac{\partial\,u_i}{\partial\,x_p}\frac{\partial\,v_p}{\partial\,x_j} + \frac{\partial\,u_j}{\partial\,x_p}\frac{\partial\,v_p}{\partial\,x_i}\right)$ $= D_{ij} - \frac{1}{2}\left(\frac{\partial\,u_i}{\partial\,x_p}\frac{\partial\,v_p}{\partial\,x_j} + \frac{\partial\,u_j}{\partial\,x_p}\frac{\partial\,v_p}{\partial\,x_i}\right)$
Finite Strain Rates	
Green Strain Rate Tensor	$\dot{\mathbf{E}} = \frac{1}{2}\left(\dot{\mathbf{F}}^T\,\mathbf{F} + \mathbf{F}^T\,\dot{\mathbf{F}}\right)$ $= \mathbf{F}^T\,\mathbf{D}\,\mathbf{F}$
Almansi Strain Rate Tensor	$\dot{\mathbf{A}} = \frac{1}{2}\,(\mathbf{L} + \mathbf{L}^T - 2\,\mathbf{A}\,\mathbf{L} - 2\,\mathbf{L}^T\,\mathbf{A})$ $= \mathbf{D} - \mathbf{A}\,\mathbf{L} + \mathbf{L}^T\,\mathbf{A}$
Oldroyd Strain Rate Tensor	$\breve{\mathbf{A}} = \dot{\mathbf{A}} + \mathbf{L}^T\,\mathbf{A} + \mathbf{A}\,\mathbf{L}$ $= \mathbf{F}^{-T}\left\{\mathbf{F}^{-1}\,\dot{\mathbf{A}}\,\mathbf{F}^{-T}\right\}\,\mathbf{F}^{-1}$ $= \mathbf{F}^{-T}\left\{\dot{\mathbf{E}}\right\}\,\mathbf{F}^{-1}$ $\equiv \mathbf{D}$

and since

$$\left(\mathbf{F}^{-1}\mathbf{A}\,\mathbf{F}^{-T}\right) = \mathbf{E} \tag{2.88}$$

$\breve{\mathbf{A}}$ is equal to a push-forward operation applied to the Green strain rate tensor:

$$\breve{\mathbf{A}} = \mathbf{F}^{-T}\left\{\dot{\mathbf{E}}\right\}\,\mathbf{F}^{-1} \tag{2.89}$$

Therefore, the relation between the spatial strain rate tensor $\mathbf{D}$, the material derivative of the Euler-Almansi strains $\dot{\mathbf{A}}$ and the material derivative of Green-Lagrange strains $\dot{\mathbf{E}}$ is then defined as follows:

$$\breve{\mathbf{A}} = \mathbf{F}^{-T} \left\{ \dot{\mathbf{E}} \right\} \mathbf{F}^{-1} = \dot{\mathbf{A}} + \mathbf{L}^T \mathbf{A} + \mathbf{A}\,\mathbf{L} = \mathbf{D} \tag{2.90}$$

Recalling (2.65), we find that the Oldroyd rate (2.90) of the Euler-Almansi strains $\breve{\mathbf{A}}$ equals the spatial strain rates $\mathbf{D}$

$$\breve{\mathbf{A}} = \dot{\mathbf{A}} + \mathbf{L}^T \mathbf{A} + \mathbf{A}\,\mathbf{L} = \mathbf{D} \tag{2.91}$$

Equation (2.91) gives a new meaning to the spatial strain rate tensor $\mathbf{D}$ being a finite strain measure and equivalent to the Oldroyd rate of the Euler-Almansi strain tensor[8].

An overview on the derived velocity and strain rate measures is collected in table 2.3.

2.1.6 Compatibility Conditions

Strain measures, as derived in the last two subsections, represent relations between components of strain tensors, e.g. the linear Lagrangean strains ε_{ij} and components of the displacement vector u_i. Since we have six independent strain components and three independent displacement components, there is no unique description of a displacement state, given the strain state is known. Thus, a relation

$$\varepsilon_{ij} = \frac{1}{2} \left(\frac{\partial u_i}{\partial x_j} \right) \tag{2.92}$$

is over-determined with respect to the displacements. A unique ε-$\mathbf{u}$ relation would guarantee that a continuum that is smooth in its initial configuration, stays smooth in a deformed state. To guarantee this, the *compatibility conditions* are formulated by elimination of the displacements, i.e. through differentiation and successive change of indices:

$$\frac{\partial^2 \varepsilon_{ij}}{\partial x_k\, \partial x_l} + \frac{\partial^2 \varepsilon_{kl}}{\partial x_i\, \partial x} - \frac{\partial^2 \varepsilon_{ik}}{\partial x_j\, \partial x_l} - \frac{\partial^2 \varepsilon_{jl}}{\partial x_i\, \partial x_k} = 0 \tag{2.93}$$

leading to a total of 81 equations, most of which are either identical or merge due to symmetry conditions. There are, however, six equations remaining, called the essential compatibility equations:

$$\begin{aligned}
\frac{\partial^2 \varepsilon_{xx}}{\partial x_y\, \partial x_z} &= \frac{\partial}{\partial x_x} \left(-\frac{\partial \varepsilon_{yz}}{\partial x_x} + \frac{\partial \varepsilon_{zx}}{\partial x_y} + \frac{\partial \varepsilon_{xy}}{\partial x_z} \right) \\
\frac{\partial^2 \varepsilon_{yy}}{\partial x_z\, \partial x_x} &= \frac{\partial}{\partial x_y} \left(-\frac{\partial \varepsilon_{zx}}{\partial x_y} + \frac{\partial \varepsilon_{xy}}{\partial x_z} + \frac{\partial \varepsilon_{yz}}{\partial x_x} \right) \\
\frac{\partial^2 \varepsilon_{zz}}{\partial x_x\, \partial x_y} &= \frac{\partial}{\partial x_z} \left(-\frac{\partial \varepsilon_{xy}}{\partial x_z} + \frac{\partial \varepsilon_{yz}}{\partial x_x} + \frac{\partial \varepsilon_{zx}}{\partial x_y} \right)
\end{aligned}$$

[8] See for example Haupt [165], p. 49, for proof of the relations between the positive tensor $\mathbf{D}$ as well as the negative tensor $-\mathbf{D}$ to the co- and contravariant Oldroyd rates of the Euler-Almansi strain tensor $\mathbf{A}$ and the Finger tensor $\mathbf{f}$, respectively.

$$2\,\frac{\partial^2\,\varepsilon_{xy}}{\partial\,x_x\,\partial\,x_y} = \frac{\partial^2\,\varepsilon_{xx}}{\partial\,x_y^2} + \frac{\partial^2\,\varepsilon_{yy}}{\partial\,x_x^2}$$

$$2\,\frac{\partial^2\,\varepsilon_{yz}}{\partial\,x_y\,\partial\,x_z} = \frac{\partial^2\,\varepsilon_{yy}}{\partial\,x_z^2} + \frac{\partial^2\,\varepsilon_{zz}}{\partial\,x_y^2}$$

$$2\,\frac{\partial^2\,\varepsilon_{zx}}{\partial\,x_z\,\partial\,x_x} = \frac{\partial^2\,\varepsilon_{zz}}{\partial\,x_x^2} + \frac{\partial^2\,\varepsilon_{xx}}{\partial\,x_z^2}$$

In a two-dimensional system the six equations are reduced to a single one:

$$2\,\frac{\partial^2\,\varepsilon_{xy}}{\partial\,x_x\,\partial\,x_y} = \frac{\partial^2\,\varepsilon_{xx}}{\partial\,x_y^2} + \frac{\partial^2\,\varepsilon_{yy}}{\partial\,x_x^2} \tag{2.94}$$

2.2 Stress Measures

2.2.1 Cauchy Stresses

At any given point in a continuum, a *stress vector* $\mathbf{t}(\mathbf{x}, t, \mathbf{n})$ shall be defined as the differential surface forces $\mathbf{df}$ acting on an infinitesimal surface area element $\mathrm{d}S$ with surface normal $\mathbf{n}$

$$\mathbf{t}(\mathbf{x}, t, \mathbf{n}) = \frac{\mathbf{df}}{\mathrm{d}S} \tag{2.95}$$

where the *Cauchy principle* assumes, that the stress vector not only depends on the location $\mathbf{x}$ and the time t but in general also on the surface normal. Since there is an infinite number of planes cutting through the regarded point, a stress measure with a finite number of parameters is found in planes perpendicular to the coordinate axes. The resultant *surface traction* on each coordinate plane is decomposed into a normal and two shear components, for example in the one-direction:

$$\mathbf{t}_1 = \sigma_{11}\mathbf{e_1} + \sigma_{12}\mathbf{e_2} + \sigma_{13}\mathbf{e_3} \tag{2.96}$$

or in general

$$\mathbf{t}_i = \sigma_{ij}\mathbf{e_j} \tag{2.97}$$

With the three normal components σ_{11}, σ_{22} and σ_{33} and the six shear components σ_{12} ... σ_{23} the nine quantities of the *stress tensor* define the three stress vectors as:

$$\sigma_{ij} = \begin{bmatrix} \mathbf{t_1} \\ \mathbf{t_2} \\ \mathbf{t_2} \end{bmatrix} = \begin{bmatrix} \sigma_{11}\ \sigma_{12}\ \sigma_{13} \\ \sigma_{21}\ \sigma_{22}\ \sigma_{23} \\ \sigma_{31}\ \sigma_{32}\ \sigma_{33} \end{bmatrix} \tag{2.98}$$

The stress vector on any surface with surface normal $\mathbf{n}$ is then defined by

$$t_i = \sigma_{ji} n_j \tag{2.99}$$

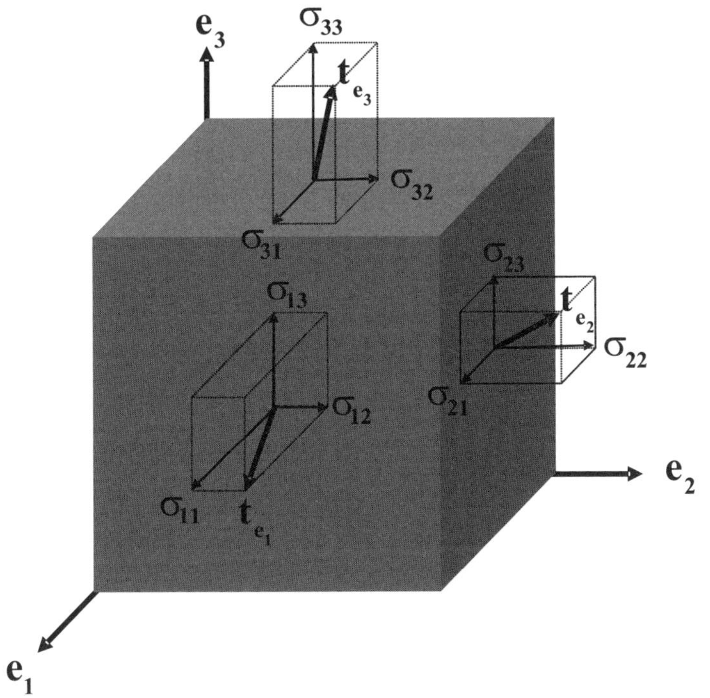

Fig. 2.7. Surface tractions t_{e_i} on the faces of a cube and related stress components σ_{ij}.

As indicated in Figure 2.7, the chosen convention[9] for indices ij of a stress component σ_{ij} identifies the first index i with the plane on which the surface traction acts and the second index j with the direction of the associated base vector e_j.

On a surface element with surface normal pointing into the positive direction, a stress component is defined positive if it points into the positive coordinate direction. This convention on the sign of a stress component implies that *tensile stresses are defined positive* and *compressive stresses are negative*, respectively.

The tensor σ_{ij} is called *Cauchy stress tensor*, refers to the current configuration and is thus a spatial or Eulerian stress measure. The conservation of angular momentum (2.197) proofs the symmetry of the Cauchy stress tensor.

[9] This index convention is in accordance with many engineering definitions (see for example Malvern [263]). The opposite definition associating the first index with the base vector's direction can accordingly be found, e.g. in Truesdell and Noll [398] or Holzapfel [194].

The so called *Kirchhoff stress tensor* or *weighted Cauchy stress tensor* takes regard of the density change between initial and current configuration by:

$$\sigma_{ij}^{W} = (det\ \mathbf{F})\sigma_{ij} = \frac{\rho_0}{\rho}\sigma_{ij} \tag{2.100}$$

2.2.2 Alternative Stress Measures

Whenever conservation equations or constitutive models shall be formulated with regard to initial configurations, and thus material strain formulations are to be combined with appropriate stress tensors, alternative stress measures are needed. One way to establish a material stress measure is found in the *first Piola-Kirchhoff stress tensor* $\mathbf{T}$, which is formulated by relating the current force vector $\mathrm{d}\mathbf{f}$ to the initial surface element $\mathrm{d}^0 S$:

$$\mathbf{T} = \frac{\mathrm{d}\mathbf{f}}{\mathrm{d}^0 S} = (det\,\mathbf{F})\ \sigma\ \mathbf{F}^{-\mathbf{T}} \tag{2.101}$$

A stress measure that is closely related to the first Piola-Kirchhoff stresses and often applied for total Lagrangean descriptions in dynamic problems[10] is the so called *nominal stress tensor*:

$$\mathbf{P} = (det\,\mathbf{F})\ \mathbf{F}^{-\mathbf{1}}\ \sigma \tag{2.102}$$

which can be understood as the transpose of the first Piola-Kirchhoff stresses (2.101).

Since the first Piola-Kirchhoff stress tensor is not symmetric ($T_{ij} \neq T_{ji}$), it is sometimes of limited usefulness in the application. To symmetrize the tensor, the differential surface force vector is also referred to the initial configuration as pseudo force vector:

$$\mathrm{d}^0\mathbf{f} = \mathbf{F}^{-\mathbf{1}}\mathrm{d}\mathbf{f} \tag{2.103}$$

That second transformation leads to the *second Piola-Kirchhoff stresses* $\hat{\mathbf{T}}$

$$\hat{\mathbf{T}} = \mathbf{F}^{-\mathbf{1}}\mathbf{T} = (det\,\mathbf{F})\ \mathbf{F}^{-\mathbf{1}}\ \sigma\ \mathbf{F}^{-\mathbf{T}} \tag{2.104}$$

which is can equally be understood as a *Piola transformation* on the Cauchy stresses σ:

$$\hat{\mathbf{T}} = J\ \mathbf{F}^{-\mathbf{1}}\ \sigma\ \mathbf{F}^{-\mathbf{T}} \tag{2.105}$$

or as a pull-back operation (2.18)

$$\hat{\mathbf{T}} = \Phi_*^{-1}[\sigma^W] \tag{2.106}$$

on the Kirchhoff stress tensor σ^W (2.100) .

[10] See for example its explanation and applications in Belytschko et al. [47].

2.2.3 Rate Dependent Stress Measures

Objectivity is an important precondition for scientific work, since it states that the laws identified for physical processes must not depend on the the observer's position in space and time. That implies that the mathematical expressions which we find, necessarily need to allow for a change of the observer without influence on its result. The calculation of time rates of changes of the stress tensor is the component of hydrocodes where the postulate for objectivity must be fulfilled. Formulations of objective stress rates will be discussed in the following. Basic considerations about objectivity, or *frame indifference*, and its meaning to non-linear continuum mechanics are described in detail by Holzapfel [194].

Objectivity or Frame Indifference

Briefly summarized, *objectivity* means that any process observed to happen in a time interval $|t - t_0|$ between two locations $|\mathbf{x} - \mathbf{x}_0|$ must be observed identically when observed from another position as happening in $|t^{\#} - t_0^{\#}|$ within $|\mathbf{x}^{\#} - \mathbf{x}_0^{\#}|$. The transformation between two observers is called *Euclidean transformation* when

$$\mathbf{x}^{\#} = \mathbf{x}_0^{\#} + \mathbf{Q}(t)\mathbf{x} - \mathbf{Q}(t)\mathbf{x}_0 \tag{2.107}$$

and

$$t^{\#} = t + t_0^{\#} - t_0 \tag{2.108}$$

are related through the orthogonal tensor $\mathbf{Q}(t)$. Objectivity with respect to vectors $\mathbf{v}$ and tensors $\mathbf{A}$ means that they are transformed according to (2.109) and (2.110), respectively, whereas scalars α remain unchanged (2.111):

$$\mathbf{v}^{\#}(\mathbf{x}^{\#},\, t^{\#}) = \mathbf{Q}(t)\mathbf{v}(\mathbf{x},\, t) \tag{2.109}$$

$$\mathbf{A}^{\#}(\mathbf{x}^{\#},\, t^{\#}) = \mathbf{Q}(t)\mathbf{A}(\mathbf{x},\, t)\mathbf{Q}^T(t) \tag{2.110}$$

$$\alpha^{\#}(\mathbf{x}^{\#},\, t^{\#}) = \alpha(\mathbf{x},\, t) \tag{2.111}$$

Objective Stress Rates

Basically, the time rate of change of stresses for isotropic elastic behaviour (3.46) can be quantified by the material time derivative of the Cauchy stress tensor:

$$\dot{\sigma}_{ij} = 2\,\mu\,\dot{\varepsilon}_{ij} + \lambda\,\dot{\varepsilon}_{kk}\,\delta_{ij} \tag{2.112}$$

Since the material time derivatives of objective spatial tensors, and the Cauchy stress tensor is a spatial tensor, are not necessarily objective[11], alternative

[11] For instance see Holzapfel [194], chapter 5.3, for further explanations

stress rates are needed for rate formulations of constitutive equations. For example, application of (2.112) would lead to changes of the stress state in the course of rigid body rotations.

An example for objective stress rates is the *Jaumann-Zaremba rate* which is often found in explicit codes:

$$\hat{\sigma}_{ij}^{J} = \dot{\sigma}_{ij} - W_{ik}\,\sigma_{kj} + \sigma_{ik}\,W_{kj} \tag{2.113}$$

with the spin tensor W_{ij} defined by 2.72. A possible disadvantage of the Jaumann rate is found in strong oscillations under pure shear when combined with kinematic hardening (see Benson [53]). For such situations, an alternative formulation is found in the *Green-Naghdi* rate:

$$\hat{\sigma}_{ij}^{GN} = \dot{\sigma}_{ij} - \Omega_{ik}\,\sigma_{kj} + \sigma_{ik}\,\Omega_{kj} \tag{2.114}$$

with the angular velocity tensor $\mathbf{\Omega} = \dot{\mathbf{R}} \cdot \mathbf{R}$ and the orthogonal rotation tensor $\mathbf{R}$. That means the only difference between the Jaumann rate (2.113) and the Green-Naghdi rate (2.114) is found in the description of rotation. Both stress rates coincide if $\mathbf{W} = \dot{\mathbf{R}} \cdot \mathbf{R}$ which describes rigid-body rotation. Furthermore it can be shown that both rates are special cases of the Lie derivatives (2.64) of σ [12].

Another often applied stress rate measure is the *Truesdell rate*:

$$\hat{\sigma}_{ij}^{T} = \dot{\sigma}_{ij} + \frac{\partial v_k}{\partial x_k}\,\sigma_{ij} - \frac{\partial v_i}{\partial x_k}\,\sigma_{kj} - \sigma_{ik}\,\frac{\partial v_j}{\partial x_k} \tag{2.115}$$

which can also be seen as a Piola transformation (2.105) of the time derivative of the second Piola-Kirchhoff stresses $\hat{\mathbf{T}}$:

$$\hat{\sigma}_{ij}^{T} = J^{-1}\,\mathbf{F}\,\frac{\partial}{\partial t}\left(\hat{\mathbf{T}}\right)\,\mathbf{F}^{T} \tag{2.116}$$

Principal Stresses and Stress Invariants

The arbitrary normal direction of surface elements on which the stress vectors were formulated made a decomposition into normal and shear components necessary. If the stress tensor is rotated in a way that its normal stress components are in line with the surface normal, the resulting rotated tensor is free of shear stresses. This operation is possible for any stress tensor and leads to the normal directions called *principal directions* and the three *principle stress* components σ_1 , σ_2 and σ_3. To identify the principal stress system, we demand:

$$\sigma_{ij}\,n_j = \sigma\,n_i \tag{2.117}$$

[12] see for example Holzapfel [194] chapter 5

or

$$(\sigma_{ij} - \sigma\, \delta_{ij})\, n_j = 0 \tag{2.118}$$

together with

$$|\sigma_{ij} - \sigma\, \delta_{ij}| = 0 \tag{2.119}$$

leading to the *characteristic equation*

$$\sigma^3 - I_1\sigma^2 - I_2\sigma - I_3 = 0 \tag{2.120}$$

The roots of the characteristic equations are called the *invariants of the stress tensor*:

$$I_1 = \sigma_{ii} = \sigma_{xx} + \sigma_{yy} + \sigma_{zz} \tag{2.121}$$

$$I_2 = \frac{1}{2}\sigma_{ij}\sigma_{ij} = \sigma_{xx}\sigma_{yy} + \sigma_{yy}\sigma_{zz} + \sigma_{xx}\sigma_{zz} - \sigma_{xy}^2 - \sigma_{xz}^2 - \sigma_{yz}^2 \tag{2.122}$$

$$I_3 = \frac{1}{3}\sigma_{ij}\sigma_{jk}\sigma_{ki} = det\left[\sigma_{ij}\right] \tag{2.123}$$

In terms of principal stresses we obtain for the invariants:

$$I_1 = \sigma_1 + \sigma_2 + \sigma_3 \tag{2.124}$$

$$I_2 = \sigma_1\sigma_2 + \sigma_2\sigma_3 + \sigma_1\sigma_3 \tag{2.125}$$

$$I_3 = \sigma_1\sigma_2\sigma_3 \tag{2.126}$$

Stress Tensor Decomposition

As any other tensor, stress tensors can be decomposed into a *spherical tensor* and a *deviator*. For example applied to the Cauchy stress tensor, this decomposition yields:

$$\sigma_{ij} = S_{ij} + \frac{1}{3}\left(\sigma_{11} + \sigma_{22} + \sigma_{33}\right)\, \delta_{ij} = S_{ij} - p\, \delta_{ij} \tag{2.127}$$

where the spherical tensor $-p\, \delta_{ij}$ represents the *hydrostatic pressure* that can be expressed through the first invariant of the total stress tensor:

$$-p = \frac{1}{3}\, \sigma_{kk} = \frac{1}{3}\, I_1 \tag{2.128}$$

The invariants J_i of the stress deviator S_{ij} are formulated accordingly to the original stress tensor invariants as:

$$J_1 = S_1 + S_2 + S_3 = 0$$

$$J_2 = \frac{1}{2}\left(S_1^2 + S_2^2 + S^2\right) \tag{2.129}$$

$$J_3 = S_1 S_2 S_3 \tag{2.130}$$

Later we will see that the decomposition of the stress tensor into a hydrostatic and a deviatoric part is useful with regard to the following two aspects:

- The formulation of plasticity and failure models when (hydrostatic) pressure dependency or Independence is of interest
- Modelling of shock wave effects where the bulk compressive material behaviour necessarily needs to be modelled by hydrostatic stress contributions in the form of specific nonlinear equations of states.

2.3 Descriptions of Static Equilibrium

2.3.1 Direct Formulation of Equilibrium

Equilibrium of mechanical systems can be formulated in various ways. A direct formulation is achieved through application of acting forces to the boundaries of an infinitesimal element. Since equilibrium is described by acting forces, however, we are interested in stresses, the forces are related to the boundary surfaces. Figure 2.8 illustrates the direct formulation of equilibrium with a two-dimensional example.

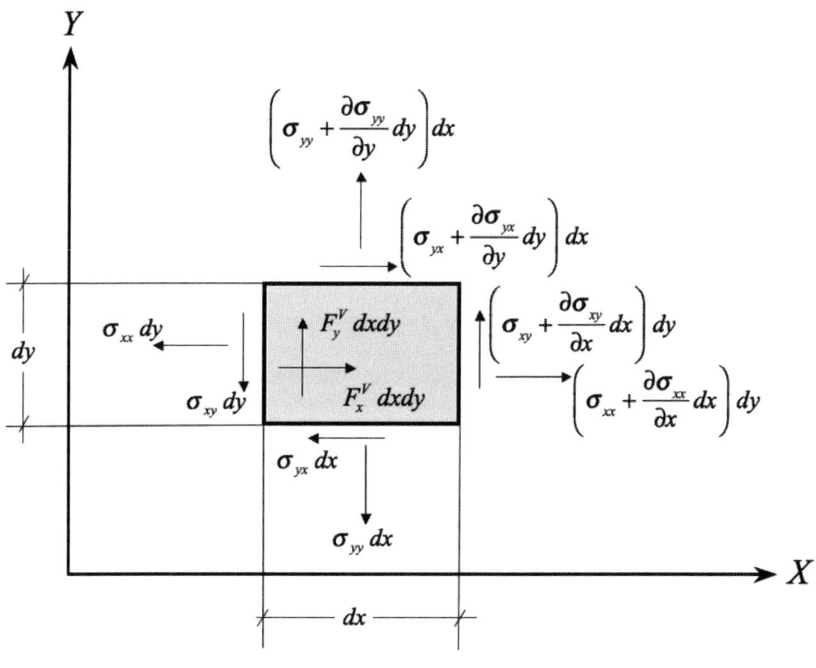

Fig. 2.8. Direct formulation of equilibrium for a two-dimensional infinitesimal element.

Setting the surface and volume forces illustrated in Figure 2.8 into equilibrium delivers the equilibrium conditions in matrix form:

$$\begin{bmatrix} \frac{\partial}{\partial x} & 0 & \frac{\partial}{\partial y} \\ 0 & \frac{\partial}{\partial y} & \frac{\partial}{\partial x} \end{bmatrix} \begin{bmatrix} \sigma_{xx} \\ \sigma_{yy} \\ \sigma_{xy} \end{bmatrix} = - \begin{bmatrix} F_x^V \\ F_y^V \end{bmatrix} \tag{2.131}$$

A hanging bar with constant cross sectional area A, constant Young's modulus E and loaded by its own weight is an example that reduces the complexity of the system to a one-dimensional problem. The related equilibrium for the deformed configuration is described by:

$$\frac{\partial \sigma_{xx}}{\partial x_x} = -F_x \tag{2.132}$$

For the distributed gravitational load in longitudinal direction we can write:

$$P(x) = m\,g\,L\left(1 - \frac{x}{L}\right) \tag{2.133}$$

leading to the areal forces:

$$F_x = \frac{P(x)}{A} \tag{2.134}$$

Assuming linear elastic behaviour $\sigma_{xx} = E\,\varepsilon_{xx} = E\,\frac{\partial u_x}{\partial x_x}$ equation (2.132) turns into

$$E\,A\,\frac{\partial^2 u_x}{\partial x^2} = -P(x) \tag{2.135}$$

relating the deformation state to the applied load at equilibrium.

2.3.2 Calculus of Variations

Before we can derive an alternative approach to formulate equilibrium, a necessary and extremely useful tool needs to be briefly introduced which is the *calculus of variations*. Later we will formulate functionals, i.e. functions of functions, which under certain conditions describe equilibrium states. These particular conditions are stationary values of the functionals. Accordingly, we will try to find these stationary, or more precisely minimum values, by formulating minimization problems which in turn can be solved with the calculus of variation.

Functional Form to be Solved

Minimization problems in continuous systems involve functionals $F(x) = F\,(x,\, f_i(x))$, often in integral forms, which can for example be of the type:

$$\Pi = \int_{x_0}^{x_1} F\,(x,\, f(x),\, f'(x))\,\mathrm{d}\,x = \int_{x_0}^{x_1} F\,(x,\, y,\, y')\,\mathrm{d}\,x \tag{2.136}$$

or, more general:

$$\Pi = \int_{x_0}^{x_1} F\,(x,\, y_i(x))\,\mathrm{d}\,x \qquad (2.137)$$

The formulation of the minimization problem in the interval $(x_0,\, x_1)$

$$\min_{y_i} = \min_{y_i} \int_{x_0}^{x_1} F\,(x,\, y_i(x))\;\mathrm{d}\,x \qquad (2.138)$$

shall involve the boundary conditions

$$y_i(x_0) = y_0 \qquad y_i(x_1) = y_1 \qquad (2.139)$$

If the functional F contains a function y and one or more of its derivatives, as for example in (2.136), the minimization problem (2.138) reduces to finding the specific function $\bar{y}$ for which the functional takes a minimum. In accordance to the classical differential calculus, the minimizing function $\bar{y}$ describes a local minimum if any slightly different function y in a vicinity U of $\bar{y}$ results in a larger functional. Mathematically expressed, we look for a solution $\bar{y}$ for which:

$$\Pi(\bar{y}) \leq \Pi y \quad \forall\, y \in U \qquad (2.140)$$

Basic Concept of the Variational Calculus

There are mathematically more profound descriptions of the variational calculus compared to what can be provided in the context of this brief introduction, e.g. in Lanczos [243]. However, what is needed to understand the engineering applications of variations in the framework of finite element methods to be discussed in section 5.7, is the following basic strategy of the calculus:

- First, we assume a certain function $\bar{y}$ to minimize the function, in other words to be a solution.
- Then a *variation* of this function is established by adding a virtual perturbation $\delta\, y$. This leads to the varied function $y = \bar{y} + \delta\, y$
- The perturbation $\delta\, y = \epsilon\, \eta(x)$ is arbitrary in its shape, expressed by a function $\eta(x)$ and size, expressed by ϵ.
- No matter how small the variation may be, a minimum of the functional form Π is always characterized by $\bar{\Pi}\,(\bar{y} + \epsilon\, \eta(x)) \geq \Pi\,(\bar{y})$, where $\bar{\Pi}$ is the varied functional.
- In the framework of differential calculus we demand for a local minimum of a function its first derivative to become zero and its second derivative to be positive. A related formulation for functionals is used as an alternative to the statement given in the last bullet: It is to demand for the first derivative, now called *first variation* of the functional $\delta\, \bar{\Pi}\,(\bar{y} + \epsilon\, \eta(x))$ to become zero for the minimizing function $\bar{y}$ which is characterized by $\epsilon = 0$.

- With the formulation :

$$\delta \bar{\Pi}\left(\bar{y}+\epsilon\,\eta(x)\right)\big|_{\epsilon=0}=0 \tag{2.141}$$

a criterion for the solution $\bar{y}$ can be found. Further development of (2.141), e.g. by Taylor series expansion, leads to a direct criterion for the function $\bar{y}$.

Application of the Variational Calculus

This procedure shall be followed now performing an example application. Let us consider that a minimum shall be found for the functional form:

$$\Pi=\int_{x_0}^{x_1} F\left(x,\,y,\,y'\right)\,\mathrm{d}\,x \tag{2.142}$$

and assume that $\bar{y}(x)$ is a minimizing solution.

Variation of the Assumed Solution

A virtual perturbation or *variation* $\delta\,y$ changing the minimizing function $\bar{y}$ to get:

$$y=\bar{y}+\delta\,y \tag{2.143}$$

shall be described by

$$\delta\,y=\epsilon\,\eta(x) \tag{2.144}$$

This variation is assumed to be of an arbitrarily small magnitude ϵ multiplied by a function $\eta_i(x)$ which is of arbitrary shape but fulfilling the essential boundary condition (2.139). Thus, we demand for the variation that

$$\delta\,y(x_0)=\delta\,y(x_1)=0 \tag{2.145}$$

With (2.144) it can be shown that

$$\delta\left(\frac{\partial\,y}{\partial\,x}\right)=\frac{\partial}{\partial\,x}\,\delta\,y \tag{2.146}$$

Figure 2.9 illustrates the nature of a variation.

In the same manner the first derivative is varied to become:

$$y'=\bar{y}'+\delta\,y' \tag{2.147}$$

with

$$\delta\,y'=\epsilon\,\eta'(x) \tag{2.148}$$

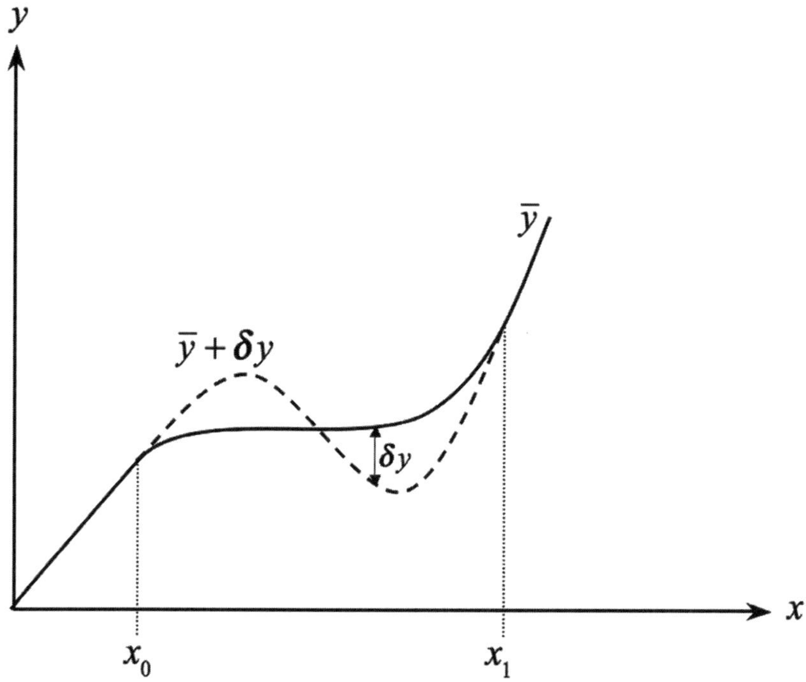

Fig. 2.9. Presumably known function $\bar{y}$ and its varied form $y = \bar{y}+\delta\, y$. The variation $\delta\, y$ reflects the size of the arbitrary modification in $\bar{y}$.

Variation of the Functional

Implementing the varied function $\delta\, y$ into the integrand of the functional form (2.142) yields a varied functional:

$$\bar{F} = F\left(x,\, \bar{y} + \delta\, y,\, \bar{y}' + \delta\, y'\right) \tag{2.149}$$

and thus a variation of the functional form:

$$\bar{\Pi} = \int_{x_0}^{x_1} F\left(x,\, \bar{y} + \delta\, y,\, \bar{y}' + \delta\, y'\right) \mathrm{d}\, x \tag{2.150}$$

Condition for an Extremum - Euler-Lagrange Equation

The varied functional is now a function of the perturbation size ϵ since:

$$\bar{\Pi}\left(\epsilon\right) = \Pi\left(\bar{y} + \epsilon\, \eta(x)\right) = \int_{x_0}^{x_1} F\left(x,\, \bar{y} + \epsilon\, \eta(x),\, \bar{y}' + \epsilon\, \eta'(x)\right) \mathrm{d}\, x \tag{2.151}$$

For the function $\bar{y}$, i.e. at $\bar{\Pi}\left(\epsilon = 0\right)$, to be a minimum of the the functional, we have to demand that

$$\delta \bar{\Pi}\,(\epsilon = 0) = 0 \tag{2.152}$$

and, thus, that

$$\begin{aligned}
\delta \bar{\Pi}\,(\epsilon) &= \int_{x_0}^{x_1} \frac{\mathrm{d}\,F}{\mathrm{d}\,\epsilon}\,\mathrm{d}\,x = \int_{x_0}^{x_1} \left[\frac{\partial\,F}{\partial\,y}\,\eta(x) + \frac{\partial\,F}{\partial\,y'}\,\eta'(x)\right]\,\mathrm{d}\,x \\
&= \int_{x_0}^{x_1} \left[\frac{\partial\,F}{\partial\,y}\,\eta(x) - \frac{\mathrm{d}}{\mathrm{d}\,x}\left(\frac{\partial\,F}{\partial\,y'}\right)\,\eta(x)\right]\,\mathrm{d}\,x + \eta(x)\,\frac{\partial\,F}{\partial\,y'}\bigg|_{x_0}^{x_1} \\
&= \int_{x_0}^{x_1} \eta(x)\left[\frac{\partial\,F}{\partial\,y} - \frac{\mathrm{d}}{\mathrm{d}\,x}\left(\frac{\partial\,F}{\partial\,y'}\right)\right]\,\mathrm{d}\,x
\end{aligned} \tag{2.153}$$

is fulfilled for $\epsilon = 0$ if

$$\frac{\partial\,F}{\partial\,y} - \frac{\mathrm{d}}{\mathrm{d}\,x}\left(\frac{\partial\,F}{\partial\,y'}\right) = 0 \tag{2.154}$$

Equation (2.154) is called *Euler-Lagrange equation* for the functional form (2.142). Generally, functions y_i that fulfill the Euler-Lagrange equation related to the respective functional form are at the same time a solutions for a minimization problem of the type (2.138)[13].

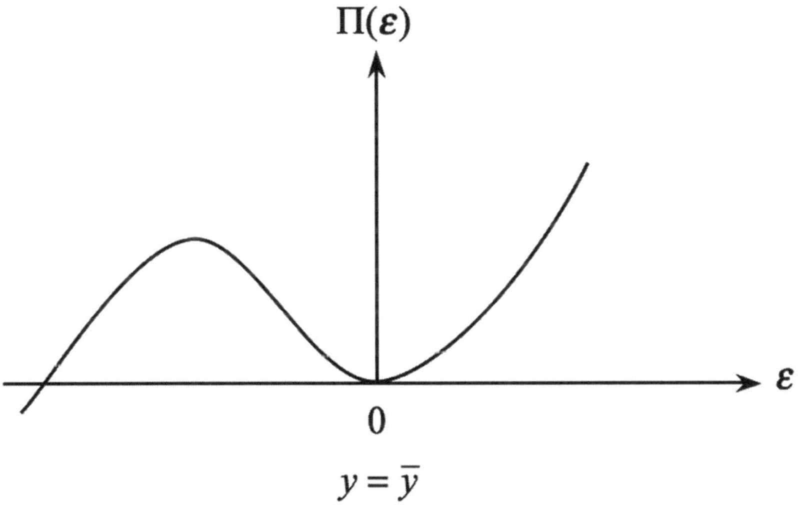

Fig. 2.10. Condition for a local minimum of the functional Π at the minimizing $y = \bar{y}$ which is represented in the $\Pi - \epsilon$ plane by $\epsilon = 0$.

[13] Further Euler-Lagrange equations for various minimization problems and related functional forms can for example be found in Dow [114].

Alternative Approach to Find the Euler-Lagrange Equation

Often, an alternative approach to find the Euler-Lagrange equation is given. It illustrates the nature of the solution by comparison of the integrands in the original and the varied functionals. To quantify that difference, we expand a Taylor series about the solution $\bar{y}$ and $\bar{y}'$:

$$\begin{aligned} F(x, \bar{y}+\delta y, \bar{y}'+\delta y') = F(x, \bar{y}, \bar{y}') & + \frac{\partial F(x, \bar{y}, \bar{y}')}{\partial y}\delta y + \frac{\partial F(x, \bar{y}, \bar{y}')}{\partial y'}\delta y' \\ & + \frac{1}{2}\left[\frac{\partial^2 F(x, \bar{y}, \bar{y}')}{\partial y^2}\delta y^2 + \frac{\partial^2 F(x, \bar{y}, \bar{y}')}{\partial y'^2}\delta y'^2 \right. \\ & \left. + \frac{\partial^2 F(x, \bar{y}, \bar{y}')}{\partial y\, \partial y'}\delta y\, \delta y'\right] \dots \end{aligned} \tag{2.155}$$

Since the variations in δy and $\delta y'$ shall become infinitely small in the vicinity of the solution, we can neglect the quadratic terms in (2.155). Thus, the difference

$$\Delta F = \bar{F} - F = \frac{\partial F}{\partial y}\delta y + \frac{\partial F}{\partial y'}\delta y' + \frac{1}{2}\left[\frac{\partial^2 F}{\partial y^2}\delta y^2 + \dots\right] + \dots \tag{2.156}$$

can be approximated by

$$\Delta F = \frac{\partial F}{\partial y}\delta y + \frac{\partial F}{\partial y'}\delta y' = \delta F \tag{2.157}$$

where δF is called the *first variation* of the functional leading to the first variation in our functional form:

$$\delta \Pi = \int_{x_0}^{x_1} \delta F \,\mathrm{d}x = \int_{x_0}^{x_1} \left(\frac{\partial F}{\partial y}\delta y + \frac{\partial F}{\partial y'}\delta y'\right) \mathrm{d}x \tag{2.158}$$

Integration by parts yields for the second term in the integrand of (2.158):

$$\int_{x_0}^{x_1} \left(\frac{\partial F}{\partial y'}\delta y'\right) \mathrm{d}x = \delta y \left.\frac{\partial F}{\partial y'}\right|_{x_0}^{x_1} - \int_{x_0}^{x_1} \delta y \frac{\mathrm{d}}{\mathrm{d}x}\left(\frac{\partial F}{\partial y'}\right) \mathrm{d}x \tag{2.159}$$

The variation δy vanishes at the boundaries x_0 and x_1 according to (2.145) and, therefore, we find:

$$\delta \Pi = \int_{x_0}^{x_1} \delta y \left[\frac{\partial F}{\partial y} - \frac{\mathrm{d}}{\mathrm{d}x}\left(\frac{\partial F}{\partial y'}\right)\right] \mathrm{d}x = 0 \tag{2.160}$$

where the minimum requirement is now expressed by the equality with zero. The variations δy and $\delta y'$ are arbitrary and, thus, we can again write as condition for the minimum in the functional form the *Euler-Lagrange equation* already found in (2.154):

$$\frac{\partial F}{\partial y} - \frac{\mathrm{d}}{\mathrm{d}\,x}\left(\frac{\partial F}{\partial y'}\right) = 0 \tag{2.161}$$

Up to here, we found a partial differential equation called Euler-Lagrange equation that describes a criterion for possible solutions $y(x)$ of the minimization problem (2.138). This finding enables us to describe a criterion for static equilibrium which will be topic of the next subsection.

2.3.3 Equilibrium Formulated as Variational Problem

With the calculus of variations we can now follow a more general way to find equilibrium states for continuous systems. Usually, for that purpose principles of mechanics are formulated[14] , e.g the principle of virtual work or the principle of minimum potential energy. Inherent to that approach is the formulation of a minimization problem. In case of the *principle of minimum potential energy* the sum of internal energy Π_{int} and external energy Π_{ext} of a system is requested to take a minimum as condition for a deformation $\mathbf{u}$ state that represents equilibrium:

$$\min_{\mathbf{u}} \Pi = \min_{\mathbf{u}} (\Pi_{int}) \tag{2.162}$$

For a discrete system, e.g. a combination of m masses and n springs, the total potential energy is described by a function, whereas in case of a continuous system it is formulated by a functional.

With a constitutive equation $\sigma = \varepsilon^T \mathbf{E}$ the internal component of the total potential energy we formulate the strain energy function integrated over a bodies domain Ω :

$$\Pi_{int} = \int_{Omega} \sigma\, \varepsilon \,\mathrm{d}\Omega = \int_{Omega} \varepsilon^T\, \mathbf{E}\, \varepsilon \,\mathrm{d}\Omega \tag{2.163}$$

whereas the external contribution is expressed by the negative product of acting forces $\mathcal{F}$ and related deformations:

$$\Pi_{int} = -\int_{\Omega} \mathcal{F}\,\mathbf{u}\,\mathrm{d}\Omega \tag{2.164}$$

In a particular application, the forces are formulated more precisely in terms of surface and volume forces.

Coming back to the example of a hanging bar of length L under its own weight, the internal energy is expressed by the strain energy:

$$\Pi_{int} = \frac{1}{2}\int_0^L A\,E\left(\frac{\partial u_x}{\partial x}\right)^2 \mathrm{d}\,x \tag{2.165}$$

---

[14] A comprehensive introduction to the principles of mechanics and the related calculus of variations can be found in Lanczos [243].

On the other hand, the work of the gravitational load to the deformation yields the external energy

$$\Pi_{ext} = \int_0^L P(x)\, u_x \mathrm{d}\, x = \int_0^L m\, g\, L \left(1 - \frac{x}{L}\right) u_x \mathrm{d}\, x \tag{2.166}$$

Thus, the principle demands for the total energy Π to become a minimum:

$$\min_u \Pi = \min_u \int_0^L \left[\frac{1}{2} A\, E \left(\frac{\partial\, u_x}{\partial\, x}\right)^2 - P(x)\, u_x\right] \mathrm{d}\, x \tag{2.167}$$

Using the variational calculus means to apply the Euler-Lagrange equation (2.154) to the functional in the integrand

$$F = \frac{1}{2} A\, E \left(\frac{\partial\, u_x}{\partial\, x}\right)^2 - P(x)\, u_x \tag{2.168}$$

With the partial differentials

$$\frac{\partial\, F}{\partial\, u} = -P(x) \quad \text{and} \quad \frac{\partial\, F}{\partial\, u'} = A\, E\, \frac{\partial\, u_x}{\partial\, x} \tag{2.169}$$

the Euler-Lagrange equation becomes:

$$\frac{\partial\, F}{\partial\, u} - \frac{\mathrm{d}}{\mathrm{d}\, x}\left(\frac{\partial\, F}{\partial\, u'}\right) = -P(x) - A\, E\, \frac{\mathrm{d}^2\, u}{\mathrm{d}\, x^2} = 0 \tag{2.170}$$

and thus:

$$A\, E\, \frac{\mathrm{d}^2\, u}{\mathrm{d}\, x^2} + P(x) = 0 \tag{2.171}$$

which is identical with the result in equation (2.135) achieved by the direct equilibrium formulation.

The Euler-Lagrange equation, generalized using a differential operator L as

$$L\, [u(x)] - f(x) = 0 \tag{2.172}$$

together with conditions on the boundaries of a domain U

$$L|_U\, [u(x)] = u(x)|_U \tag{2.173}$$

forms a boundary value problem. Since these are generally not approachable to closed form solutions, approximative solutions are needed. Numerical methods, specifically finite element methods, providing approximations for the minimization problems formulated with the variational calculus will be introduced in section 5.7.

With respect to dynamic processes, a certain form of static equilibrium will always be achieved in a final deformation state. However, investigation of transient events is primarily focused on the process path which finally leads to that equilibrium. To describe the process rather than solely the equilibrium, all intermediate states of loading and related deformation states need to be described. For high dynamic processes this approach necessarily includes effects of wave propagation. The physical way of describing this path is the application of the conservation equations for mass, momentum and energy (section 2.4) which lead to an equation of motion instead of equilibrium.

In order to find an approximation for the equation of motion, *Hamilton's principle of stationary action* will be applied (section 2.5.3) equivalently to utilizing the principal of minimum potential energy for equilibrium.

2.4 Conservation Equations

Conservation of mass, momentum and energy is an empirical observation that is mathematically stated in the material independent universal laws of physics called *conservation equations* or *first principals.* For an arbitrary conserved volume specific variable u, it means that in a closed system u stays constant. Formulated for an open control volume V, conservation of u is fulfilled when the net flux of u through the surface of V equals the time rate of change of u inside the control volume. The condition can either be formulated for finite control volume V or in a differential description for an infinitesimal volume element $\mathrm{d}V$. Additionally, both versions may be formulated in a material or in a spatial description providing four individual equations for each conserved quantity u. The difference in material and spatial descriptions results from observing the flow field of a fluid by either using a moving material volume of the fluid or by the use of a control volume that is fixed in space. The first option is achieved by the use of Lagrangean coordinates (see chapter 2.1.1) leading to a discretizing grid that contains constant parcels of the fluid and deforms with it. A spatial description, on the other hand, leads to an Eulerian grid fixed in space in which the fluid material is continually changing. Formulated in a material description, the resulting conservation equations are called *non-conservative type.* In a spatial description they are of the so called *conservative type.*

2.4.1 Four Ways of Describing Conservation

Demanding conservation of a volume specific variable in Lagrangean coordinates $U(^0\mathbf{X},\ t)$ within a finite volume leads to the *integral non-conservative* formulation:

$$\frac{\mathrm{D}\,U}{\mathrm{D}\,t} = \int_{^0V} \frac{\mathrm{d}\,U}{\mathrm{d}\,t} + U\ \nabla\mathbf{v}\ \mathrm{d}^0V = 0 \tag{2.174}$$

whereas in an Eulerian configuration, demanded conservation leads to the *integral conservative* form:

$$\frac{\mathrm{D}\,u}{\mathrm{D}\,t} = \int_{^tV} \frac{\mathrm{d}\,u}{\mathrm{d}\,t} + \nabla(u\,\mathbf{v})\ \mathrm{d}^tV = 0 \tag{2.175}$$

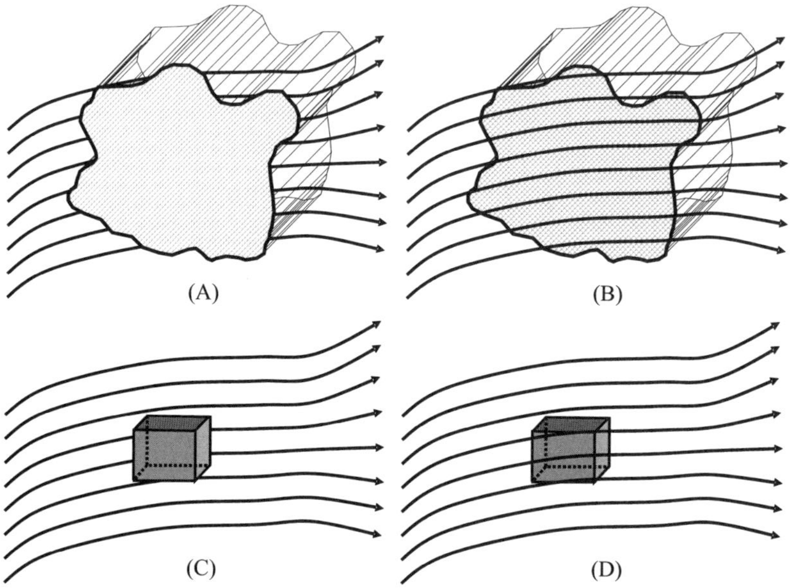

Fig. 2.11. Options of defining a control volume: (A) Body fixed (Lagrangean) finite, (B) Space fixed (Eulerian) finite, (C) Space fixed (Eulerian) infinitesimal, (D) Body fixed (Lagrangean) infinitesimal.

Equation (2.174) can equally be demanded for an infinitesimal volume dV leading to the *differential non-conservative* form:

$$\frac{\mathrm{D}\,U}{\mathrm{D}\,t} + U\,\frac{\partial v_i}{\partial X_i} = \frac{\partial\,U}{\partial t} + U\,\frac{\partial v_i}{\partial X_i} = 0 \tag{2.176}$$

In case of an Eulerian configuration, the substantial derivative (2.63) leads to the differential conservative form of (2.176) as:

$$\frac{\mathrm{d}\,u}{\mathrm{d}\,t} + u\,\frac{\partial v_i}{\partial x_i} = \frac{\partial u}{\partial t} + \frac{\partial u}{\partial x_i}\,\frac{\partial x_i}{\partial t} + u\frac{\partial v_i}{\partial x_i} = \frac{\partial u}{\partial t} + \frac{\partial(u\,v_i)}{\partial x_i} = 0 \tag{2.177}$$

giving the *differential conservative* formulation:

$$\frac{\partial u}{\partial t} + \nabla(u\mathbf{v}) = 0 \tag{2.178}$$

The resulting four definitions are collected in table 2.4 and illustrate in Fig. 2.11.

Table 2.4. Formulations of a conservation equation for a mass specific quantity u

	Finite control volume (Global form)	**Infinites. control volume** (Local form)
Spatial (Eulerian) Configuration	Integral, conservative $\int_V \frac{\partial u}{\partial t} + \frac{\partial u\, v_i}{\partial x_j}\,\mathrm{d}V = 0$	Differential, conservative $\frac{\partial u}{\partial t} + \frac{\partial(u\ v_i)}{\partial x_j} = 0$
Material (Lagrangean) Configuration	Integral, non-conservative $\int_V \frac{\partial u}{\partial t} + u\,\frac{\partial\, v_i}{\partial X_j}\,\mathrm{d}V = 0$	Differential, non-conservative $\frac{\partial u}{\partial t} + u\frac{\partial v_i}{\partial X_j} = 0$

Mathematically the four formulations can be transformed into each other without loss of accuracy. However, the discretized versions of the equations, as implemented in finite methods codes, may deliver different results. Depending on the type of problem that is to be solved, e.g. fluid flow or structural deformation, an adequate formulation needs to be chosen.

2.4.2 Conservation of Mass

In order to guarantee mass conservation, the net flux of mass through the surface of the control volume V must equal the time rate of change of mass inside V. The integral equations of mass conservation are in conservation and in non-conservation form, respectively:

$$\int_{^tV} \frac{\partial \rho}{\partial t} + \frac{\partial \rho v_i}{\partial^t x_j}\mathrm{d}^tV = 0 \tag{2.179}$$

$$\int_{^0V} \frac{\partial \rho}{\partial t} + \rho\frac{\partial v_i}{\partial^0 X_j}\mathrm{d}^0V = 0 \tag{2.180}$$

The respective differential conservative form of the *continuity equation* is:

$$\frac{\partial \rho}{\partial t} + \frac{\partial \rho v_i}{\partial^t x_j} = 0 \tag{2.181}$$

and its non-conservative version:

$$\frac{\partial \rho}{\partial t} + \rho \frac{\partial v_i}{\partial^0 X_j} = 0 \tag{2.182}$$

2.4.3 Conservation of Momentum

Newton's second law is the basis to formulate the criterion for the conservation of momentum. The resulting *equation of motion* can be formulated with regard to linear momentum $\mathbf{I}(t)$

$$\mathbf{I}(t) = \int_V \rho \mathbf{v} \mathrm{d}V \tag{2.183}$$

or angular momentum $\mathbf{D}(t)$

$$\mathbf{D}(t) = \int_V ({}^0\mathbf{X} \otimes \rho \mathbf{v}) \mathrm{d}^0 V \tag{2.184}$$

where the latter is also called *moment of momentum.*

Linear Momentum

Conservation of linear momentum balances the time rate of change in momentum of an arbitrary body to the resultant forces:

$$\frac{\partial}{\partial t} \int_V \rho \mathbf{v} \mathrm{d}V = \int_S \mathbf{f} \mathrm{d}S + \int_V \rho\, \mathbf{b} \mathrm{d}V \tag{2.185}$$

where $\mathbf{f}$ are forces acting on differential surface elements $\mathrm{d}S$ and $\mathbf{b}$ are body forces on differential volumes. Using the Gaussian theorem, the surface integral over forces $\mathbf{f}$ can be written as volume integral

$$\int_S \mathbf{f} \mathrm{d}S = \int_S \sigma \mathbf{n} \mathrm{d}S = \int_V \frac{\partial \sigma}{\partial^{\mathbf{t}} \mathbf{x}} \mathrm{d}^t V \tag{2.186}$$

The *integral conservative* formulation of the *linear momentum* conservation reads:

$$\int_{{}^tV} \left\{ \frac{\partial \rho \mathbf{v}}{\partial t} + \frac{\partial \rho \mathbf{v}}{\partial^{\mathbf{t}} \mathbf{x}} \mathbf{v} - \frac{\partial \sigma}{\partial^{\mathbf{t}} \mathbf{x}} - \rho\, \mathbf{b} \right\} \mathrm{d}\, {}^tV = 0 \tag{2.187}$$

To find the material description for the conservation of linear momentum, equation (2.187) must be formulated with respect to initial conditions. With (2.25), the body forces $\mathcal{B}$ in material description are:

$$\mathcal{B} = \mathbf{b} \frac{d\, {}^tV}{d\, {}^0V} = \mathbf{bJ} \tag{2.188}$$

Transformation of the Cauchy stress tensor σ to the reference condition is achieved through the nominal stresses $\mathbf{P}$ (2.102):

$$\int_V \frac{\partial \sigma}{\partial^{\mathbf{t}} \mathbf{x}} \mathrm{d}^t V = \int_V \frac{\mathbf{J}^{-1}\mathbf{F}\,\partial \mathbf{P}}{\mathbf{F}\partial^{\mathbf{0}}\mathbf{X}} \mathbf{J} \mathrm{d}^0 V = \int_V \frac{\partial \mathbf{P}}{\partial^{\mathbf{0}}\mathbf{X}} \mathrm{d}^0 V \tag{2.189}$$

which then leads to the *integral non-conservative* form:

$$\int_{{}^0V} \left\{ \rho_0 \frac{\mathrm{d}\mathbf{v}}{\mathrm{d}t} - \frac{\partial \mathbf{P}}{\partial^{\mathbf{0}}\mathbf{X}} - \rho_0\, \mathcal{B} \right\} \mathrm{d}\, {}^0V = 0 \tag{2.190}$$

The differential non-conservative formulation of linear momentum is:

$$\rho_0 \frac{\mathrm{d}\mathbf{v}}{\mathrm{d}t} - \frac{\partial \mathbf{P}}{\partial^{\mathbf{0}}\mathbf{X}} - \rho_0\, \mathcal{B} = 0 \tag{2.191}$$

or

$$\rho_0 \dot{\mathbf{v}} - \nabla \cdot \mathbf{P} - \rho_0\, \mathcal{B} = 0 \tag{2.192}$$

and the differential conservative form:

$$\frac{\partial \rho \mathbf{v}}{\partial t} + \frac{\partial \rho \mathbf{v}}{\partial^t \mathbf{x}} \mathbf{v} - \frac{\partial \sigma}{\partial^t \mathbf{x}} - \rho\, \mathbf{b} = 0 \tag{2.193}$$

Equation (2.193) represents the conservative form of the *Navier-Stokes Equations* and can be expressed in index writing using the Levi-Civita symbol ϵ^{ijk} (2.3) and the body forces $b_i = \rho\, f_i$:

$$\frac{\partial \rho v_i}{\partial t} + \frac{\partial \epsilon^{ijk} v_j \rho v_k}{\partial^t x_j} - \frac{\partial \sigma_{ij}}{\partial x_i} - \rho\, f_i = 0 \tag{2.194}$$

where

$$\epsilon^{ijk} v_j = \begin{pmatrix} 0 & -v_z & v_y \\ v_z & 0 & -v_z \\ -v_y & v_z & 0 \end{pmatrix} \tag{2.195}$$

Angular Momentum

To conserve the moment of momentum, the time rate of change in the angular momentum with regard to an arbitrary reference point ${}^t\mathbf{x_i}$ has to balance the resultant moments. The *integral Eulerian* form of angular momentum conservation is:

$$\int_{{}^tV} \left\{ \epsilon_{ijk}\, {}^t x_j \left(\rho \frac{\partial v_j}{\partial t} - \frac{\partial \sigma_{jk}}{\partial^t x_j} - \rho\, f_j \right) \right\} + \epsilon_{ijk}\, \sigma_{jk}\, \mathrm{d}\, {}^tV = 0 \tag{2.196}$$

For the *differential conservative* formulation follows that:

$$\epsilon_{ijk}\, x_j \left(\frac{\partial \rho \mathbf{v}}{\partial t} + \frac{\partial \rho \mathbf{v}}{\partial^t \mathbf{x}} \mathbf{v} - \frac{\partial \sigma_{jk}}{\partial x_j} - b_j \right) + \epsilon_{ijk}\, \sigma_{jk} = 0 \tag{2.197}$$

As a consequence of the linear momentum conservation (2.193) the part of equation (2.197) set in brackets must equal zero and, therefore, an equivalent form of the differential non-conservative can be written as:

$$\epsilon_{ijk}\,\sigma_{jk} = 0 \tag{2.198}$$

which, written in full length, means

$$(\sigma_{23} - \sigma_{32}) + (\sigma_{31} - \sigma_{13}) + (\sigma_{12} - \sigma_{21}) = 0 \tag{2.199}$$

and thus is demanding for the symmetry of the Cauchy stress tensor σ :

$$\sigma_{ij} = \sigma_{ji} \tag{2.200}$$

An enlightening and much more fundamental combined derivation of both stress tensors and the balance equations of mechanics can be found in Haupt [165].

2.4.4 Conservation of Energy

In the context of this book, energy terms are denoted by the letters E and e for absolute and mass specific energies, respectively. Therefore, we write

$$E_{tot} = \int \rho e_{tot} \mathrm{d}V \tag{2.201}$$

where the index *tot* accounts for total energy consisting of internal energy E and kinetic energy E_{kin}:

$$E_{tot} = E + E_{kin} \tag{2.202}$$

Accordingly the time rate of change in the total specific energy can be written as:

$$\rho \frac{\mathrm{d}e_{tot}}{\mathrm{d}t} = \rho \frac{\mathrm{d}e}{\mathrm{d}t} + \frac{\mathrm{d}}{\mathrm{d}t}\left(\frac{1}{2}\rho \mathbf{v}^2\right) = \rho \frac{\mathrm{d}e}{\mathrm{d}t} + \rho \mathbf{v} \frac{\mathrm{d}\mathbf{v}}{\mathrm{d}t} \tag{2.203}$$

The first law of thermodynamics demands the sum of mechanical and thermal energy to be constant in a closed system. Accordingly, any change in total specific energy e_{tot} needs to be balanced by the power performed by external and body forces $\nabla \cdot (\sigma \mathbf{v}) + \mathcal{B}\,\mathbf{v}$ and the rate of heat supply $\dot{q}$ which consists of the specific heat flux $\mathbf{h}$ and radiation induced heat supply $\dot{\tilde{q}}$ as:

$$\dot{q} = -\nabla \cdot \mathbf{h} + \rho \dot{\tilde{q}} \tag{2.204}$$

Thus, the integral version of the non-conservative form of energy conservation is:

$$\rho \frac{\partial}{\partial t} \int_{{}^0V} e_{tot}\, \mathrm{d}^0V - \int_{{}^0V} \frac{\partial(\mathbf{P}\mathbf{v})}{\partial\, {}^0\mathbf{X}}\, \mathrm{d}^0V - \int_{{}^0V} \mathcal{B}\,\mathbf{v} + \rho \dot{q}\, \mathrm{d}^0V = 0 \tag{2.205}$$

and the integral conservative formulation reads:

$$\int_{{}^tV} \frac{\partial \rho e_{tot}}{\partial t} + \frac{\partial(\rho e_{tot}\mathbf{v})}{\partial\, {}^t\mathbf{x}}\, \mathrm{d}\,{}^tV - \int_{{}^tV} \frac{\partial(\sigma \mathbf{v})}{\partial\, {}^t\mathbf{x}}\, \mathrm{d}\,{}^tV - \int_{{}^tV} \mathbf{b}\,\mathbf{v} + \rho \dot{q}\, \mathrm{d}^tV = 0 \tag{2.206}$$

The differential Lagrangean version yields:

$$\rho \frac{\mathrm{d}\, e_{tot}}{\mathrm{d}\, t} - \nabla \cdot (\mathbf{P}\mathbf{v}) - \mathcal{B}\, \mathbf{v} - \rho \dot{q} = 0 \tag{2.207}$$

which can be formulated with the total energy split into kinetic and internal energy according to equation (2.203) as:

$$\rho \frac{\mathrm{d}\, e}{\mathrm{d}\, t} + \rho \mathbf{v} \frac{\mathrm{d}\mathbf{v}}{\mathrm{d}t} - \nabla \cdot (\mathbf{P}\mathbf{v}) - \mathcal{B}\, \mathbf{v} - \rho \dot{q} = 0 \tag{2.208}$$

If the chain rule is applied to $\nabla \cdot (\mathbf{P}\mathbf{v})$:

$$\nabla \cdot (\mathbf{P}\mathbf{v}) = \nabla \cdot \mathbf{P} \cdot \mathbf{v} + \mathbf{P} : \nabla \otimes \mathbf{v} \tag{2.209}$$

and the equation of motion (2.192) is used to replace

$$\nabla \cdot \mathbf{P} \cdot \mathbf{v} = \rho \mathbf{v} \frac{\mathrm{d}\mathbf{v}}{\mathrm{d}t} - \mathcal{B}\, \mathbf{v} \tag{2.210}$$

another version of (2.207) can be found in:

$$\rho \frac{\mathrm{d}\, e}{\mathrm{d}\, t} - \mathbf{P} : \nabla \otimes \mathbf{v} - \rho \frac{\mathrm{d}\, q}{\mathrm{d}\, t} = 0 \tag{2.211}$$

where equation (2.211) in contrast to (2.207) contains specific internal energy only! It is the formulation of energy conservation which is usually implemented in wave propagation codes.

The Eulerian formulations of energy conservation are:

$$\frac{\partial\, \rho\, e_{tot}}{\partial\, t} + \nabla \cdot (\rho\, e_{tot} \mathbf{v}) - \nabla \cdot (\sigma \mathbf{v}) - \mathbf{b}\mathbf{v} - \rho \dot{q} = 0 \tag{2.212}$$

2.4.5 Compressed Formulation of the Conservation Equations

To write the conservation equations for mass, momentum and energy in a compressed formulation, we can define a density matrix $\mathbf{u}({}^t\mathbf{x}, t)$ containing mass densities ρ, momentum densities $\rho\, \mathbf{v}$ and energy densities $\rho\, e_{tot}$:

$$\mathbf{u}({}^t\mathbf{x}, t) = \begin{pmatrix} \rho \\ \rho\, \mathbf{v} \\ \rho\, e_{tot} \end{pmatrix} \tag{2.213}$$

as well as a flux matrix $\psi(\mathbf{u})$

$$\psi(\mathbf{u}) = \begin{pmatrix} \rho\, \mathbf{v} \\ \rho\, \mathbf{v} \otimes \mathbf{v} - \sigma \\ \rho\, e_{tot} \mathbf{v} - \sigma \mathbf{v} \end{pmatrix} \tag{2.214}$$

and a matrix $\mathbf{Q}$ containing source terms :

$$\mathbf{Q} = \begin{pmatrix} 0 \\ \rho\, f \\ \rho\, f\ \mathbf{v} + \rho \dot{q} \end{pmatrix} \tag{2.215}$$

to obtain the integral equations in a more compact way:

$$\frac{\partial}{\partial t} \int_{{}^tV} \mathbf{u}\, \mathrm{d}\,{}^tV + \int_{{}^tV} \nabla\psi\, \mathrm{d}\,{}^tV = \rho \int_{{}^tV} \mathbf{Q}\, \mathrm{d}\,{}^tV \tag{2.216}$$

and, in the same manner, the differential version:

$$\frac{\partial \mathbf{u}}{\partial t} + \nabla\psi = \mathbf{Q} \tag{2.217}$$

2.5 Variational Solutions of the Balance Equations

With the equation of motion in the form (2.191) a time dependent partial differential equation was derived that describes conservation of linear momentum. In the regime of static problems we found a direct and a variational way to formulate equilibrium. Similarly, we can describe the time dependent motion of structural continua in different ways:

i) Direct formulation of the equations in their differential form and local approximation at discrete locations for example by finite difference schemes.

ii) Construction of a so called *weak form* of the differential equations, including related conditions for their validity, and approximation by finite element methods.

iii) Formulation of variational problems, i.e. in the case of dynamic problems application of Hamilton's principle of least action, and solution by finite elements.

iv) A fundamental alternative to formulating the conservation equations as performed in i)-iii) is provided by the use of energy methods. Though there is still only limited experience with *Hamiltonian descriptions* in fluid and structural dynamics codes, existing implementations[15] give first promising insight to the potential of that approach.

Solution strategy iii) of applying a variational technique is often used for oscillation problems and modal analyses. An example for its application to derive the equations of motion in an alternative way is found in Gourma [138].

[15] Hamiltonian descriptions for simulations of viscous compressible fluid dynamics as well as hypervelocity impact processes can be found in Fahrenthold and Koo [124] and [124], respectively.

Incremental solutions in time are usually performed with the first two options mentioned above. Transformation of differential equations to difference equations i) will be discussed in section 5.5. The second approach ii), formulation of weak forms, is used for approximative solution techniques in many finite element codes. Some necessary theoretical basics will be introduced next.

2.5.1 What are Weak Forms?

The basic task for the approaches to be discussed next will be to provide solutions for a boundary value problem of the general form:

$$L\,[u(x)] - f(x) = 0 \tag{2.218}$$

with Dirichlet boundary conditions (see equation (5.11)) on the boundary Γ of a domain Ω:

$$u(x)|_{\Gamma} = u_0(x) \tag{2.219}$$

and Neumann boundary (equation (5.12)) conditions:

$$\left.\frac{\partial\, u()x}{\partial\, x}\right|_{\Gamma} = 0 \tag{2.220}$$

Equation (2.218) together with its boundary conditions (2.219) and (2.220) is called the *strong form* of the partial differential equation.

A so called *weak form* is found if equation (2.218) is multiplied by an arbitrary smooth function $w(x)$ and integrate over the domain:

$$\int_{\Omega} w(x)\,\{L\,[u(x)] - f(x)\}\,\mathrm{d}\Omega = 0 \tag{2.221}$$

Integration by parts delivers an integral form in which the differential $L\,[u(x)]$ is reduced in its order by one.

Both strong form (2.218) and weak form (2.221) shall be briefly illustrated with the example of the hanging bar under its weight for which we found the strong form of its related Euler-Lagrange equation (2.171) to be

$$A\,E\,\frac{\mathrm{d}^2\,u}{\mathrm{d}\,x^2} + P(x) = 0 \tag{2.222}$$

A weak form of (2.222) is formulated if we first multiply by a smooth function $w(x)$ and integrate over the length L of the bar:

$$\int_0^L w(x)\left\{A\,E\,\frac{\mathrm{d}^2\,u}{\mathrm{d}\,x^2} + P(x)\right\}\mathrm{d}x = 0 \tag{2.223}$$

After separation of the integrand

$$\int_0^L w(x) \left\{ A\,E\,\frac{\mathrm{d}^2\,u}{\mathrm{d}\,x^2} \right\} \mathrm{d}x + \int_0^L w(x)\,P(x)\,\mathrm{d}x = 0 \tag{2.224}$$

and integration by parts it can be shown that the weak form of (2.222) is:

$$\int_0^L A\,E\,\frac{\mathrm{d}\,u}{\mathrm{d}\,x}\frac{\mathrm{d}\,w}{\mathrm{d}\,x}\,\mathrm{d}x - \int_0^L w(x)\,P(x)\,\mathrm{d}x = 0 \tag{2.225}$$

Its advantage over the strong form is the reduced order in the derivative of u and the related bigger class of possible solutions.

Various types of weak forms will be introduced in the context of finite element approximations with weighted residual in section 5.7.1.

2.5.2 Weak Forms of the Equation of Motion

For a solution of the conservation equations for mass, momentum and energy with a weak form, specific attention is often reduced to the momentum balance. This is sufficient when mass and energy conservation are solved formulating the velocity divergence $\nabla \cdot \mathbf{v}$ and gradient $\nabla \otimes \mathbf{v}$ as well as the divergence of the stress tensor $\nabla \cdot \sigma$, respectively. Accordingly, weak forms will be derived for the momentum balance in the following. In chapter 5 discrete versions for the momentum equation as well as for mass and energy conservation will be derived.

Weak forms of the differential non-conservative momentum balance (2.191) can be written in two basic types of formulations, both of which are Lagrangean. They are distinguished with respect to the configurations which the derivatives and the weak form integrals relate to:

- A *total Lagrangean* formulation is what in classical continuum mechanics is called material or simply Lagrangean system. The dependent variables as wells as their derivatives and the integrals in the weak forms are taken with respect to the initial undeformed configuration.
- *Updated Lagrangean* formulations, on the other hand, also describe the dependent variables referenced to the initial configuration. The derivatives and the integrals, however, are formulated in spatial coordinates.

The derivation of both versions employed here follows the one presented in Belytschko et al [46] and leads to weak forms in terms of the *principle of virtual work* and the *principle of virtual power* for the total and updated Lagrangean formulations, respectively.

Total Lagrangean Formulation

In its total Lagrange version, mass and energy conservation are simply written as:

$$\rho = \frac{1}{J}\,\rho_0 \tag{2.226}$$

and

$$\dot{e} = \frac{1}{\rho_0}\,\dot{\mathbf{F}} : \mathbf{P}^T \tag{2.227}$$

The strong form of the linear momentum equation (2.191) in Lagrangean coordinates $\mathbf{X}$ is facilitated by the nominal stresses $\mathbf{P}$ and the density in the initial configuration ρ_0 as:

$$\frac{\partial\,\mathbf{P}}{\partial\,\mathbf{X}} + \rho_0\,\mathcal{B} - \rho_0\,\frac{\partial^2\,\mathbf{u}\,(\mathbf{X},\,t)}{\partial\,t^2} = 0 \tag{2.228}$$

where the body forces are again denoted by $\mathcal{B}$. Application of the principle of virtual work is realized in multiplying (2.228) with the virtual displacements $\delta\,\mathbf{u}$ and integration over the domain Ω_0 in its initial configuration:

$$\int_{\Omega_0} \delta\,\mathbf{u}\left(\frac{\partial\,\mathbf{P}}{\partial\,\mathbf{X}} + \rho_0\,\mathcal{B} - \rho_0\,\frac{\partial^2\,\mathbf{u}\,(\mathbf{X},\,t)}{\partial\,t^2}\right)\,\mathrm{d}\,\Omega_0 = 0 \tag{2.229}$$

A disadvantage of (2.229) in the context of finite element approximations with shape functions is the stress gradient $\frac{\partial\,\mathbf{P}}{\partial\,\mathbf{X}}$. Reformulation of the stress gradient term by:

$$\int_{\Omega_0} \delta\,\mathbf{u}\frac{\partial\,\mathbf{P}}{\partial\,\mathbf{X}}\,\mathrm{d}\,\Omega_0 = \int_{\Omega_0} \frac{\partial}{\partial\,\mathbf{X}}\,(\delta\,\mathbf{u}\,\mathbf{P})\,\,\mathrm{d}\,\Omega_0 - \int_{\Omega_0} \frac{\partial\,\delta\mathbf{u}}{\partial\,\mathbf{X}}\,\mathbf{P}\,\mathrm{d}\,\Omega_0 \tag{2.230}$$

and application of the Gaussian theorem on the first term of the right hand side delivers the sum of two integrals over the initial boundary Γ_0 of the domain:

$$\int_{\Omega_0} \frac{\partial}{\partial\,\mathbf{X}}\,(\delta\,\mathbf{u}\,\mathbf{P})\,\,\mathrm{d}\,\Omega_0 = \int_{\Gamma_0} \delta\,\mathbf{u}\,\mathbf{n}_0\,\mathbf{P}\,\mathrm{d}\,\Gamma_0 + \int_{\Gamma_0} \delta\,\mathbf{u}\,\,(\mathbf{n}_0\,\mathbf{P})\,\,\mathrm{d}\,\Gamma_0 \tag{2.231}$$

where the second integral equals zero due to traction continuity and the first can be expressed by the virtual displacement gradient $\dfrac{\partial\,(\delta\,\mathbf{u})}{\partial\,\mathbf{X}}$ on the traction boundary (see Belytschko et al [46]) to deliver the *weak form of momentum conservation in the total Lagrangean formalism*:

$$\int_{\Omega_0} \delta\,\frac{\partial\,(\delta\,\mathbf{u})}{\partial\,\mathbf{X}}\,\mathbf{P}\,\mathrm{d}\,\Omega_0 - \int_{\Omega_0} \rho_0\,\delta\,\mathbf{u}\,\mathcal{B}\,\mathrm{d}\,\Omega_0 + \int_{\Omega_0} \rho_0\,\delta\,\mathbf{u}\,\frac{\partial^2\,\mathbf{u}}{\partial\,t^2}\,\mathrm{d}\,\Omega_0 - \int_{\Gamma_0} \delta\,\mathbf{u}\,\,(\mathbf{n}_0\,\mathbf{P})\,\,\mathrm{d}\,\Gamma_0 = 0 \tag{2.232}$$

Updated Lagrangean Formulation

To derive the Updated Lagrangean form of the balance equations, the kinematics are formulated in the current configuration and the Cauchy stresses σ

are used along with the current density ρ to yield the related strong forms of mass conservation (2.182):

$$\dot{\rho} = -\rho \nabla \cdot \mathbf{v} \tag{2.233}$$

energy conservation (2.211):

$$\dot{e} = \frac{1}{\rho} \sigma : \nabla \otimes \mathbf{v} \tag{2.234}$$

and linear momentum balance (2.192):

$$\dot{\mathbf{v}} = \frac{1}{\rho} \nabla \cdot \sigma + \rho \mathcal{B} \tag{2.235}$$

This time the principle of virtual power is employed to find the weak form of momentum conservation to be:

$$\int_{\Omega} \rho \, \delta \mathbf{v} \frac{\partial \mathbf{v}}{\partial t} \, \mathrm{d}\Omega + \int_{\Omega} \frac{\partial \, \delta \mathbf{v}}{\partial \mathbf{x}} \sigma \, \mathrm{d}\Omega - \int_{\Omega} \rho \, \delta \mathbf{v} \, \mathcal{B} \, \mathrm{d}\Omega - \int_{\Omega} \delta \mathbf{v} \, (\mathbf{n} \, \sigma) \, \mathrm{d}\Gamma = 0 \tag{2.236}$$

2.5.3 Hamilton's Principle of Least Action

The equivalent approach to the principle of stationary potential energy which was used to describe static equilibrium is represented in dynamic systems by *Hamilton's principle of least action*. The functional that is to be solved by a variational approach is again the equation of motion (2.228):

$$\frac{\partial \mathbf{P}}{\partial \mathbf{X}} + \rho_0 \, \mathcal{B} - \rho_0 \, \frac{\partial^2 \mathbf{u} \, (\mathbf{X}, t)}{\partial t^2} = 0 \tag{2.237}$$

To find a solution $\mathbf{v}$, the specific kinetic energy of the body is formulated by integration over the domain:

$$e_{kin}(\mathbf{v}) = \frac{1}{2} \int_{\Omega_0} \rho_0 \, v^2 \, \mathrm{d}\Omega_0 \tag{2.238}$$

The *Lagrange function* L describes the difference between kinetic energy and total potential energy:

$$L = e_{kin}(\mathbf{v}) - \Pi \tag{2.239}$$

where the total potential energy is described by the internal and external force components given in the equation of motion (2.237) are:

$$\Pi_{int} = \int_{\Omega_0} \frac{\partial \mathbf{P}}{\partial^0 \mathbf{X}} \, \mathrm{d}\Omega_0 \tag{2.240}$$

and

$$\Pi_{ext} = - \int_{\Omega_0} \rho_0 \, \mathcal{B} \, \mathrm{d}\Omega_0 \tag{2.241}$$

leading to the Lagrange function:

$$L = \frac{1}{2} \int_{\Omega_0} \rho_0 \, v^2 \, \mathrm{d}\Omega_0 + \int_{\Omega_0} \frac{\partial \mathbf{P}}{\partial^0 \mathbf{X}} - \mathcal{B} \, \mathrm{d}\Omega_0 \tag{2.242}$$

Hamilton's principle demands a stationary value for the time integral over the Lagrange function (2.242) :

$$\delta \int_{t_0}^{t_1} L \, \mathrm{d}t = \int_{t_0}^{t_1} \delta \, L \, \mathrm{d}t = 0 \tag{2.243}$$

To obtain a weak form, virtual displacements $\delta \, \mathbf{u}$ are used. Multiplication with the integrand and subsequent integration by parts yields the weak form for the principle of least action:

$$\int_{\Omega_0} \rho_0 \, \delta \, \mathbf{u} \, \frac{\partial^2 \, \mathbf{u}}{\partial \, t^2} \, \mathrm{d}\Omega_0 + \int_{\Omega_0} \delta\varepsilon^T \, \mathbf{P} \, \varepsilon \, \mathrm{d}\Omega_0 - \int_{\Omega_0} \rho_0 \delta \, \mathbf{u}^T \, \mathcal{B} \, \mathrm{d}\Omega_0 - \int_{\Gamma_0} \mathbf{n} \, \mathbf{P} \, \mathrm{d}\Gamma_0 = 0 \tag{2.244}$$

A motion ($\mathbf{u}$, $\mathbf{v}$) that satisfies (2.244) fulfils the related *Euler-Lagrange equation*:

$$\frac{\mathrm{d}}{\mathrm{d}t} \frac{\partial \, L}{\partial \, \mathbf{v}} - \frac{\partial \, L}{\partial \, \mathbf{u}} = 0 \tag{2.245}$$

and is a solution to the equation of motion.

2.6 Thermodynamic Basics

To describe the thermodynamics of processes, state variables and relations between them are needed. *State variables* are e.g. the pressure p, volume V and density ρ, total energy E_{tot}, internal energy E and temperature T, the enthalpy H, the entropy S and the Helmholtz and Gibbs free energies Ψ and G, respectively. Any two of the state variables are independent.

2.6.1 Energy Is Conserved - The First Law

As already formulated in chapter 2.4.4, the *first law of thermodynamics* demands the conservation of total energy. This is guaranteed if the increase in energy $\mathrm{d}E_{tot}$ is equaled by thermal energy entering the system as heat flow $\mathrm{d}Q$ and mechanical work $\mathrm{d}W$ added to the same system:

$$\mathrm{d}E_{tot} = \mathrm{d}Q - \mathrm{d}W \tag{2.246}$$

where the negative sign of $\mathrm{d}W$ is based on the convention that work *done* by a system, in other words leaving the system, is understood as positive. The mechanical work is expressed by the change in volume due to the stress applied $\mathrm{d}W = -\sigma \mathrm{d}V$ or, if only hydrostatic pressure is considered $\mathrm{d}W = p\mathrm{d}V$. A local form using mass specific terms yields:

$$de_{tot} = dq + \frac{1}{\rho}\sigma d\varepsilon \tag{2.247}$$

Equation (2.211) already represented a combination of energy conservation and the equation of motion (2.191) to give a local time rate formulation in material description as:

$$\rho\dot{e} - \mathbf{T}\nabla \cdot \mathbf{v} - \rho\dot{q} = 0 \tag{2.248}$$

It should be mentioned that equation (2.248) applies equally for quasi-static, dynamic, reversible and irreversible processes[16] . Restrictions or specifications are possible, e.g. by application of the Cauchy stresses instead of the first Piola-Kirchhoff tensor $\mathbf{T} = J\mathbf{F}^{-1}\sigma$ in combination with the Eulerian strain rate tensor $\dot{\mathbf{e}} = \mathbf{F}\dot{\varepsilon}$ to find a material description for small strains:

$$\rho\dot{e} - J\,\sigma \cdot \dot{\mathbf{e}} - \rho\dot{q} = 0 \tag{2.249}$$

or by the treatment of finite strains through the Green-Lagrangean strain rate tensor $\dot{\mathbf{E}}$:

$$\rho\dot{e} - \mathbf{T} \cdot \dot{\mathbf{E}} - \rho\dot{q} = 0 \tag{2.250}$$

which in literature is sometimes also formulated Cauchy stresses σ in combination with $\mathbf{J} = \rho_0/\rho$ and with the time rate of the right Cauchy-Green tensor $\dot{\mathbf{C}} = 2\dot{\mathbf{E}}$ like:

$$\rho\dot{e} - \frac{1}{2}\frac{\rho_0}{\rho}\,\mathbf{F}^{-1}\,\sigma \cdot \dot{\mathbf{C}} - \rho\dot{q} = 0 \tag{2.251}$$

There are manifold ways to add or remove forms of energy to and from a system. Surface, body or contact forces leading to deformation, heat flow, friction or chemical reactions are only some examples. In chapter 2.4 on conservation equations different kinds of thermal and mechanical contributions to total energy were already noted to formulate energy conservation in the context of continuum mechanical problems.

2.6.2 Entropy Increases - The Second Law

The observation that heat always flows from regions of higher to regions of lower temperature and that the amount of thermal energy needed to provide mechanical work is alway larger than the gained external work lead to the formulation of the second law of thermodynamics. To guarantee the validity of the first law, a new state variable had to be postulated, which is called *entropy*. In the course of experimental or hypothetical investigations to establish thermodynamic relations the behaviour of closed systems is examined. Thermodynamic *processes* are forms of changes in state variable initiated through exchange of the system with its environment. The process itself can be described as path in the space of state variables. Any natural process starting

[16] See also Valanis [406] and [407]

from a condition of thermodynamic equilibrium follows subsequent conditions of non-equilibrium until a new equilibrium condition is reached.

A *reversible* process is a path of continuous equilibrium conditions resulting in a minimum change in entropy. Physically, this is only possible for infinitesimal small changes and, thus, for infinitely long lasting processes. However, we will later on assume reversibility of processes provided the change applied to the system is fast compared to the response of the system. Both reversible and irreversible processes must of course obey the first law.

In the limit case of a reversible process, the heat $\mathrm{d}Q$ entering the system at an absolute temperature T leads to the following change in *entropy*:

$$\mathrm{d}S = \frac{\mathrm{d}Q}{T} \tag{2.252}$$

An *irreversible process* is characterized by an increase in entropy which is always larger than the increase observed in the reversible case:

$$\mathrm{d}S > \frac{\mathrm{d}Q}{T} \tag{2.253}$$

Thus, generally the time rate of change in specific entropy is expressed by the following inequality

$$\dot{s} \geq \frac{\dot{q}}{T} \tag{2.254}$$

called *Clausius-Duhem inequality*. In combination with the first law formulated by equation (2.249) we find the following time rate relation between specific entropy s, specific internal energy e and the first Piola-Kirchhoff stresses $\mathbf{T}$:

$$\rho T \dot{s} \geq \rho \dot{e} - \mathbf{T}\dot{\varepsilon} \tag{2.255}$$

Equation (2.254) is one formulation to express the second law of thermodynamics. Further ways to express that inequality will be introduced in the course of section 2.6.4 on Clausius-Duhem formulations.
In combination with the first law (2.246) and equation (2.247) the change in internal energy due to mechanical work and thermal energy in a reversible process can be noted with the *Gibbs fundamental equation*:

$$\mathrm{d}E = \mathrm{d}Q - \mathrm{d}W \leq T\mathrm{d}S - p\mathrm{d}V \tag{2.256}$$

where the equal sign represents the reversible case. Hence, if the volume is kept constant ($\mathrm{d}V = 0$) in a reversible *isochoric* process, the temperature is a measure for the change of internal energy with respect to entropy:

$$\left.\frac{\mathrm{d}E}{\mathrm{d}S}\right|_V = T \tag{2.257}$$

For an *isentropic* process, i.e. reversible and *adiabatic* ($dQ = 0$), entropy is constant $dS = 0$ and we find

$$\left.\frac{dE}{dV}\right|_S = -p \tag{2.258}$$

2.6.3 Thermodynamic Potentials

Four of the above mentioned state variables are so called thermodynamic potentials:

- internal energy
- enthalpy
- Helmholtz free energy
- Gibbs free energy

Enthalpy

Having the internal energy already described before, the next potential *enthalpy* H is defined as sum of internal energy plus the work done by the system to the environment and is therefore also called *heat content*:

$$H = E + pV \tag{2.259}$$

or

$$dH = TdS + Vdp \tag{2.260}$$

Helmholtz Free Energy

In order to describe the amount of energy which is available to perform mechanical work at constant temperature, the thermal contributions TS are subtracted from the internal energy. The resulting thermodynamic potential is called *Helmholtz free energy* or *Helmholtz potential* Ψ :

$$\Psi = E - T\,S \tag{2.261}$$

With the free energy per unit mass ψ

$$\Psi = \rho \int \psi \, dV \tag{2.262}$$

we can use mass specific terms.

$$\psi = e - T\,s \tag{2.263}$$

Applying the Gibbs fundamental equation (2.256) and for the reversible limit case we get:

$$d\psi = dp - Tds - sdT = -pdV - sdT \tag{2.264}$$

where the equality accounts for the reversible limit case. Taking into regard the full Cauchy stress tensor σ instead of its hydrostatic component p together with the linear Lagrangean strain tensor ε and postulating an isothermal ($\mathrm{d}T = 0$) process, the change in free energy per unit mass ψ of a solid material is described by:

$$\mathrm{d}\psi = \frac{1}{\rho}\sigma_{ij}\mathrm{d}\varepsilon_{ij} \tag{2.265}$$

and hence the Cauchy stresses are related to the Helmholtz free energy per unit volume ψ as:

$$\sigma_{ij} = \rho_0 \, \frac{\partial\psi}{\partial\varepsilon_{ij}} \tag{2.266}$$

Again, the premise for equation (2.266) is the assumption of a reversible isothermal process.

The Helmholtz free energy represents that portion of internal energy that can potentially be transformed into mechanical work. In which way the mechanical work performs in a specific material, i.e. whether elastic, plastic, ductile, brittle, damage or failure occurs, is described by a set of additional *internal state variables* $\aleph^{\mathbf{i}}$. These internal state variables, e.g. a plastic strain tensor or damage tensors and scalars, reflect mechanisms on the micro-scopic level of the materials' structure which are not directly observable. Thus, the internal state variables are used to describe the influence of effects like dislocation, slip or micro-defects on the macro-scopically observable behaviour of materials in a phenomenological way. Therefore, the state functions are defined in terms of internal state variables. If we assume a number of n internal state variables $\aleph^i_{kl}$, the Helmholtz free energy can be formulated as $\Psi = \Psi(\varepsilon_{ij}, T, \aleph^i_{kl})$ and hence its total derivative yields:

$$\mathrm{d}\Psi = \frac{\partial\Psi}{\partial\varepsilon_{ij}}\mathrm{d}\varepsilon_{ij} + \frac{\partial\Psi}{\partial T}\mathrm{d}T + \sum_i \left(\frac{\partial\Psi}{\partial\aleph^i_{kl}}\mathrm{d}\aleph^i_{kl}\right) \tag{2.267}$$

In an isothermal process, the Helmholtz potential decreases to a minimum value consistent with the temperature until an equilibrium state is approached where no more mechanical work can be done.

Gibbs Free Energy

For processes at constant temperature and constant pressure, the *Gibbs free energy* is preferably used to describe the available energy:

$$G = H - T\,S = E + p\,V - T\,S \tag{2.268}$$

leading to

$$\mathrm{d}G = \mathrm{d}E + p\mathrm{d}V + V\mathrm{d}p - T\mathrm{d}S - S\mathrm{d}T \tag{2.269}$$

and thus for $T = const.$ and $p = const.$:

$$\mathrm{d}G = \mathrm{d}E + p\mathrm{d}V - T\mathrm{d}S \tag{2.270}$$

2.6.4 Formulations of the Clausius-Duhem Inequality

There are various formulations of the Clausius-Duhem inequality (2.254) to be found in literature. Depending on the focus and the relevant type of thermodynamic processes they may look quite different at first. In the following a derivation will be used that bases on formulations derived by Coleman and Gurtin [95] and Valanis [405] which is presented in a similar way by Wu [429].

With the help of the equation of motion (2.191) in a material frame referencing to an undeformed state, the first law of thermodynamics was formulated by (2.211) in terms of specific internal energy e as:

$$\rho\dot{e} - \mathbf{P}\nabla \cdot \mathbf{v} - \rho\dot{q} = 0 \tag{2.271}$$

Replacing the change in specific internal energy de by the change in specific Helmholtz free energy $\mathrm{d}\psi(\varepsilon, T, \aleph)$ according to equation (2.264)

$$\mathrm{d}e = \mathrm{d}\psi(\varepsilon, T, \aleph) + T\mathrm{d}s + s\mathrm{d}T \tag{2.272}$$

a rate formulation of the first law reads:

$$\dot{\psi} + T\dot{s} + s\dot{T} - \frac{1}{\rho}\mathbf{P}\dot{\varepsilon} - \dot{q} = 0 \tag{2.273}$$

Since the second law formulated in the Clausius-Duhem inequality (2.254) demands that $T\mathrm{d}s - \mathrm{d}q \geq 0$, we get:

$$\dot{\psi} + s\dot{T} - \frac{1}{\rho}\mathbf{P}\dot{\varepsilon} = \dot{q} - T\dot{s} \leq 0 \tag{2.274}$$

leading to another version of the *Clausius-Duhem-inequality*:

$$\dot{\psi} + s\dot{T} - \frac{1}{\rho}\mathbf{P}\dot{\varepsilon} \leq 0 \tag{2.275}$$

If we introduce the *rate of entropy production per unit volume* γ:

$$\gamma = \dot{s} - \dot{q}/T \tag{2.276}$$

multiplied by the absolute temperature T, equation (2.274) can as well be transformed into a version of the Clausius-Duhem inequality:

$$\dot{\psi} + s\dot{T} - \frac{1}{\rho}\mathbf{P}\dot{\varepsilon} = T\,\gamma \geq 0 \tag{2.277}$$

that is also often found in derivations of constitutive equations for reversible ($\gamma = 0$) isothermal ($\dot{T} = 0$) processes as:

$$\dot{\psi} - \frac{1}{\rho}\mathbf{P}\dot{\varepsilon} = T\,\gamma = 0 \tag{2.278}$$

In material sciences, inequality (2.275) is useful to find thermo-mechanically consistent formulations of constitutive models, i.e. for descriptions of elasticity as well as for yield, failure and damage processes. Therefore, equation (2.274) shall be developed a little further. First, the total derivative $\mathrm{d}\psi$ from (2.267) will be inserted into (2.273) to get

$$\frac{\partial \psi}{\partial \varepsilon_{ij}} \dot{\varepsilon}_{ij} + \frac{\partial \psi}{\partial T} \dot{T} + \frac{\partial \psi}{\partial \aleph^i_{kl}} \dot{\aleph}^i_{kl} + T\dot{s} + s\dot{T} - \frac{1}{\rho} P_{ij} \dot{\varepsilon}_{ij} - \dot{q} = 0 \tag{2.279}$$

Taking into regard relation (2.266) between the free energy ψ and the nominal stresses $\mathbf{P}$

$$\frac{\partial \psi}{\partial \varepsilon_{ij}} = \frac{1}{\rho} P_{ij} \tag{2.280}$$

and equation (2.263) in the form of

$$s = -\frac{\partial \psi}{\partial T} \tag{2.281}$$

equation (2.279) becomes:

$$\frac{\partial \psi}{\partial \aleph^i_{kl}} \dot{\aleph}^i_{kl} + T\dot{s} - \dot{q} = 0 \tag{2.282}$$

that is with $T\mathrm{d}s - \mathrm{d}q \geq 0$ leading to a third form of the Clausius-Duhem-inequality:

$$\frac{\partial \psi}{\partial \aleph^i_{kl}} \dot{\aleph}^i_{kl} \leq 0 \tag{2.283}$$

and which by use of the *rate of entropy production* γ (2.276) can also be written as

$$T\gamma \geq 0 \tag{2.284}$$

or equally as

$$\dot{s} \geq \frac{\dot{q}}{T} \tag{2.285}$$

2.6.5 Consequences for Constitutive Equations

With the thermo-mechanical derivations given in the above sections a set of equations was established that need to be obeyed in the description of material behaviour in order to be thermodynamically consistent. To define such restrictions, the formulation of the Helmholtz free energy (2.265) for isothermal processes is expanded to irreversible components including plasticity and damage, both decoupled from each other and from elasticity. Lemaitre [247], for example, writes:

$$\psi = \frac{1}{\rho} \left\{ \frac{1}{2} \varepsilon^{el}_{ij} \, E_{ijkl} \, \varepsilon^{el}_{kl} \, (1 - d) + R^{\infty} \left(r + \frac{1}{b} e^{-b\, r} \right) + \frac{X^{\infty}}{3} \gamma \, \alpha_{ij} \, \alpha_{ij} \right\} \tag{2.286}$$

where R^{∞} and b are material parameters describing isotropic hardening, X^{∞} and γ the respective ones for kinematic hardening. The scalar r denotes the accumulated equivalent strain associated to isotropic strain hardening, whereas the back stress tensor associated strain tensor α_{ij} is again related to kinematic hardening.

Elastic Stresses

In case of small deformations, the linear strain tensor can be split additively into an elastic and a plastic component by $\varepsilon = \varepsilon^{el} + \varepsilon^{pl}$. In equation (2.266) the reversible, thus elastic, components of the Cauchy stress under small deformations was already derived from the specific Helmholtz free energy. The origin of equation (2.266) is that for a non-dissipative isothermal process we demand according to the Clausius-Duhem equation (2.278) :

$$\mathrm{d}\,\psi - \frac{1}{\rho}\sigma\,\mathrm{d}\,\varepsilon = 0 \tag{2.287}$$

leading to:

$$\sigma_{ij}^{el} = \rho \frac{\partial \psi}{\partial \varepsilon_{ij}^{el}} = E_{ijkl}\,\varepsilon_{kl}^{el} \tag{2.288}$$

Under large deformations the elastic components of the Green strain rate tensor $\dot{E}_{ij}^{el}$ are derived by the Kröner-Lee decomposition according to (2.53). Referring to the undeformed condition, the nominal stresses P_{ij} are then related to the free energy density as follows:

$$P_{ij}^{el} = \rho \frac{\partial \psi}{\partial E_{ij}^{el}} \tag{2.289}$$

Inelastic Stresses

Inelastic stress components under small deformations are calculated from the free energy and the plastic linear strain ε_{ij}^{pl} by

$$\sigma_{ij}^{pl} = \rho \frac{\partial \psi}{\partial \varepsilon_{ij}^{pl}} \tag{2.290}$$

whereas large deformations again are governed by the plastic Green strains derived from equation (2.54) and thus leading to Lagrangean first Piola-Kirchhoff stresses:

$$T_{ij}^{pl} = \rho \frac{\partial \psi}{\partial E_{ij}^{pl}} \tag{2.291}$$

Table 2.5. State variables and conjugate thermodynamic forces.

	External State Variable	Internal State Variable	Conjugate Force
	ε_{ij}		$\sigma_{ij} = \rho \frac{\partial \psi}{\partial \varepsilon_{ij}}$
Elasticity		ε_{ij}^{el}	$\sigma_{ij}^{el} = \rho \frac{\partial \psi}{\partial \varepsilon_{ij}}{}^{el}$
			$= E_{ijkl}\, \varepsilon_{kl}^{el}\,(1-d)$
Plasticity		ε_{ij}^{pl}	$\sigma_{ij}^{pl} = \rho \frac{\partial \psi}{\partial \varepsilon_{ij}}{}^{pl}$
Isotropic		r	$R = \rho \frac{\partial \psi}{\partial r}$
Hardening			$= R^{\infty}\left(1 - e^{-br}\right)$
Kinematic		α_{ij}	$X_{ij}^{D} = \rho \frac{\partial \psi}{\partial \alpha_{ij}}$
Hardening			$= \frac{2}{3} X^{\infty}\, \gamma\, \alpha_{ij}$
Damage		d	$Y = -\rho \frac{\partial \psi}{\partial d_{ij}}$
			$= \frac{1}{2} \varepsilon_{ij}^{el}\, E_{ijkl}\, \varepsilon_{kl}^{el}$

Damage

Further so called *thermodynamic tensions* associated to internal state variables $\aleph_{ij}$, e.g. a damage tensor, are described similarly with the Helmholtz potential. The strain energy density release rate is related to the evolution of a damage tensor d_{ij} according to equation (2.286) through:

$$Y = -\rho \frac{\partial \psi}{\partial d_{ij}} = \frac{1}{2} \varepsilon_{ij}^{el}\, E_{ijkl}\, \varepsilon_{kl}^{el} \tag{2.292}$$

Isotropic and Kinematic Strain Hardening

According to the free energy formulation in equation (2.286) a description of isotropic and kinematic strain hardening follows:

$$R = \rho \frac{\partial \psi}{\partial r} = R^{\infty}\left(1 - e^{-br}\right) \tag{2.293}$$

and

$$X_{ij}^{D} = \rho \frac{\partial \psi}{\partial \alpha_{ij}} = \frac{2}{3} X^{\infty}\, \gamma\, \alpha_{ij}. \tag{2.294}$$

Thermodynamic Conjugate Forces

With an isothermal formulation of the Helmholtz potential depending on externally observable state variables and additional internal state variables a set of thermodynamic conjugate forces for plasticity and damage was established above and is collected in table 2.5.

3

Constitutive Equations

With the basic formulations collected in the previous sections, the thermodynamics and deformation processes of solids, fluids and gases can be described mathematically. These equations are, however, material independent, i.e. properties which are characteristic for a specific material like its stiffness, ductility, strength, compressibility or viscosity and their influence on the deformation behavior under a given load cannot be quantified yet. To be able to describe material specific behaviour, we need so called material laws, i.e. relations between stress and strain measures formulated by *constitutive equations*.

Dynamic deformation processes, specifically when shock wave formation is involved, are usually modelled by a decomposed stress tensor. The decomposition spits the stress tensor into a deviatoric tensor S_{ij} and a spherical hydrostat $p\,\delta_{ij}$. As already introduced in equation (2.127), the usual decomposition is defined by

$$\sigma_{ij} = S_{ij} - p\,\delta_{ij} \tag{3.1}$$

Accordingly, in the course of this chapter constitutive equations will separately be described, starting with basic formulations for equations of state. The usefulness of the decomposition results from the needed non-linear character of equations of state to describe shock waves. In the absence of shock waves, the stress tensor can most often be desccribed without decomposition. This close relation to shock waves as well as the specific methodology of deriving nonlinear equations of state for shock wave applications is the reason why the more specific introduction to equations of state will be covered in chapter 4. In the course of its application to anisotropic materials we will then also see that the usual decomposition is limitted to isotropic materials and alternative approaches are needed.

Concerning the stress deviator, elastic, hyperelastic and rate dependent formulations will be derived. As one particular aspect of deriving constitutive equations from micro-mechanical mehods, the use of statistical mechanics to

to formulate elastic potentials for polymers will be illustrated.
Of specific interest for time dependent soutions are objective stress rate formulations. They will be discussed at the end of this chapter.

3.1 Equations of State

As stated before, out of the set of thermodynamic state variables any two are independent. A formulation of hydrostatic pressure p depending on density ρ and specific internal energy e, or equivalently on volume V and temperature T, delivers an equation of state (EOS):

$$p = p(\rho,\, e) = p(V,\, T) = p(V,\, e) \tag{3.2}$$

However, (3.2) does not cover heat conduction and other effects that would require temperature. Equations of state of that kind are therefore often called *incomplete EOS*. Complete formulations, on the other hand, take into account temperature T and entropy s and are therefore formulated using the specific Helmholtz free energy ψ as:

$$\psi(V,\, T) = e - T\, s \tag{3.3}$$

From the Gibbs fundamental equation (2.256) and for the reversible limit case we find as derived before in equation (2.264) :

$$\mathrm{d}e = T\,\mathrm{d}s - p\,\mathrm{d}V \tag{3.4}$$

and thus

$$p(V,\, T) = -\frac{\partial\,\psi}{\partial\,V} \tag{3.5}$$

and

$$s(V,\, T) = -\frac{\partial\,\psi}{\partial\,T} \tag{3.6}$$

for a complete equation of state. Usually, the applications addressed in the context of this book are not affected by heat conduction effects and, thus, incomplete equations of state can be used. For detailed derivations of complete EOS, interested readers may be referred to Menikoff [273] to find examples of Helmholtz free energies and derived equations of state. As examples for complete equations of state, porous materials will be described with the Menikoff-Kober EOS and the Hayes EOS in section 4.5.2.

Equilibrium surfaces described by (3.2), e.g. in the p-V-e-space, define all possible conditions of thermodynamic equilibrium of a specific material including solid, fluid and gaseous states. For a pure material, the EOS describes a state surface like the schematic one displayed in Figure 3.1.

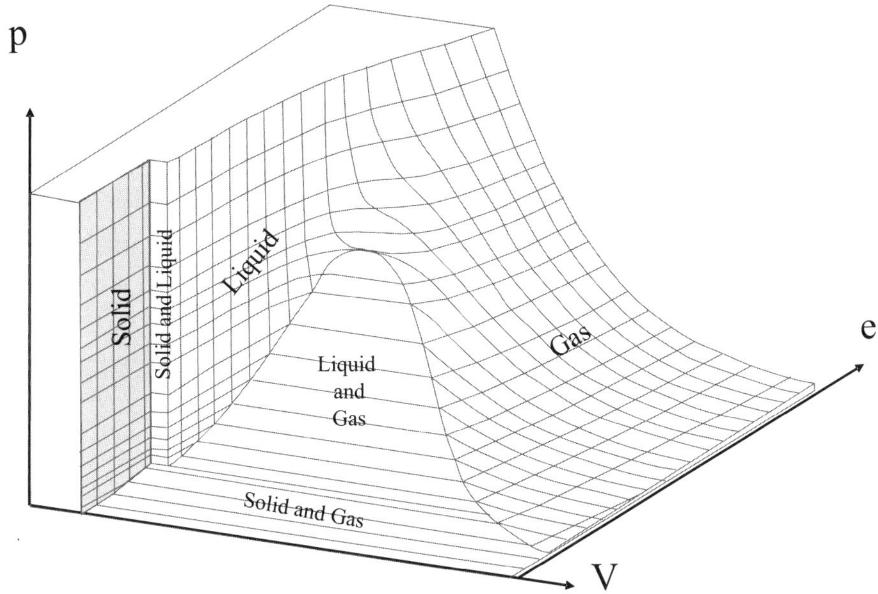

Fig. 3.1. Schematic state surface for a pure material representing all conditions of thermodynamic equilibrium.

To describe the local bulk sound speed c_B in a material at a specific state $(p,\ \rho,\ e)$, the slope of the state surface is calculated from the equation of state as:

$$c_B^2 = \left.\frac{\partial p}{\partial \rho}\right|_e + \frac{p}{\rho^2}\left.\frac{\partial p}{\partial e}\right|_\rho \tag{3.7}$$

3.1.1 Axiomatic Equations of State

Depending on the medium and scale of interest, different paths lead to constitutive formulations and parameters of an equation of state (3.2). Kinetic gas theory is an example for an axiomatic derivation of constitutive equations. It observes a gas as statistical distribution of molecules. The temperature of the gas depends on the distributions of velocities and densities of the molecules. Using the Maxwell-Boltzmann distribution, for example, one can observe a perfect monatomic gas as an assembly of N molecules having n degrees of freedom in motion each. Via the Maxwell-Boltzmann distribution, i.e. the probability of finding a specific molecule in a specific energy state, the *Maxwell distribution of velocities* for molecules of mass m in a gas at Temperature T is formulated. Using the Boltzmann constant k, the maximum of the distribution is described by the velocity:

$$v = \sqrt{\frac{2\,k\,T}{m}} \tag{3.8}$$

This allows for an estimation of force components acting on the surface of an infinitesimal volume[1] finally leading to the equation of state for N molecules of a *perfect gas*:

$$p = \frac{1}{V} N\, k\, T = \rho\, N\, k\, T = \rho\, R\, T \tag{3.9}$$

with the specific gas constant $R = k\, N$.

For ideal gases, according to Joule's law $\left(\frac{\partial e}{\partial V}\right)_T = 0$, the change in internal energy is described by the *caloric equation of state*:

$$\mathrm{d}e = c_v\, \mathrm{d}T \tag{3.10}$$

This change de can equally be expressed through the first law for an isentropic $(\mathrm{d}\, s = 0)$ process:

$$\mathrm{d}e = -p\, \mathrm{d}V \tag{3.11}$$

and the specific heat at constant volume c_v can be expressed using the adiabatic exponent κ for ideal gas as:

$$\kappa = 1 + \frac{R}{c_v} \tag{3.12}$$

leading to

$$\frac{R}{\kappa - 1}\, \mathrm{d}T + p\, \mathrm{d}\, V = 0 \tag{3.13}$$

which can be rearranged using the equation of state (3.9) in the form $p\, V = R\, T$ to

$$\frac{p\, \mathrm{d}\, V + V\, \mathrm{d}\, p}{\kappa - 1} + p\, \mathrm{d}\, V = 0 \tag{3.14}$$

Integration for constant κ leads to

$$ln\, p + \kappa\, ln\, V = 0 \tag{3.15}$$

and thus in non-logarithmic form:

$$p\, V^{\kappa} = const. \tag{3.16}$$

leading to an isentropic version of the ideal gas equation of state:

$$p = (\kappa - 1)\rho\, e \tag{3.17}$$

The limitations of the perfect gas theory are to be found in high densities and high temperatures. Alternative statistical approaches for real gases are the Fermi-Dirac or the Einstein-Bose distribution. Since it would lead to far to discuss this in the context of this book, the reader may be referred to Slater

[1] A detailed derivation can be found in Slater [359].

[359] or other text books on gas dynamics or chemical physics.

Investigation of regular crystalline solids delivered options to derive axiomatic equations of states for solids through statistical mechanics in a similar way as in the case of gas dynamics. Interatomic or intermolecular binding and repulsive forces are now additional factors in the description of potential energy. Describing the dynamics of a crystal of N atoms through a system of $3N$ harmonic oscillators is a standard approach to find equations of state for solids[2]. If the Helmholtz free energy is regarded as sum of a volume dependent cold potential $\psi_{T=0K}(V)$ term, a vibration induced thermal component $\psi_{th}(V,\ T)$ and a potential term due to thermally excited electrons $\psi_e(V,\ T)$:

$$\psi = \psi_{0K}(V) + \psi_{th}(V,\ T) + +\psi_e(V,\ T) \tag{3.18}$$

The three contributing terms can be estimated from theoretical assumptions on interatomic or intermolecular forces and statistical mechanics on the motion of crystal-lattice atoms or molecules, respectively.

In Bushman et al. [72] formulations for the cold curve, the thermal component and the electron contribution are derived. The same approach will be part of the derivation leading to the so called Mie-Grüneisen equation of state. Since that is one way to describe a nonlinear equation of state for shock wave descriptions it will be discussed in section 4.5.

3.1.2 Empirical Equations of State

Linear Isothermal EOS

For most engineering applications involving equations of states, empirical relations with experimentally derived data are used. Its most simple representation is the so called *linear equation of state* which assumes isothermal processes ($dT = 0$) and a linear pressure-volume or pressure-density relation. Via the bulk modulus K:

$$K = -V \left(\frac{\partial p}{\partial V}\right)_S = \rho\, c^2 \tag{3.19}$$

the linear equation of state is formulated as:

$$p = K\,\varepsilon_{kk} = K\,\frac{\rho - \rho_0}{\rho_0} = K\,\left(\frac{\rho}{\rho_0} - 1\right) = K\,\mu \tag{3.20}$$

with the compression term μ describing the ratio of change in volume and density, respectively:

$$\mu = -\frac{\Delta V}{V} = \frac{\rho}{\rho_0} - 1 \tag{3.21}$$

[2] Literature with detailed derivations on that topic can be found in Landau and Lifshitz [244], Zel'dovich and Raizer [433], Slater [359] and Bushman et al. [72].

For isotropic materials, the bulk modulus K is linked to the Young's modulus E and the shear modulus G via the Poisson ratio μ by:

$$K = \frac{E}{3\,(1 - 2\,\mu)} = \frac{2\,G\,(1 + \mu)}{3\,(1 - 2\,\mu)} \tag{3.22}$$

meaning that the knowledge of any two other elastic constants provides the needed material dependent input for the linear equation of state.

Nonlinear EOS

Whenever the linear elastic region described in equation (3.20) is left, which is for example the case when a wide spectrum of pressure and energy shall be covered by the EOS, nonlinear relations are needed. A polynomial description of an equation of state can for instance be written as:

$$p = K_1\mu + K_2\mu^2 + K_3\mu^3 + (B_0 + B_1\mu)\;\rho_0\,e \tag{3.23}$$

where K_i and B_i are material constants usually defined separately for compression and expansion, respectively. An important difference to the linear equation (3.20) is marked by the energy dependency of the last term in (3.23). Whereas the linear equation is only a compression curve along an isotherm, the latter can really be called equation of state in the sense of (3.2).

Later we will see how experimental data resulting from the observation of shock wave propagation can provide information to identify the material parameters in (3.23). The underlying theory is composed of

- the thermomechanics of shock waves, i.e. essentially the Rankine-Hugoniot equations providing a line of reference configurations on the state surface, used to identify the parameters K_i
- and an assumption on the pressure change off the Hugoniot-line along isochores, defining the constants B_i.

Since the basic theory of shock waves in solids and specifically its thermomechanic aspects need to be discussed before, related formulations for equations of state will be derived in section 4.5.

3.2 Constitutive Equations for Total Stresses

In section 3.1 we already introduced first constitutive formulations for the spherical component of the total stress tensor. This would be sufficient, if we only wanted to describe a fluid or if material strengths were negligible. To derive constitutive equations for solids, the deviatoric material behavior needs to be formulated also. With the decomposition (2.127) of the stress tensor, both

- a separated calculation of the deviatoric and volumetric components ($\mathbf{S}_{ij}$, p)
- as well as a direct constitutive formulation for the total stresses

are possible. As mentioned before, the relevancy of the decomposition becomes apparent when shock wave propagation and related dissipative effects are of interest.

The constitutive formulations collected in the following sections are applicable to total stress tensor formulations and to stress deviator descriptions.

3.2.1 Cauchy Elasticity

For many engineering materials the assumption of an initial linear reversible relation between components of stress and strain is valid. Therefore, an introduction to elastic material models typically starts with the one-dimensional linear elastic *Hooke's law*:

$$\sigma_{xx} = E\,\varepsilon_{xx} \tag{3.24}$$

with the material specific *Young's modulus* E. The material behavior is called *elastic* as long as its stress-strain relation is reversible and path-independent, i.e.

- the current stress state is fully determined by the current state of deformation
- the deformation process involves no dissipation.

When the first condition describes path independence of the stress state, then the second guarantees path independence of the energy. If both conditions are fulfilled in a one-dimensional case described by equation (3.24), the existence of a potential function $w(\varepsilon_{xx})$ with:

$$\sigma_{xx} = \frac{\mathrm{d}\,w(\varepsilon_{xx})}{\mathrm{d}\varepsilon_{xx}} \tag{3.25}$$

is ensured simultaneously. The potential function $w(\varepsilon_{xx})$ is identical to the strain-energy density and in the linear case of Hooke's law (3.24) it is expressed by the quadratic strain formulation:

$$w(\varepsilon_{xx}) = \frac{1}{2}\,E\,\varepsilon_{xx}^2 \tag{3.26}$$

Transformation of the one-dimensional Hooke's law to the general three-dimensional case does not necessarily guarantee the existence of an associated potential function in the hyperspace of six independent strain components. As mentioned before by means of the quadratic function (3.26) describing a parabolic line in the $w - \varepsilon_{xx}$ plane, path independence of both stress state and energy is the precondition for a potential and vice versa. If, in the three-dimensional state, only path independence of the stress state is guaranteed through a constitutive equation, the material behavior is called *Cauchy elastic.* Only if a related hypersurface $W(\varepsilon)$ exists, i.e. the path independence of energy is guaranteed as well, the derived constitutive model describes *hyperelasticity*, also called *Green elasticity.* The latter will be discussed after a brief introduction to various forms of Cauchy elasticity.

For Cauchy elastic materials the Hookean law (3.24) is generalized to the three-dimensional form:

$$\sigma_{ij} = E_{ijkl}\, \varepsilon_{kl} \tag{3.27}$$

valid for infinitesimal strains and involving a fourth order material *elasticity tensor* E_{ijkl}.

3.2.2 General Elastic Anisotropy

Isotropic elasticity is most simple, but not the most common material behavior. In order to distinguish it from various forms of anisotropy, a derivation of constitutive formulations, starting from the general anisotropic case, shall be performed next. For that purpose, the scalar strain energy shall be formulated as quadratic expansion in terms of the linearized strain tensor ε_{ij}:

$$\rho\, \psi = \alpha_0 + \alpha_{ij}\, \varepsilon_{ij} + \frac{1}{2}\, E_{ijkl}\, \varepsilon_{ij}\, \varepsilon_{kl} \tag{3.28}$$

using a scalar material parameter α_0, a second order material tensor α_{ij} and a forth order material tensor E_{ijkl}. Since the undeformed material shall store no free energy $(\psi(\varepsilon_{ij} = 0) = 0)$, the scalar parameter must vanish $\alpha_0 = 0$. The same is concluded for the second order tensor α_{ij} if the resulting Cauchy stresses:

$$\sigma_{ij} = \rho\, \frac{\partial\, \psi}{\partial\, \varepsilon_{ij}} = \alpha_{ij} + \frac{1}{2}\, E_{ijkl}\, \varepsilon_{kl} \tag{3.29}$$

in the undeformed state are investigated:

$$\sigma_{ij}(\varepsilon = 0) = \alpha_{ij} \tag{3.30}$$

leading to the generalized Hooke's law:

$$\sigma_{ij} = \rho \frac{\partial \psi}{\partial \varepsilon_{ij}} = E_{ijkl}\, \varepsilon_{kl} \tag{3.31}$$

Equation (3.31) describes a general anisotropic elasticity with 81 independent components of E_{ijkl}. A first step of reducing the complexity of the material tensor is found in the symmetry of the stress tensor (2.200):

$$\sigma_{ij} = \sigma_{ji} \tag{3.32}$$

which leads to 36 independent physical components. According to the generalized Hooke's law (3.31), the material tensor can be expressed as second derivative of the free energy

$$E_{ijkl} = \rho \frac{\partial^2 \psi}{\partial \varepsilon_{ij}\, \partial \varepsilon_{kl}} \tag{3.33}$$

and since the result is independent on the order of the derivatives, we find a reduction to 21 independent components of the material tensor through

$$E_{ijkl} = E_{klij} \tag{3.34}$$

describing general anisotropic behaviour.

The material symmetry (3.34) allows for a simplification in writing the fourth order $3x3x3x3$ material tensor E_{ijkl} by a compressed notation. The so called *Voigt notation*[3] is one option to write E_{ijkl} in a $6x6$ format through the transformation:

$$\begin{bmatrix} E_{1111} & E_{1122} & E_{1133} & E_{1123} & E_{1113} & E_{1112} \\ E_{2211} & E_{2222} & E_{2233} & E_{2223} & E_{2213} & E_{2212} \\ E_{3311} & E_{3322} & E_{3333} & E_{3323} & E_{3313} & E_{3312} \\ E_{2311} & E_{2322} & E_{2333} & E_{2323} & E_{2313} & E_{2312} \\ E_{1311} & E_{1322} & E_{1333} & E_{1323} & E_{1313} & E_{1312} \\ E_{1211} & E_{1222} & E_{1233} & E_{1223} & E_{1213} & E_{1212} \end{bmatrix} \Longrightarrow \begin{bmatrix} E_{11} & E_{12} & E_{13} & E_{14} & E_{15} & E_{16} \\ E_{21} & E_{22} & E_{23} & E_{24} & E_{25} & E_{26} \\ E_{31} & E_{32} & E_{33} & E_{34} & E_{35} & E_{36} \\ E_{41} & E_{42} & E_{43} & E_{44} & E_{45} & E_{46} \\ E_{51} & E_{52} & E_{53} & E_{54} & E_{55} & E_{56} \\ E_{61} & E_{62} & E_{63} & E_{64} & E_{65} & E_{66} \end{bmatrix} \tag{3.35}$$

With the Voigt notation used further on in the context of this book, E_{ij} is equivalent to E_{ijkl} according to the transformation (3.35) whereas a single indexed E_i describes values of Young's moduli in one of three orthogonal directions i.

Dellinger et al. [106] pointed out that a certain disadvantage of the Voigt notation should be noted in the fact that it misses to represent a sometimes important norm since

$$\sum_{i,j} E_{ij}^2 \neq \sum_{i,j,k,l} E_{ijkl}^2 \tag{3.36}$$

That is why they prefer the so called *Kelvin notation* as alternative in which the elements of the material tensor with compressed indices 4, 5 and 6 are scaled by a factor $\sqrt{2}$. Performed for each of the two indices, the scaling leads, for example, to an element $E_{35} = \sqrt{2}\, E_{3313}$ as well as to $E_{66} = 2\, E_{1212}$.

3.2.3 Elasticity with Symmetry Planes

General anisotropy in the elastic behaviour cannot be characterized for mathematical descriptions with reasonable experimental efforts. Therefore, for

[3] See for example Auld [25].

practical application formulations of reduced generality though not necessarily reduced accuracy are needed. And most materials show one or several aspects allowing for simpler descriptions of their elastic deformation behaviour.

Monoclinic Elasticity

Material behaviour that is symmetric with respect to specific planes leads to further reductions in the number of independent elastic constants. If there is one single plane of that kind, the material behaviour is called *monoclinic* and the resulting stress strain relation can be satisfied with 13 parameters. If $z = 0$ is the plane of symmetry, equation (3.27) is written in contracted notation as:

$$\begin{bmatrix} \sigma_{xx} \\ \sigma_{yy} \\ \sigma_{zz} \\ \sigma_{yz} \\ \sigma_{xz} \\ \sigma_{xy} \end{bmatrix} = \begin{bmatrix} E_{11} & E_{12} & E_{13} & 0 & 0 & E_{16} \\ E_{21} & E_{22} & E_{23} & 0 & 0 & E_{26} \\ E_{31} & E_{32} & E_{33} & 0 & 0 & E_{36} \\ 0 & 0 & 0 & E_{44} & E_{45} & 0 \\ 0 & 0 & 0 & E_{54} & E_{55} & 0 \\ E_{61} & E_{62} & E_{63} & 0 & 0 & E_{66} \end{bmatrix} \begin{bmatrix} \varepsilon_{xx} \\ \varepsilon_{yy} \\ \varepsilon_{zz} \\ \varepsilon_{yz} \\ \varepsilon_{xz} \\ \varepsilon_{xy} \end{bmatrix} \tag{3.37}$$

Allocations of the 13 independent parameters of elasticity in the material tensor for the three symmetry planes $x = 0$, $y = 0$ and $z = 0$ are illustrated symbolically in Figure 3.2. Typical fields of applications for monoclinic elastic behaviour are fabric or composite materials.

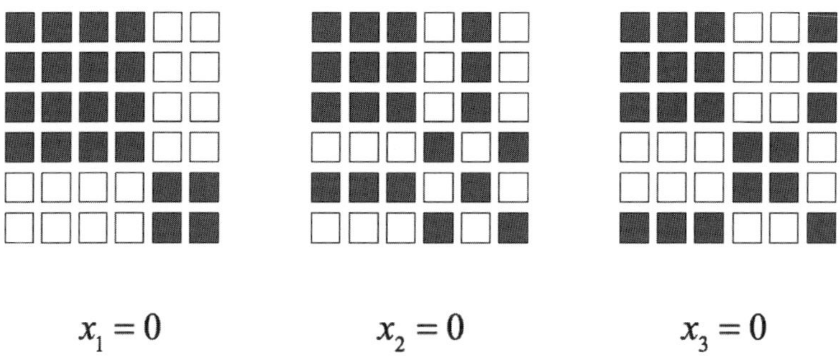

Fig. 3.2. Symbolic allocation of elements in a monoclinic material's elasticity tensor for three different planes of symmetry written in contracted notation.

Orthotropic Elastic Behaviour

Whenever a second plane of symmetry exists, the resulting material behaviour is *orthotropic* and nine individual parameters are sufficient to describe elastic material response:

$$\begin{bmatrix} \sigma_{xx} \\ \sigma_{yy} \\ \sigma_{zz} \\ \sigma_{yz} \\ \sigma_{xz} \\ \sigma_{xy} \end{bmatrix} = \begin{bmatrix} \frac{1-\nu_{yz}\nu_{zy}}{E_y E_z\ \Delta} & \frac{\nu_{yx}-\nu_{zx}\nu_{yz}}{E_y E_z\ \Delta} & \frac{\nu_{zx}-\nu_{yx}\nu_{zy}}{E_y E_z\ \Delta} & 0 & 0 & 0 \\ \frac{\nu_{yx}-\nu_{zx}\nu_{yz}}{E_y E_z\ \Delta} & \frac{1-\nu_{xz}\nu_{zx}}{E_x E_z\ \Delta} & \frac{\nu_{zy}-\nu_{xy}\nu_{zx}}{E_x E_z\ \Delta} & 0 & 0 & 0 \\ \frac{\nu_{zx}-\nu_{yx}\nu_{zy}}{E_y E_z\ \Delta} & \frac{\nu_{zy}-\nu_{xy}\nu_{zx}}{E_x E_z\ \Delta} & \frac{1-\nu_{xy}\nu_{yx}}{E_x E_y\ \Delta} & 0 & 0 & 0 \\ 0 & 0 & 0 & 2G_{yz} & 0 & 0 \\ 0 & 0 & 0 & 0 & 2G_{xz} & 0 \\ 0 & 0 & 0 & 0 & 0 & 2G_{xy} \end{bmatrix} \begin{bmatrix} \varepsilon_{xx} \\ \varepsilon_{yy} \\ \varepsilon_{zz} \\ \varepsilon_{yz} \\ \varepsilon_{xz} \\ \varepsilon_{xy} \end{bmatrix} \tag{3.38}$$

with Young's moduli in principal directions E_i, Poisson ratios ν_{ij}, shear moduli G_{ij} and

$$\Delta = \frac{1 - \nu_{xy}\nu_{yx} - \nu_{yz}\nu_{zy} - \nu_{xz}\nu_{zx} - 2\nu_{yx}\nu_{zy}\nu_{xz}}{E_x E_y E_z} \tag{3.39}$$

Note that the number of independent parameters is nine since

$$\frac{\nu_{ij}}{E_i} = \frac{\nu_{ji}}{E_j} \tag{3.40}$$

Laminated composites as well as short fiber reinforced plastics are typical representatives showing orthotropic elasticity.

Transverse Isotropic Behaviour

Individual uni-directional lamina and metal sheets show isotropic behaviour in one plane, which is the through-thickness plane, and deviating properties in the direction normal to that plane. In case of uni-directional lamina the deviating direction goes along with the fiber orientation whereas for sheet metal it is the rolling direction. That kind of behaviour is called *transverse isotropic*. The resulting five independent parameters are identified for a material with isotropic behaviour in the $x = 0$ plane as:

$$\begin{bmatrix} \sigma_{xx} \\ \sigma_{yy} \\ \sigma_{zz} \\ \sigma_{yz} \\ \sigma_{xz} \\ \sigma_{xy} \end{bmatrix} = \begin{bmatrix} E_{11} & E_{12} & E_{12} & 0 & 0 & 0 \\ E_{12} & E_{22} & E_{23} & 0 & 0 & 0 \\ E_{12} & E_{23} & E_{22} & 0 & 0 & 0 \\ 0 & 0 & 0 & \frac{E_{22}-E_{23}}{2} & 0 & 0 \\ 0 & 0 & 0 & 0 & E_{55} & 0 \\ 0 & 0 & 0 & 0 & 0 & E_{55} \end{bmatrix} \begin{bmatrix} \varepsilon_{xx} \\ \varepsilon_{yy} \\ \varepsilon_{zz} \\ \varepsilon_{yz} \\ \varepsilon_{xz} \\ \varepsilon_{xy} \end{bmatrix} \tag{3.41}$$

Isotropic Behaviour

Complete independence of the elastic material response on the loading direction is called *isotropic*. The stiffness matrix of an isotropic material remains invariant under any change of orthonormal basis. It is formulated with only two parameters, e.g. using the Lamé constants λ and μ:

$$E_{ijkl} = \lambda\, \delta_{ij}\delta_{kl} + \mu\, (\delta_{ik}\delta_{jl} + \delta_{il}\delta_{jk}) \tag{3.42}$$

A more illustrative depiction of the isotropic constitutive law is reached using the Voigt notation and the two components E_{11} and E_{12} of the material matrix:

$$\begin{bmatrix} \sigma_{xx} \\ \sigma_{yy} \\ \sigma_{zz} \\ \sigma_{yz} \\ \sigma_{xz} \\ \sigma_{xy} \end{bmatrix} = \begin{bmatrix} E_{11} & E_{12} & E_{12} & 0 & 0 & 0 \\ E_{12} & E_{11} & E_{12} & 0 & 0 & 0 \\ E_{12} & E_{12} & E_{11} & 0 & 0 & 0 \\ 0 & 0 & 0 & \frac{1}{2}(E_{11}-E_{12}) & 0 & 0 \\ 0 & 0 & 0 & 0 & \frac{1}{2}(E_{11}-E_{12}) & 0 \\ 0 & 0 & 0 & 0 & 0 & \frac{1}{2}(E_{11}-E_{12}) \end{bmatrix} \begin{bmatrix} \varepsilon_{xx} \\ \varepsilon_{yy} \\ \varepsilon_{zz} \\ \varepsilon_{yz} \\ \varepsilon_{xz} \\ \varepsilon_{xy} \end{bmatrix} \tag{3.43}$$

which are expressed by the Lamé constants as follows:

$$E_{11} = \lambda + 2\,\mu \tag{3.44}$$

and

$$E_{12} = \lambda \tag{3.45}$$

Accordingly, Hooke's law now reduces to an equation with the two independent constants λ and μ :

$$\sigma_{ij} = 2\,\mu\,\varepsilon_{ij} + \lambda\,\varepsilon_{kk}\,\delta_{ij} \tag{3.46}$$

Elastic parameters that can be directly characterized by material tests are for example the Young's modulus E, the Poisson ratio ν and the shear modulus G through which the Lamé constants are defined as:

$$\lambda = \frac{\nu\,E}{(1+\nu)(1-2\,\nu)} \qquad \mu = G = \frac{E}{2(1+\nu)} \tag{3.47}$$

The hydrostatic component of a stress state in an isotropic material is expressed through the Lamé constants by:

$$p = -\frac{1}{3}\,\sigma_{kk} = -\frac{1}{3}\,tr\sigma = -\frac{1}{3}\,(3\,\lambda + 2\,\mu)\varepsilon_{kk} = -K\,tr\varepsilon \tag{3.48}$$

with the *bulk modulus*

$$K = \lambda + \frac{2}{3}\,\mu \tag{3.49}$$

3.2.4 Green Elasticity - Hyperelastic Behavior

When finite strains and nonlinear elasticity are discussed, elasticity is usually formulated in a thermodynamical context using the free energy Ψ. Elastic behavior is originally defined as a material's stress-strain response for which a strain energy function exists. *Hyperelastic* material behavior is then often defined via the existence of a specific Helmholtz potential that is solely depending on the deformation gradient[4] $\psi(\mathbf{F})$. As a consequence of this reduction of the free energy it is then equal to the *strain energy* $\psi(\mathbf{F}) = 1/\rho\, W = 1/\rho\, \int \sigma_{ij}\, \mathrm{d}\varepsilon_{ij}$.

Having introduced the Helmholtz potential, e.g. in the form (2.265), and several consequences of thermo-mechanics for constitutive equations already, we can describe elastic deformations involving infinitesimal strains ε as isothermal reversible processes. These assumptions reduce the free energy, e.g. in the form (2.286), to

$$\psi = \frac{1}{2\,\rho}\, \varepsilon_{ij}\, E_{ijkl}\, \varepsilon_{kl} \tag{3.50}$$

Stress Tensors for Hyperelasticity

The difference between the definitions of elastic and hyperlastic material behavior is that the latter demands for a strain energy which, as Malvern [263] describes it precisely, is 'a scalar function of one of the strain or deformation tensors, whose derivatives with respect to a strain component determines the corresponding stress component'.

If nonlinear hyperelasticity shall be described, the specific free energy ψ is formulated as function of the deformation gradient $\psi(\mathbf{F})$ or of any other nonlinear strain measure, e.g. the Green-Lagrange strains $\psi(\mathbf{E})$ or the right Cauchy-Green tensor $\psi(\mathbf{C})$. The related nonlinear continuum mechanics are fundamentally derived in Truesdell and Noll [398], in Malvern [263] and more recently in a very modern terminology by Holzapfel [194]. Constitutive laws for hyperelastic materials can now be formulated using the various stress and strain measures, e.g. the Cauchy stress tensor:

$$\sigma = \rho\, J^{-1}\mathbf{F} \left(\frac{\partial\, \psi(\mathbf{F})}{\partial\, \mathbf{F}} \right)^T = 2\,\rho\, J^{-1}\mathbf{F} \frac{\partial\, \psi(\mathbf{C})}{\partial\, \mathbf{C}} \mathbf{F}^T = \rho\, J^{-1}\, \mathbf{F} \frac{\partial\, \psi(\mathbf{E})}{\partial\, \mathbf{E}}\, \mathbf{F}^T \tag{3.51}$$

Application of the first Piola-Kirchhoff stresses yields:

$$\mathbf{T} = \rho\, \frac{\partial\, \psi(\mathbf{F})}{\partial\, \mathbf{F}} \tag{3.52}$$

[4] This is also the definition of a *homogeneous material*.

which turns into

$$\mathbf{T} = \rho \, \frac{\partial \, \psi(\mathbf{C})}{\partial \, \mathbf{C}} \, \frac{\partial \, \mathbf{C}}{\partial \, \mathbf{F}} = 2 \, \rho \, \mathbf{F} \frac{\partial \, \psi(\mathbf{C})}{\partial \, \mathbf{C}} \tag{3.53}$$

if we recall the identities

$$\mathbf{C} = \mathbf{F}^T \mathbf{F} = 2 \, \mathbf{E} + \mathbf{1} \tag{3.54}$$

Using the second Piola-Kirchhoff stress tensor we find:

$$\hat{\mathbf{T}} = \rho \, \frac{\partial \, \psi(\mathbf{E})}{\partial \, \mathbf{E}} = \rho \, \frac{\partial \, \psi(\mathbf{C})}{\partial \, \mathbf{C}} \, \frac{\partial \, \mathbf{C}}{\partial \, \mathbf{E}} = 2 \, \rho \, \frac{\partial \, \psi(\mathbf{C})}{\partial \, \mathbf{C}} \tag{3.55}$$

Hyperelasticity in Isotropic Materials

As any other second order tensor, the right Cauchy-Green tensor $\mathbf{C}$ can be expressed in terms of its invariants I_1, I_2, I_3:

$$I_1(\mathbf{C}) = tr\,\mathbf{C} \quad I_2(\mathbf{C}) = \frac{1}{2} \left[(tr\,\mathbf{C})^2 - tr\,\mathbf{C}^2 \right] \quad I_3(\mathbf{C}) = \det \mathbf{C} \tag{3.56}$$

For isotropic materials, the free energy potential can equivalently be written in terms of these invariants

$$\psi(I_1, I_2, I_3) = \psi(\mathbf{C}) \tag{3.57}$$

The derivative of free energy in terms of $\mathbf{C}$ can then be expressed by partial derivatives with respect to the invariants I_i through the chain rule:

$$\frac{\partial \, \psi(\mathbf{C})}{\partial \, \mathbf{C}} = \frac{\partial \, \psi}{\partial \, I_1} \, \frac{\partial \, I_1}{\partial \, \mathbf{C}} + \frac{\partial \, \psi}{\partial \, I_2} \, \frac{\partial \, I_2}{\partial \, \mathbf{C}} + \frac{\partial \, \psi}{\partial \, I_3} \, \frac{\partial \, I_3}{\partial \, \mathbf{C}} \tag{3.58}$$

where the partial derivatives of the three individual invariants with respect to the right Cauchy Green tensor C can be expressed as

$$\frac{\partial \, I_1}{\partial \, \mathbf{C}} = \mathbf{I} \qquad \frac{\partial \, I_2}{\partial \, \mathbf{C}} = I_1 \, \mathbf{I} - \mathbf{C} \qquad \frac{\partial \, I_3}{\partial \, \mathbf{C}} = I_3 \, \mathbf{C}^{-1} \tag{3.59}$$

Applying these partial derivatives, *isotropic hyperelastic* material behavior can be expressed with the second Piola-Kirchhoff stresses and a Helmholtz potential depending on the strain invariants $\psi(I_1, I_2, I_3)$ as:

$$\hat{\mathbf{T}} = 2 \, \rho_0 \, \frac{\partial \, \psi(\mathbf{C})}{\partial \, \mathbf{C}} = 2 \, \rho_0 \, \left[\left(\frac{\partial \, \psi}{\partial \, I_1} + I_1 \frac{\partial \, \psi}{\partial \, I_2} \right) \mathbf{I} - \frac{\partial \, \psi}{\partial \, I_2} \, \mathbf{C} + I_3 \, \frac{\partial \, \psi}{\partial \, I_3} \, \mathbf{C}^{-1} \right] \tag{3.60}$$

Alternatively to the formulation of free energy potentials with the invariants of the Cauchy-Green strain tensor, the related principal stretches λ_1, λ_2, λ_3, i.e. the eigenvalues of the left stretch tensor $\mathbf{V}$ (2.27), can be used for isotropic materials instead:

$$\psi(\lambda_1, \lambda_2, \lambda_3) = \psi(\mathbf{C}) \tag{3.61}$$

Under the given assumption of isotropic material behavior, the principal Cauchy stresses σ_i can be expressed through the principal stretch depending potential as:

$$\sigma_i = J^{-1}\, \lambda_i\, \frac{\partial \psi}{\partial \lambda_i} \tag{3.62}$$

Hyperelasticity in Anisotropic Materials

Transversely Isotropic Materials

Anisotropy of any kind changes in a way that the stress at a material point no longer depends solely on he deformation gradient $\mathbf{F}$ but also on one or more directions. In the case of transversely isotropic materials we have one single direction showing different behavior. Unidirectional composite materials are a typical representative for transversely isotropic behavior. The direction of the fibers, let it be denoted in the referential configuration by $\mathbf{a}_0$, is a preferred direction of material behavior. This directional dependency also leads to an additional dependency of the Helmholtz free energy on the preferred direction compared to (3.57), i.e. in case of a unidirectional composite

$$\psi = \psi(\mathbf{C}, \mathbf{a}_0 \otimes \mathbf{a}_0) \tag{3.63}$$

Following Spencer [361] and Holzapfel [194] this leads to a description of the free energy through

- three invariants of the right Cauchy-Green tensor $I_1(\mathbf{C})$, $I_2(\mathbf{C})$, $I_3(\mathbf{C})$ plus
- two *pseudo-invariants* depending on the preferred direction $I_4(\mathbf{C}, \mathbf{a}_0)$ and $I_5(\mathbf{C}, \mathbf{a}_0)$

where

$$I_1 = tr\mathbf{C} \qquad I_2 = \frac{1}{2}\left\{(tr\, C)^2 - tr\, C^2\right\} \qquad I_3 = det\, \mathbf{C} = \left(\frac{\rho_0}{\rho}\right)^2$$
$$I_4 = \mathbf{a}_0\, \mathbf{C}\, \mathbf{a}_0 = \lambda^2 \qquad I_5 = \mathbf{a}_0\, \mathbf{C}^2\, \mathbf{a}_0 \tag{3.64}$$

where λ is the stretch of the material in the preferred direction $\mathbf{a}_0$.

To find a constitutive relation, e.g. in terms of the second Piola-Kirchhoff tensor, the derivative of free energy with respect to the Cauchy-Green tensor is needed. Using the chain rule again, the derivatives of the now five invariants with respect to $\mathbf{C}$ partly defined in (3.59) are completed by

$$\frac{\partial I_4}{\partial \mathbf{C}} = \mathbf{a}_0 \otimes \mathbf{a}_0 \qquad \frac{\partial I_5}{\partial \mathbf{C}} = \mathbf{a}_0 \otimes \mathbf{C}\, \mathbf{a}_0 + \mathbf{a}_0\, \mathbf{C} \otimes \mathbf{a}_0 \tag{3.65}$$

leading to a second Piola-Kirchhoff tensor for transverse isotropic materials:

$$\hat{\mathbf{T}} = 2\,\rho_0\,\Bigg[\left(\frac{\partial\psi}{\partial I_1} + I_1\frac{\partial\psi}{\partial I_2}\right)\mathbf{I} - \frac{\partial\psi}{\partial I_2}\,\mathbf{C} + I_3\,\frac{\partial\psi}{\partial I_3}\,\mathbf{C}^{-1} + \frac{\partial\psi}{\partial I_4}\,\mathbf{a}_0\otimes\mathbf{a}_0 + \frac{\partial\psi}{\partial I_5}\,(\mathbf{a}_0\otimes\mathbf{C}\,\mathbf{a}_0 + \mathbf{a}_0\,\mathbf{C}\otimes\mathbf{a}_0)\Bigg] \tag{3.66}$$

Spencer [361] shows that the specific choice of the pseudo-invariants (3.64) leaves the free energy unchanged for any change in the reference configuration expressed by any proper orthogonal tensor $\mathbf{Q}$ as

$$\psi = \psi(\mathbf{C},\,\mathbf{a}_0\otimes\mathbf{a}_0) = \psi(\mathbf{Q}\,\mathbf{C}\,\mathbf{Q}^T,\mathbf{Q}\,\mathbf{a}_0\otimes\mathbf{a}_0\,\mathbf{Q}^T) \tag{3.67}$$

and, thus, guarantees objectivity.

Orthotropic Materials

If two orthogonal directions with individual elastic properties exist, in terms of composites also called a material with two families of fibers, the material is orthotropic in the reference configuration. The second preferred direction of that material type shall be denoted in the reference configuration by $\mathbf{g}_0$, where the orthogonality demands for $\mathbf{a}_0\cdot\mathbf{g}_0 = 0$. The now two field vector products

$$\mathbf{a}_0\otimes\mathbf{a}_0 = \mathbf{A}_0 \quad \text{and} \quad \mathbf{g}_0\otimes\mathbf{g}_0 = \mathbf{G}_0 \tag{3.68}$$

which are also called *structural tensors*, lead to two more pseudo-invariants

$$I_6 = \mathbf{g}_0\cdot\mathbf{C}\mathbf{g}_0 \quad \text{and} \quad I_7 = \mathbf{g}_0\cdot\mathbf{C}^2\mathbf{g}_0 \tag{3.69}$$

The again expanded formulation for the second Piola-Kirchhoff tensor yields:

$$\begin{aligned}\hat{\mathbf{T}} = 2\,\rho_0\,\Bigg[&\left(\frac{\partial\psi}{\partial I_1} + I_1\frac{\partial\psi}{\partial I_2}\right)\mathbf{I} - \frac{\partial\psi}{\partial I_2}\,\mathbf{C} + I_3\,\frac{\partial\psi}{\partial I_3}\,\mathbf{C}^{-1}\\ &+ \frac{\partial\psi}{\partial I_4}\,\mathbf{A}_0 + \frac{\partial\psi}{\partial I_5}\,(\mathbf{a}_0\otimes\mathbf{C}\,\mathbf{a}_0 + \mathbf{a}_0\,\mathbf{C}\otimes\mathbf{a}_0) + \frac{\partial\psi}{\partial I_6}\,\mathbf{G}_0\\ &+ \frac{\partial\psi}{\partial I_7}\,(\mathbf{g}_0\otimes\mathbf{C}\,\mathbf{g}_0 + \mathbf{g}_0\,\mathbf{C}\otimes\mathbf{g}_0) + \frac{1}{2}\,\frac{\partial\psi}{\partial I_8}\,(\mathbf{a}_0\cdot\mathbf{g}_0)\,(\mathbf{a}_0\otimes\mathbf{g}_0 + \mathbf{g}_0\otimes\mathbf{a}_0)\Bigg]\end{aligned} \tag{3.70}$$

Further derivations of even more complex relations can be found in literature. It would, however, lead to far in the context of this book to introduce them also[5].

[5] For example, Holzapfel [194] expands the spectrum of anisotropic constitutive equations to materials with two families of fibers that are not orthogonal to each other, i.e. $\mathbf{a}_0\cdot\mathbf{g}_0 \neq 0$. Again two pseudo invariants are needed for that class of materials, introduced as $I_8 = (\mathbf{a}_0\cdot\mathbf{g}_0)\mathbf{a}_0\cdot\mathbf{C}\mathbf{g}_0$ and $I_9 = (\mathbf{a}_0\cdot\mathbf{g}_0)^2$. Furthermore incompressible matrix materials in composites with two families of fibers are investigated in [194].

3.2.5 Some Examples of Hyperelastic Formulations

Simulation of rubber-like materials is a typical application of hyperelastic constitutive material models. Fundamental introduction to rubber like material behavior as well as often cited experimental data are found in Ogden [301], Treloar [393], Treloar [394] and in Jones and Treloar [219].

Incompressible Formulations

Concerning the theoretical description of rubber behavior, there are both compressible formulations as well as models assuming incompressible behavior. The latter is characterized through:

$$J = \lambda_1 \, \lambda_2 \, \lambda_3 = 1 \tag{3.71}$$

Ogden and Valanis-Landel Models

A basic model for *incompressible material behavior* that is often found in literature as well as in commercial codes was proposed by Ogden [299]. The corresponding strain energy is formulated in a combination of n summands:

$$\psi = \sum_{i=1}^{n} \frac{\mu_i}{\alpha_i} \left\{ \lambda_1^{\alpha_i} + \lambda_2^{\alpha_i} + \lambda_3^{\alpha_i} - 3 \right\} \tag{3.72}$$

representing a ground-state shear modulus by:

$$G = \frac{1}{2} \sum_{i=1}^{n} \mu_i \, \alpha_i \tag{3.73}$$

Implementation in commercial codes usually limits the number n of applicable parameters couples μ_i, α_i in the Ogden model since sufficient accuracy is achieved between $n = 3$ (see Holzapfel [194], p. 236, or Treloar [394]) and $n = 6$ (as in LS-DYNA, Livermore Software Corporation, and Timmel et al. [391]).

For an *incompressible material* the constraint condition (3.71) is fulfilled through

$$\lambda_3 = (\lambda_1 \, \lambda_2)^{-1} \tag{3.74}$$

leading to a modified formulation of the potential:

$$\tilde{\psi} = \sum_{i=1}^{n} \frac{\mu_i}{\alpha_i} \left\{ \lambda_1^{\alpha_i} + \lambda_2^{\alpha_i} + (\lambda_1 \, \lambda_2)^{-\alpha_i} - 3 \right\} \tag{3.75}$$

Furthermore, compared to the compressible formulation (3.62), the corresponding principal Cauchy stresses under incompressibility (3.74) are:

$$\sigma_i = \lambda_i \, \frac{\partial \tilde{\psi}}{\partial \lambda_i} - p \tag{3.76}$$

where the terms $\lambda_i \, \partial \tilde{\psi} \, / \, \partial \lambda_i$, $(i = 1, 2)$, for the Ogden model result in:

$$\lambda_1 \frac{\partial \tilde{\psi}}{\partial \lambda_1} = \sum_{j=1}^{n} \mu_j \, \left\{ \lambda_1^{\alpha_j} - \lambda_1^{-\alpha_j} \, \lambda_2^{-\alpha_j} \right\} \tag{3.77}$$

and:

$$\lambda_2 \frac{\partial \tilde{\psi}}{\partial \lambda_2} = \sum_{j=1}^{n} \mu_j \, \left\{ \lambda_2^{\alpha_j} - \lambda_1^{-\alpha_j} \, \lambda_2^{-\alpha_j} \right\} \tag{3.78}$$

In Ogden [301] an application of (3.75) is given that is well suited to investigate the validity of theoretical values calculated from constitutive laws through comparison with biaxial experiments. Simulated was a load case that was performed experimentally before by Jones and Treloar [219]: Rectangular sheets are exposed to biaxial deformation keeping one principal stretch constant (λ_2). The resulting data were principal Cauchy stress differences ($\sigma_1 - \sigma_2$) depending on the varied principal stretch (λ_1). According to (3.77), (3.78) and (3.76) the principal Cauchy stress differences are calculated for the incompressible Ogden model as:

$$\sigma_1 - \sigma_2 = \sum_{j=1}^{n} \mu_j \, \{ \lambda_1^{\alpha_j} - \lambda_2^{\alpha_j} \} \tag{3.79}$$

In Figure 3.3 some of the cited experimental data are compared to theoretical results with the Ogden, Mooney-Rivlin and neo-Hookean models using parameters according to Table 3.1.

Table 3.1. Typical parameters for incompressible $n = 3$ Ogden models as given in Ogden [301].

i	μ_i	α_i
	10^4 [N/mm^2]	–
1	0.69	1.3
2	0.01	4.0
3	-0.0122	-2.0

As illustrated in Figure 3.3, the shape of the principal Cauchy stress differences graphs for the Ogden models does not change with varying values for λ_2. The same would be achieved if instead of the incompressible Ogden model (3.75) a *Valanis-Landel* formulation[6] was used:

[6] Valanis and Landel [408] suggested in general a decomposition of the strain energy into $\psi = \omega(\lambda_1) + \omega(\lambda_2) + \omega(\lambda_3)$.

$$\psi = \sum_{j=1}^{3} \omega(\lambda_j) \tag{3.80}$$

with

$$\omega(\lambda_j) = \sum_{i=1}^{n} \frac{\mu_i}{\alpha_i} \left\{ \lambda_j^{\alpha_i} - 1 \right\} \tag{3.81}$$

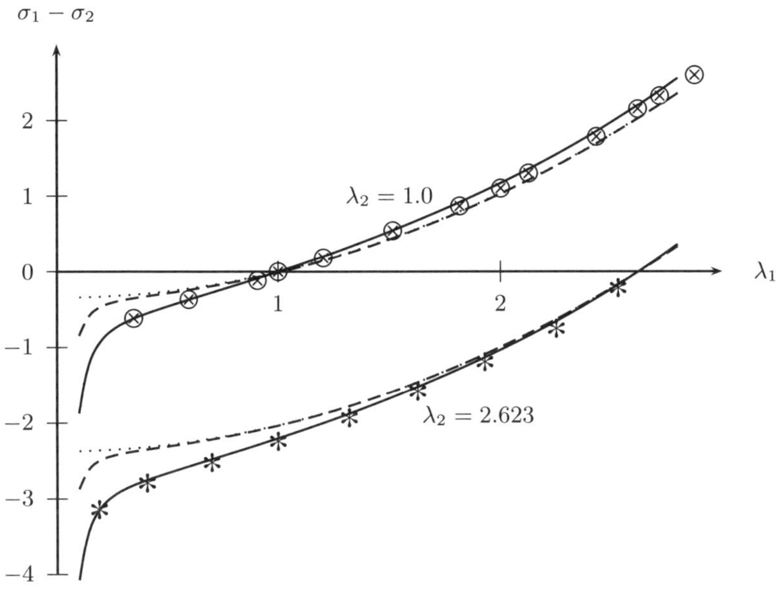

Fig. 3.3. Experimental data ($*$: $\lambda_2 = 2.623$, $\otimes$: $\lambda_2 = 1.0$) by Jones and Treloar [219] and theoretical results for biaxially loaded sheets of almost incompressible material using the material data provided by Ogden [301] as shown in Table 3.1. The principal Cauchy stress differences $\sigma_1 - \sigma_2$ are calculated with equations (3.79), (3.86) and (3.89) resulting from the Ogden (solid lines), Mooney-Rivlin (dashed lines) and neo-Hookean (dotted lines) potentials, respectively.

Mooney-Rivlin and neo-Hookean Models

Basically, the *Mooney-Rivlin* potential represent sub-groups or reduced formulations of the Ogden model which becomes evident if the Mooney-Rivlin strain energy potential is expressed using $n = 2$ parameter couples (3.75) with Ogden parameters $\alpha_1 = 2$ and $\alpha_2 = -2$. That leads to a potential of the form:

$$\begin{aligned}\psi &= \frac{c_1}{\alpha_1}\left(\lambda_1^{\alpha_1}+\lambda_2^{\alpha_1}+\lambda_3^{\alpha_1}-3\right)+\frac{c_2}{\alpha_2}\left(\lambda_1^{\alpha_2}+\lambda_2^{\alpha_2}+\lambda_3^{\alpha_2}-3\right)\\ &= \frac{c_1}{2}\left(\lambda_1^2+\lambda_2^2+\lambda_3^2-3\right)-\frac{c_2}{2}\left(\lambda_1^{-2}+\lambda_2^{-2}+\lambda_3^{-2}-3\right)\\ &= \frac{c_1}{2}\left(I_1-3\right)-\frac{c_2}{2}\left(I_2-3\right)\end{aligned} \tag{3.82}$$

with the Mooney-Rivlin parameters $c_1 = 0.5\,\mu_1$ and $c_2 = -0.5\,\mu_2$ expressed through Ogden parameters[7]:

$$\psi = \frac{\mu_1}{4}\left(\lambda_1^2+\lambda_2^2+\lambda_3^2-3\right)+\frac{\mu_2}{4}\left(\lambda_1^{-2}+\lambda_2^{-2}+\lambda_3^{-2}-3\right) \tag{3.83}$$

The Mooney-Rivlin potential leads to a shear modulus expressed through μ_2 and μ_2 by:

$$G = \mu_1 - \mu_2 \tag{3.84}$$

With the incompressibility constraint (3.74) the modified Mooney-Rivlin potential $\tilde{\psi}$ can be written using Ogden parameters as:

$$\tilde{\psi} = \frac{\mu_1}{4}\left(\lambda_1^2+\lambda_2^2+\lambda_1^{-2}\,\lambda_2^{-2}-3\right)+\frac{\mu_2}{4}\left(\lambda_1^{-2}+\lambda_2^{-2}+\lambda_1^2\,\lambda_2^2-3\right) \tag{3.85}$$

leading to a principal Cauchy stress difference of:

$$\begin{aligned}\sigma_1-\sigma_2 &= \lambda_1\,\frac{\partial\tilde{\psi}}{\partial\lambda_1}-\lambda_2\,\frac{\partial\tilde{\psi}}{\partial\lambda_2}\\ &= \frac{\mu_1}{2}\,\left(\lambda_1^2-\lambda_2^2\right)-\frac{\mu_2}{2}\,\left(\lambda_1^{-2}+\lambda_2^{-2}\right)\end{aligned} \tag{3.86}$$

The *neo-Hookean* potential reduces the number of Ogden parameter pairs to a single one and is, thus, formulated as a $n = 1$ and $\alpha_1 = 2$ model:

$$\begin{aligned}\psi &= \frac{c_1}{2}\left(\lambda_1^2+\lambda_2^2+\lambda_3^2-3\right)\\ &= \frac{\mu_1}{4}\left(\lambda_1^2+\lambda_2^2+\lambda_3^2-3\right)\\ &= \frac{c_1}{2}\left(I_1-3\right)\end{aligned} \tag{3.87}$$

with a single parameter $c_1 = 0.5\,\mu_1$ and a corresponding shear modulus

$$G = \mu_1 \tag{3.88}$$

Incompressibility and the reduction to a single parameter couple leads to a further simplification of the equation to calculate the principal Cauchy stress differences:

---

[7] At this point, a matter of confusion repeatedly found in literature ([194] as well as [301]) should be noted: Denoting the Mooney-Rivlin potential as $n = 2$ Ogden model with $\alpha_1 = 2$ and $\alpha_2 = -2$ necessarily leads to (3.82). Omitting the denominators 2 and -2, respectively, neglects either the α_i terms or half of the $\mu_i/2$ terms. Anyway, it would always produce a factor *two* in the potential as well as in derived terms. Note: the same is true for the neo-Hookean potential.

$$\sigma_1 - \sigma_2 = \frac{\mu_1}{2}\left(\lambda_1^2 - \lambda_2^2\right) \tag{3.89}$$

Compared to the experimental data of Jones and Treloar [219] there is also good agreement of neo-Hookean and Mooney-Rivlin models with respect to the vertical shift of curves for varying λ_2-values. However, both models clearly show missing capabilities to predict the measured data the further λ_1 varies from $\lambda_1 = \lambda_2$[8]. From Figure 3.3 this becomes apparent in the interval $1.5 < \lambda_1 < 3$: while both models predict good to exact values around $\lambda_1 = 2.623$ in case of $\lambda_2 = 2.623$, they tend to deviate more and more from the experimental results for $\lambda_2 = 1$.

Yeoh Potential for Incompressible Filled Rubber Materials

Significant nonlinearity of the shear modulus G with increasing deformation observed in material filled with carbon black[9] motivated the formulation of an alternative potential. The shear modulus for incompressible materials:

$$G = 2\left(\frac{\partial \psi}{\partial I_1} + \frac{\partial \psi}{\partial I_2}\right) \tag{3.90}$$

represented by models discussed so far, i.e. (3.73) for the Ogden, (3.84) for the Mooney-Rivlin and (3.88) in case of the neo-Hookean model, is independent on the deformation. Therefore, Yeoh [431] formulated a potential solely in terms of the first invariant that is cubic in $(I_1 - 3)$:

$$\psi = c_1\,(I_1 - 3) + c_2\,(I_1 - 3)^2 + c_3\,(I_1 - 3)^3 \tag{3.91}$$

leading to a shear modulus of

$$G = 2c_1 + 4c_2(I_1 - 3) + 6c_3(I_1 - 3)^2 \tag{3.92}$$

Compressible Formulations

To extend the incompressible models described above for application on compressible materials, the free energy is decomposed into a dilatational and a isochoric component. This is achieved using the modified left Cauchy-Green tensor

$$\tilde{\mathbf{B}} = J^{-2/3}\,\mathbf{B} = J^{-2/3}\,\mathbf{F}\,\mathbf{F}^T \tag{3.93}$$

to formulate the decomposition of the free energy as

$$\psi(\mathbf{B}) = \psi_{dil}(J) + \psi_{iso}(\tilde{\mathbf{B}}) \tag{3.94}$$

In terms of the invariants of the modified left Cauchy-Green tensor (3.93) :

[8] See also Ogden [301], p. 495

[9] See for example Seki et al. [350].

$$I_1(\tilde{\mathbf{B}}) = tr\,\tilde{\mathbf{B}} \quad I_2(\tilde{\mathbf{B}}) = \frac{1}{2}\left[(tr\,\tilde{\mathbf{B}})^2 - tr\,\tilde{\mathbf{B}}^2\right] \quad I_3(\tilde{\mathbf{B}}) = \det\,\tilde{\mathbf{B}} = J \tag{3.95}$$

the Helmholtz free energy can again be written as

$$\psi(\mathbf{B}) = \psi_{dil}(J) + \psi_{iso}\left(I_1(\tilde{\mathbf{B}}),\, I_2(\tilde{\mathbf{B}})\right) \tag{3.96}$$

One option to describe the dilatational component $\psi_{dil}(J)$ is applying the material bulk modulus K to get:

$$\psi_{dil}(J) = K\,\left[0.5\,(J^2 - 1) - \ln J\right] \tag{3.97}$$

Compressible Mooney-Rivlin Model

A compressible form of the Mooney-Rivlin model (3.82) can then be constructed by

$$\psi = \frac{c_1}{2}\left(I_1(\tilde{\mathbf{B}}) - 3\right) + \frac{c_2}{2}\left(I_2(\tilde{\mathbf{B}}) - 3\right) + c_3\left(J^{-1} - 1\right) + c_4\left(J^{0.5} - 1\right)^2 \tag{3.98}$$

where

$$c_3 = \frac{c_1}{4} + \frac{c_2}{2} \qquad c_4 = \frac{c_1\,(5\,\nu - 2) + c_2\,(11\,\nu - 5)}{1 - 2\,\nu} \tag{3.99}$$

Blatz-Ko Model

With the Blatz-Ko model another formulation of Helmholtz free energy for compressible materials can be written in terms of the invariants (3.95) as:

$$\psi = \frac{G}{2}\,\left[(I_1(\tilde{\mathbf{B}} - 3) + \frac{1-2\nu}{\nu}\left(J^{\frac{-2\nu}{1-2\nu}} - 1\right)\right] \tag{3.100}$$

Further phenomenological formulations of constitutive equations for compressible and incompressible materials can be found in Fu and Ogden [131], Belytschko et al. [47] and Holzapfel [194]. Examples for modelling of rubber-like materials under dynamic loading conditions using LS-DYNA have been published by Timmel, Kaliske and Kolling [391].

Elastic Potentials Based on Statistical Mechanics

Statistical mechanics is an elegant way of linking fundamental physics observed on a microscopic scale to macroscopic formulations, e.g. to constitutive laws for materials. The models discussed in the previous paragraphs for hyperelastic material behavior are all of phenomenological nature. There is, however, a number of attempts and a whole sequence of publications on modeling hyperelasticity, specifically the mechanical behavior of rubbery and glassy polymers, with statistical mechanics[10]. In that framework, statistical mechanics means formulating the macromolecular structure of elastomeric and thermoplastic polymers with chain statistics.

[10] To name only some key publications, the reader is referred to Kuhn [235], Wall [414], Flory and Rehner [128], James and Guth [207], Treloar [394], Arruda and Boyce [18], Anand [4], Anand and Gurtin [6] and Gearing [133].

Chain Statistics to Model the Thermomechanics of Polymers

Basically, chain statistics describes an individual macromolecule as a chain of N segments of identical length l where the links connect the segments under an arbitrary angle. That is where the statistical component of the theory begins, i.e. when the chain in its initial undeformed state is modeled as product of a random walk of N steps. In the reference configuration the initial chain length is described by

$$r_0 = \sqrt{N}\, l \tag{3.101}$$

In a chain with one end in the origin of a Cartesian coordinate system, the position $P(x, y, z,)$ of the other end is described through a statistical distribution. The probability $p(x, y, z)$ of an infinitesimal rectangular block $\mathrm{d}\,\tau = \mathrm{d}\,x,\ \mathrm{d}\,y,\ \mathrm{d}\,z$ around $P(x, y, z,)$, to contain the end point of the chain is proportional to the number of segment-configurations leading into this block. Initial solutions to describe the probability for the chain models, introduced by Kuhn [234] as well as Guth and Mark [156], utilized Gaussian functions

$$p(x, y, z)\mathrm{d}x\,\mathrm{d}y\,\mathrm{d}z = \frac{b^3}{\pi^{3/2}}\, e^{-b^2(x^2+y^2+z^2)}\mathrm{d}x\,\mathrm{d}y\,\mathrm{d}z \tag{3.102}$$

with $b^2 = \frac{3}{2} N\, l^2$. An important drawback of these distributions is their limitation to small chain stretches and thus total stretched chain lengths small compared to the maximum length:

$$r << N\, l \tag{3.103}$$

The distribution is further used to derive the configurational entropy of a single chain as well as of a network of interconnected chains. Boltzmann statistical mechanics describe the configurational entropy of a single chain as proportional to the logarithm of the number of available configurations, where all configurations are of equal probability for an unconfined chain. In contrast to such an isolated chain, the cross-links of entangled, inter-connected chains in a network, representing the structural polymer material that we want to model, imply constraints. That means the motion of the ends of intervening segments is restricted to a certain volume $\mathrm{d}\,\tau$ leading to a configurational entropy of a single chain of:

$$s = c - k\, b^2\, r^2 \tag{3.104}$$

with a constant parameter c and Boltzmann's constant k. The change in total entropy in a network of n deformed chains per unit volume is then:

$$\Delta S = \sum \Delta s = -\frac{1}{2} n\, k \left(\lambda_1^2 + \lambda_2^2 + \lambda_3^2 - 3\right) \tag{3.105}$$

assuming an affine deformation of the moving chain end from x_0, y_0, z_0 to x, y, z with

$$\lambda_i = \frac{x_i}{x_0} \tag{3.106}$$

With the Gaussian statistics (3.102), the corresponding strain energy of the deformed network yields:

$$W = -T\,\mathrm{d}S = \frac{1}{2} n\,k\,T\,(I_1 - 3) = \frac{1}{2} G\,(I_1 - 3) \tag{3.107}$$

with the first invariant of the left Cauchy-Green tensor I_1 and the shear modulus G.

Since the major limitation (3.103) of the Gaussian theory is that the chain must not approach its fully stretched length $r_L = N\,l$, non-Gaussian chain statistics were developed[11] inventing modifications of the distribution. Kuhn and Grün [236] derived a logarithmic probability distribution involving the *Langevin function* allowing for a corrected representation of the limiting chain extensibility.

The Langevin function is defined as

$$L(x) = \coth(x) - 1/x \tag{3.108}$$

An approximation of (3.108) is found in the series

$$\langle L(x) \rangle = \frac{1}{3}\,x - \frac{1}{45}\,x^3 + \frac{2}{945}\,x^5 - \frac{1}{4725}\,x^7 + ... \tag{3.109}$$

Accordingly, the inverse Langevin function can be approximated through the series

$$\langle L^{-1}(x) \rangle = 3\,x + \frac{9}{5}\,x^3 + \frac{297}{175}\,x^5 + \frac{1539}{875}\,x^7 + ... \tag{3.110}$$

With the Langevin statistics, the probability density (3.102) was replaced by Kuhn and Grün through a logarithmic definition:

$$\ln p(r) = c - N\left(\frac{r}{N\,l}\,\beta + \ln\frac{\beta}{\sinh\beta}\right) \tag{3.111}$$

Arruda-Boyce Eight-Chain Model

Following several other chain models, e.g. the four-chain tetrahedral model of Flory and Rehner [128], Arruda and Boyce [18] introduced the representative cubical volume of initial edge length a_0 formed by eight chains connected in the center (see Figure 3.4). With (3.101) the edge length in the so called *eight-chain-model* described in terms of initial chain lengths by

$$a_0 = \frac{2}{\sqrt{3}} r_0 \tag{3.112}$$

---

[11] see Treloar [394] for detailed descriptions of both Gaussian and non-Gaussian theories

Macroscopic normal stretches of the representative element λ_1, λ_2, λ_3 are related to the microscopical chain stretches via the root mean square:

$$\lambda_{chain} = \frac{1}{\sqrt{3}} \sqrt{\lambda_1^2 + \lambda_2^2 + \lambda_3^2} \tag{3.113}$$

With the length of the fully stretched chain $r_L = N\,l$ its maximum stretch is $\lambda_L = \frac{r_L}{r_0} = \sqrt{N}$.

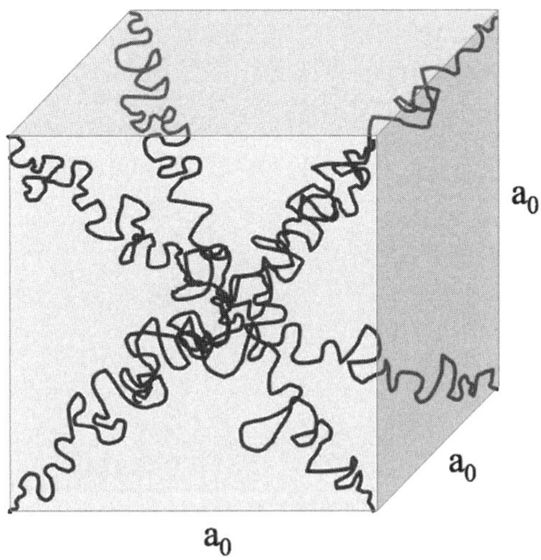

Fig. 3.4. Eight chain model for rubber constitutive equations basing on chain statistics (Arruda and Boyce [18]).

The sum of the entropies of the individual chains describes the total entropy. The non-Gaussian probability distribution function (3.111) for the length of a molecular chain allows for high chain extensions with end-to-end distances of chain ends r close to the chain's contour length L. After Kuhn and Grün [236] the configurational entropy s of a stretched single chain of current length r is

$$s = k \left[c - N \left(\frac{r}{N\,l}\,\beta + \ln \frac{\beta}{\sinh \beta} \right) \right] \tag{3.114}$$

with the parameter c, Boltzmann's constant k and the inverse Langevin function $\beta = L^{-1}(r/Nl)$. The related Helmholtz free energy is described by

$$\psi = n\,k\,T\,N \left(\frac{r}{N\,l}\,\beta + ln \frac{\beta}{\sinh \beta} \right) - T\,c' \tag{3.115}$$

with the additional parameter c' and the chain density n.

In terms of the first invariant of the Cauchy-green tensor I_1 and using the first five elements of the inverse Langevin approximation, the Helmholtz free energy for the Arruda-Boyce eight chain model can be expressed by:

$$\psi = G \left[\frac{1}{2} \left(I_1 - 3\right) + \frac{1}{20\,N} \left(I_1^2 - 9\right) + \frac{11}{1050\,N^2} \left(I_1^3 - 27\right) \right.$$
$$\left. + \frac{19}{7000\,N^3} \left(I_1^4 - 81\right) + \frac{519}{673750\,N^4} \left(I_1^5 - 243\right) + \dots \right] \quad (3.116)$$

with the shear modulus G.

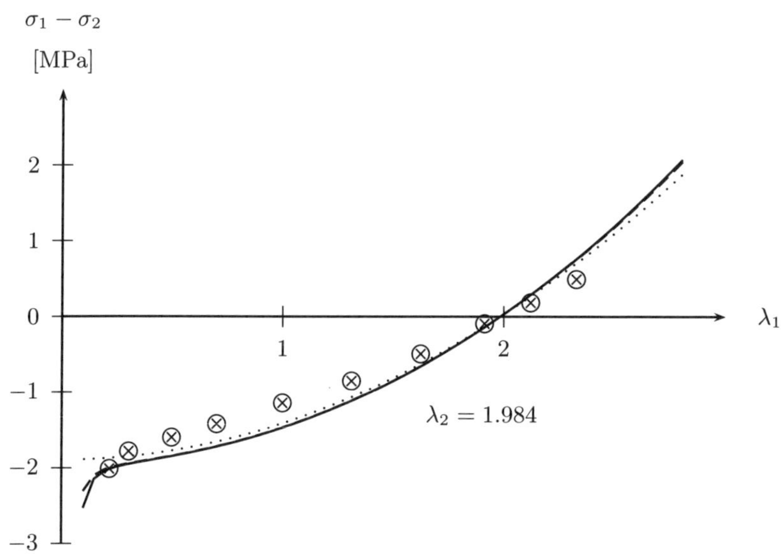

Fig. 3.5. Principal Cauchy stress differences $\sigma_1 - \sigma_2$ under biaxial loads calculated from the Arruda-Boyce potential (3.116) with one, two and five terms in the approximative inverse Langevin function. Comparison with experimental data ($\otimes$: $\lambda_2 = 1.984$) by Jones and Treloar [219] for biaxially loaded sheets of nearly incompressible rubber material using the material data given in Table 3.1.

Figure 3.5 illustrates the influence of the number of terms used in equation (3.116) to approximate the inverse Langevin function. It reflects the application of the Arruda-Boyce potential with a chain density $N = 26.5$ to describe the incompressible material investigated before in Figure 3.3 under biaxial loads. With the incompressibility constraint $\lambda_3 = 1/(\lambda_1\,\lambda_2)$ and, thus, the first invariant of the left Cauchy-Green tensor under incompressibility $\tilde{I}_1 = \lambda_1^2 + \lambda_2^2 + (\lambda_1 \lambda_2)^{-2}$ yields a related formulation of the potential (3.116) $\tilde{\psi} = \tilde{psi}(\tilde{I}_1)$. The principal stresses are calculated according to (3.76) with:

$$\sigma_i = \lambda_i \frac{\partial \tilde{\psi}}{\partial \lambda_i} - p = \lambda_i \frac{\partial \tilde{\psi}}{\partial \tilde{I}_1} \frac{\partial \tilde{I}_1}{\partial \lambda_i} - p \tag{3.117}$$

where

$$\frac{\partial \tilde{\psi}}{\partial \tilde{I}_1} = G \left[\frac{1}{2} + \frac{\tilde{I}_1}{20\,N} + \frac{33\,\tilde{I}_1^2}{1050\,N^2} + \frac{76\,\tilde{I}_1^3}{7000\,N^3} + \frac{2595\,\tilde{I}_1^4}{673750\,N^4} + ... \right] \tag{3.118}$$

and

$$\begin{aligned} \frac{\partial \tilde{I}_1}{\partial \lambda_1} &= 2 \left[\lambda_1 - (\lambda_1 \lambda_2)^{-3} \lambda_2 \right] \\ \frac{\partial \tilde{I}_1}{\partial \lambda_2} &= 2 \left[\lambda_2 - (\lambda_1 \lambda_2)^{-3} \lambda_1 \right] \end{aligned} \tag{3.119}$$

Concerning the series approximation in (3.118), the calculated results show convergence with two terms, already. Whereas the use of a single term, essentially reflecting a neo-Hookean behavior, cannot reflect the cubic behavior in λ_1. This is achieved with at least two terms. Any further enhancement in complexity yields only little improvement in the results for that load case.

Compared to the experimental data, the shape of the stress-strain curve is well reproduced with two or more terms, where at least three terms are needed to show the steep gradient around $\lambda_1 < 0.1$. In general, the accuracy of the model for that load case compared to the experimental data is lower than what was achieved with the phenomenological Ogden model discussed in Figure 3.3 but better than the Mooney-Rivlin or neo-Hookean results. Besides that, it is superior to all other chain models (for comparison see [18]).

Anand's Extension to Compressible Solids

The incompressible and isothermal eight-chain model discussed in the last paragraph was generalized to compressible solids and thermal effects by Anand [4]. Though the compressibility of rubber materials is usually not of interest since the ratio in the typical values of shear and bulk moduli is usually in the range of $\frac{G}{K} \approx 10^{-4}$, some specific applications demand for modeling the related influence. Examples are rubber O-rings and generally all highly confined structures made from rubber.

Anand invented the application of a scalar effective distortional stretch:

$$\bar{\lambda}^* = \frac{1}{\sqrt{3}} \sqrt{\mathbf{V}^* \mathbf{V}^*} = \frac{1}{\sqrt{3}} \sqrt{\text{tr}\,(\mathbf{U}^2)} \tag{3.120}$$

using the distortional component $\mathbf{V}^*$ of the stretch tensor $\mathbf{V}$, defined by equation (2.57), to describe a related Helmholtz free energy $\psi^* = \psi^*\left(\bar{\lambda}^*, J, T\right)$ and a stretch-dependent shear modulus $\mu = \mu\left(\bar{\lambda}^*, \lambda_L\right)$. The resulting pressure and temperature dependent constitutive equation in terms of Cauchy stresses yields:

$$\mathbf{T} = J^{-1}\left[\mu\,\mathbf{B}_0^p + \kappa\left\{\ln J - 3\,\alpha\,(T - T_0)\right\}\mathbf{I}\right] \tag{3.121}$$

with temperature dependent bulk modulus $\kappa = \kappa(T) = \kappa_0 + m\,(T - T_0)$, a constant coefficient of thermal expansion α and the deviatoric component of the left Cauchy-Green tensor $\mathbf{B}_0^p$:

$$\mathbf{B}_0^p = \mathbf{B}^p - \frac{1}{3}\left(\mathrm{tr}\,\mathbf{B}^p\right)\mathbf{I} \tag{3.122}$$

where $\mathbf{B}^p$ is the left Cauchy-Green tensor corresponding to the plastic deformation gradient $\mathbf{F}^p$:

$$\mathbf{B}^p = \mathbf{F}^p\,\mathbf{F}^p \tag{3.123}$$

Comparison of the model results with experimental data for incompressible material behavior by Jones and Treloar [219] shows similar quality as the eight-chain model. Comparison with experimental data on volumetric expansion and compression as well as on thermal expansion of vulcanized natural rubbers shows acceptable results, in case of large stretches and moderate volume changes very good agreement.

Further application of chain statistics and related models will be discussed in section 3.3.7 concerned with inelastic material behavior of polymers including softening, hardening and plastic anisotropy.

3.3 Constitutive Equations for Inelastic Deformations

Irreversible deformation of structures results from fundamental changes in the micro-structure of the structural materials, e.g. dislocation movement, growth and coalescence of micro-defects or disentanglement of polymer chains. Other than elastic deformations, inelastic stress and strain states go along with fundamental modifications of the initially unloaded material, albeit on a molecular or crystalline scale. These variances in the material structure in turn involve a variation of the material properties identified for the virgin material. Prediction of stress-strain states under which these changes occur, is an indispensable task of computational codes for application in numerical simulation not only of crash and impact processes. Therefore, the continuum thermo-mechanics of yield and failure need to be understood sufficiently well to adequately choose from existing models. The following chapter shall give an introduction to the basic terminology commonly used in the formulation of both yield and failure criteria.

3.3.1 Basic Terminology in Plasticity Theory

Plasticity theory provides methods to identify criteria for the initiation of plastic yield as well as models to describe the subsequent flow behaviour in terms of stress and strain increments. In order to determine under which stress and strain conditions plastic yielding occurs, the material specific conditions for inelastic, irreversible deformation need to be identified. For most materials however, there is no precise physical threshold between pure elastic and beginning plastic deformation. In fact, the co called *yield criteria* are conventions upon which amount of permanent deformation is regarded as yield start. An example for such a commonly used convention is the $\sigma_{Y\,02}$-threshold defining beginning plastic flow in a uniaxial tension test as the stress-strain location for which after total unloading $\varepsilon^{pl} = 0.2$ % plastic strain remain in the sample. As illustrated in Figure 3.6, this threshold can be identified by a parallel shift of the initial elastic slope in the one-dimensional stress strain curve. Though a first, in the stress strain graph visible, deviation from linear elasticity occures at point A, this convention defines point B as *yield point* with an initial yield stress $\sigma^0_{Y\,02}$. Complexity in this seemingly simple identification of the plastic threshold rises when the material is non-linear elastic, e.g. to be observed in polymers, or when the material is of visco-elastic nature as well as whenever dynamic material tests involve acceleration and wave effects that superimpose the measured signals.

Also, Figure 3.6 shows a simplification in the so called *perfectly plastic* behaviour which is sometimes applied when maximum expected strains are close to the initial yield strain or when first occurrence of plastic zones shall terminate the calculation. More realistic material behaviour changes the initial yield stress that was found in point B. When a material sample with the illustrated behaviour was loaded to position C, then unloaded along the shifted line of elasticity and reloaded again, a new yield stress would be found in σ^C. In the one-dimensional case, this *strain hardening* or *work hardening* called effect shifts the yield point and leads to either *isotropic* or *kinematic hardening*. Both forms of hardening are sketched in Figure 3.7 where identical absolute values for initial yield stresses under tension and compression are assumed.

Before several selected yield conditions and flow rules can be introduced, some basic terminologies to describe them need to be defined.

Planes and Surfaces in the Principal Stress Space

Since material models for computational structural dynamics need to be formulated for three-dimensional applications, the related plasticity concepts take shapes in the six-dimensional stress and strain spaces. For isotropic materials the models can be formulated in a more plausible way in the three-dimensional principal stress space. This geometrical representation was

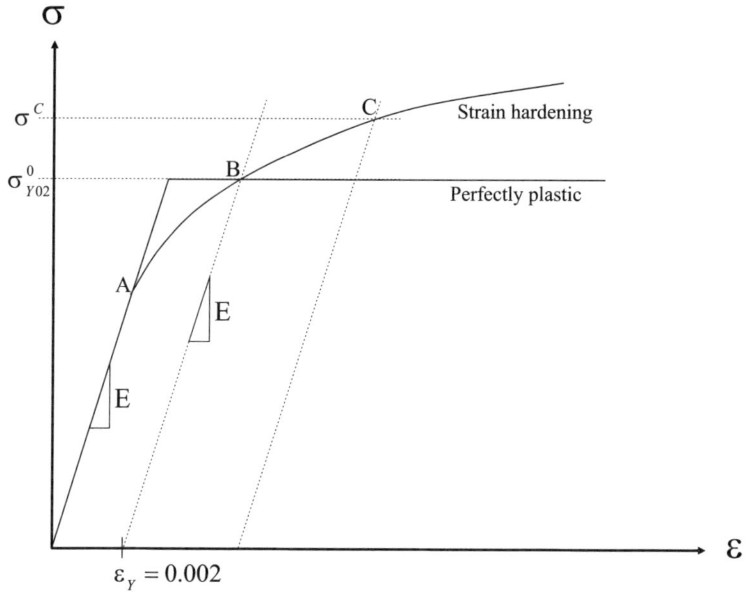

Fig. 3.6. Idealized linear perfectly plastic behaviour, realistic strain hardening and definition of an 0.2 percent plastic strain threshold for initial plastic yield .

suggested by Haigh [157] and Westergaard [418] which is the reason why the principal stress space is also called *Haigh-Westergaard space*. Its spatial diagonal

$$\sigma_1 = \sigma_2 = \sigma_3 = \frac{1}{3} J_1 \tag{3.124}$$

depicts the states of pure hydrostatic stresses ($S_1 = S_2 = S_3 = 0$) and is therefore called *hydrostatic axis*. Any plane that is normal to one of the four spatial diagonals in the principal stress space is called *octahedral plane*. For each of the eight families of octahedral planes we find

$$\pm\sigma_1 \pm \sigma_2 \pm \sigma_3 = \sqrt{3}\xi \tag{3.125}$$

where ξ is the distance between the plane and the origin of the stress space and also called the *first Haigh-Westergaard coordinate* (see the next paragraph). The normal and shear stresses on and in an octahedral plane are

$$\sigma_{oct} = \frac{1}{3}\left(\sigma_1 + \sigma_2 + \sigma_3\right) = \frac{1}{3} I_1 \tag{3.126}$$

and

$$\tau_{oct} = \frac{1}{3}\sqrt{\left(\sigma_1 - \sigma_2\right)^2 + \left(\sigma_2 - \sigma_3\right)^2 + \left(\sigma_3 - \sigma_1\right)^2} = \sqrt{\frac{2}{3} J_2} \tag{3.127}$$

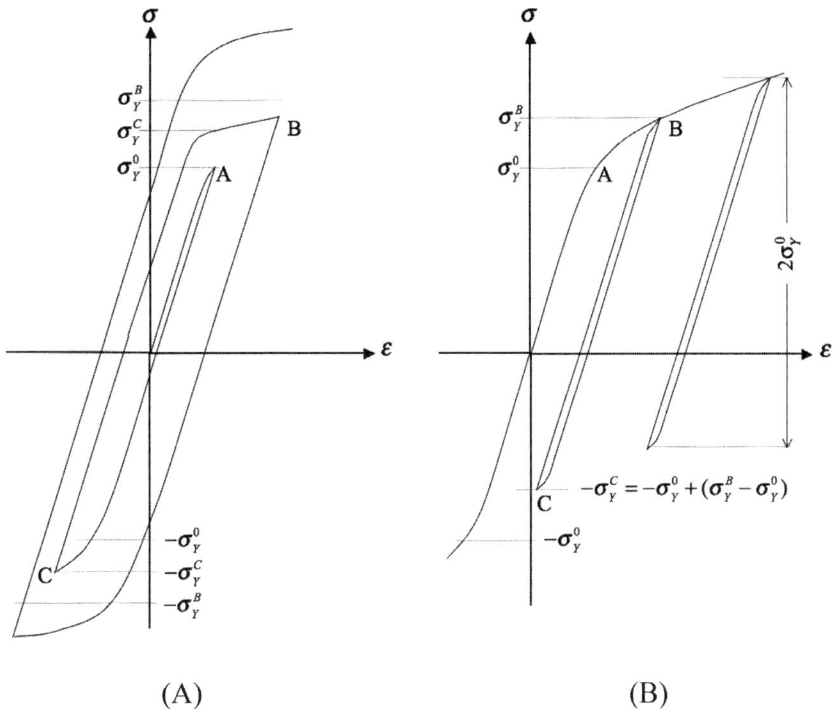

Fig. 3.7. Isotropic (A) and kinematic (B) strain hardening effect under uniaxial cyclic loading and unloading to tension and compression.

The family of octahedral planes normal to the hydrostatic axis

$$\sigma_1 + \sigma_2 + \sigma_3 = \sqrt{3}\xi \tag{3.128}$$

are called *deviatoric planes* and the one deviatoric plane that contains the origin ($\xi = 0$) is called π*-plane.*

Haigh-Westergaard Coordinates

As already mentioned before, the radial distance of a deviatoric plane to the origin of the principal stress space along the hydrostatic axis is defined with the first Haigh-Westergaard coordinate ξ defined according to (3.125) as:

$$\xi = \frac{1}{\sqrt{3}} \tag{3.129}$$

Illustrated in Figure 3.9, the second Haigh-Westergaard coordinate ρ leads from the the end point P in the deviatoric plane defined by ξ to the border

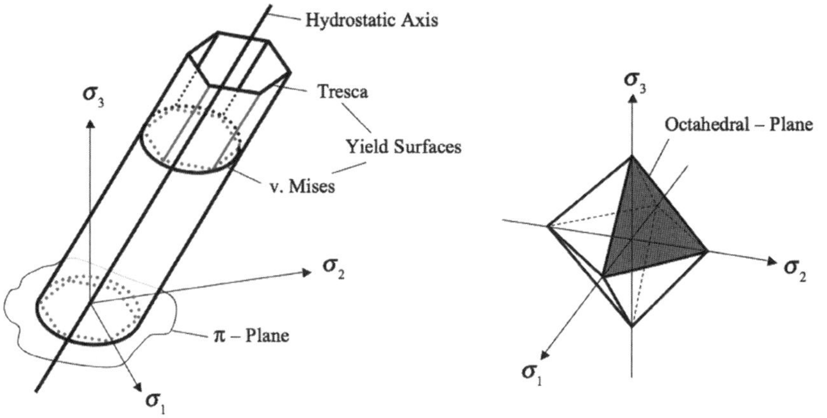

Fig. 3.8. Definitions of the hydrostatic axis, deviatoric planes, octahedral planes and π-plane in the principal stress space.

of the yield surface in that plane. As a consequence from the definition of the deviatoric planes (3.128), any stress vector pointing from the hydrostatic axis to a position in the same deviatoric plane, e.g. vector $\bar{PQ}$ in Figure 3.9 (A), can be represented by a deviator as:

$$\bar{PQ} = (S_1,\ S_2,\ S_3) \tag{3.130}$$

Accordingly, the magnitude of coordinate ρ in a given plane is expressed through the deviatoric stress components, the second invariant of the deviator or the octahedral shear stress, respectively, as:

$$\rho = \sqrt{S_1^2 + S_2^2 + S_3^2} = \sqrt{2\ J_2} = \sqrt{3}\ \tau_{oct} \tag{3.131}$$

The remaining third coordinate denotes the angle θ under which ρ points to the yield surface. Where the definition of the direction $\theta = 0°$ is such that it coincides with the direction of the projected first principal stress axis σ_1' into the deviatoric plane. Thus, the third Haigh-Westergaard coordinate can be expressed through[12] :

$$cos\,\theta = \frac{\sqrt{3}}{2}\frac{S_1}{\sqrt{J_2}} \tag{3.132}$$

or

$$cos\ 3\,\theta = \frac{3\ \sqrt{3}}{2}\frac{J_3}{J_2^{3/2}} \tag{3.133}$$

[12] For detailed geometrical derivations and further expressions for yield loci in the principal stress space the reader may be referred to Chen and Han [86] as well as Khan and Huang [223]

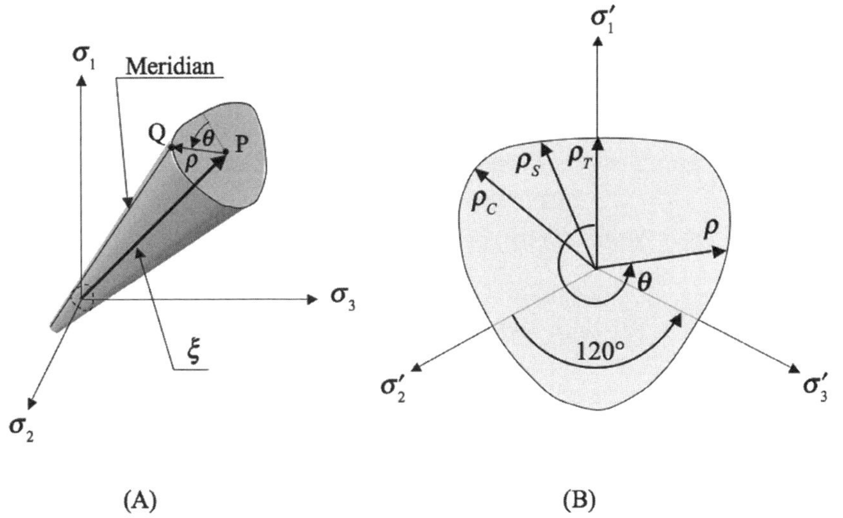

Fig. 3.9. (A) Illustration of the Haigh-Westergaard coordinates ξ, ρ and θ in the principal stress space and example meridian as intersecting line of a meridian plane $\theta = const.$ with the yield surface. (B) Projections σ_i' of the principal stress axes into the deviatoric plane with positions of the tensile meridian $\theta = 0^\circ$, ρ_T, shear meridian $\theta = 30^\circ$, ρ_S and compressive meridian $\theta = 60^\circ$, ρ_C.

Meridians

Planes perpendicular to the deviatoric planes and passing the hydrostatic axis at angles $\theta = const.$ are called *meridian planes*. The intersecting line of a meridian plane with a yield surface is called *meridian*. Magnitudes of the coordinate ρ along a certain meridian, specifically under the angles $\theta = 0^\circ$, $\theta = 30^\circ$ and $\theta = 60^\circ$, offer important information on the yield behaviour. That is due to the fact that the following stress conditions are found along these meridians:

- Positions along the 0°-meridian represent hydrostatic pressure states (depending on the coordinate ξ) superimposed by uniaxial tensile stresses: $\sigma_1 > \sigma_2 = \sigma_3$ which for pure uniaxial tension becomes: $\sigma_2 = \sigma_3 = 0$.
- The 30°-meridian represents hydrostatic pressure states superimposed by pure shear: $\sigma_1 - \sigma_2 = \sigma_2 - \sigma_3$ with $\sigma_1 > \sigma_2 > \sigma_3$
- and finally, uniaxial compression in the presence of hydrostatic pressure is represented along the 60°-meridian with $\sigma_1 = \sigma_2 > \sigma_3$.

For isotropic materials, only the section between $\theta = 0^\circ$ and $\theta = 60^\circ$ needs to be investigated experimentally since material symmetry expresses itself in the symmetric shape of the residual section.

With the Haigh-Westergaard coordinates (ξ, ρ, θ) principal stress space related yield surfaces can be described. Of specific interest is the (ξ, ρ)-plane when the positive *rho*-direction is associated with $\theta = 0°$ and the negative *rho*-direction with $\theta = 60°$ since then the tensile and compressive yield behaviour is illustrated in dependency to the superimposed hydrostatic pressure. An example for graphical illustrations of that type is shown in Figure 3.11.

3.3.2 Selected Yield Criteria

In this section, mathematical tools to model irreversible plastic deformation will be collected. For that purpose, *yield functions* will be used to identify beginning plastic deformation. Subsequent stress and strain conditions described by so called *flow rules* will be discussed in the next section.

In the general six-dimensional stress space a hypersurface describes the entirety of all possible stress states corresponding to plastic yield. The yield function $F(\sigma_{ij})$ is defined to equal zero on the yield surface

$$F(\sigma_{xx}, \sigma_{yy}, \sigma_{zz}, \sigma_{xy}, \sigma_{xz}, \sigma_{yz},) = 0 \tag{3.134}$$

and to be less than zero for elastic stress states:

$$F(\sigma_{ij}) < 0 \quad . \tag{3.135}$$

More evidently, a surface in the three-dimensional principal stress space separates elastic from plastic stress states for isotropic materials, describing yield stress states by

$$F(\sigma_1,\ \sigma_2,\ \sigma_3) = 0 \tag{3.136}$$

Utilization of the Haigh-Westergaard coordinates leads to formulations of yield conditions as

$$F(\xi,\ \rho,\ \theta) = 0 \tag{3.137}$$

Very common to the formulation of yield and fracture surfaces in the principal stress space is also the use of the invariants of the stress tensor:

$$F(I_1, I_2, I_3) = 0 \tag{3.138}$$

Pressure Independent Isotropic Formulations

Early experimental investigation of metal plasticity, e.g. Hill [184] showed that there is no pressure dependency and no purely pressure induced yielding even under hydrostatic pressure in the kilobar regime[13]. The observation of that *plastic incompressibility* leads to the reduced formulation of pressure independent yield criteria through the second and third deviator invariants:

---

[13] Bridgeman [67] tested at hydrostatic pressure of 25 [kbar]

$$F(J_2, J_3) = 0 \tag{3.139}$$

Two early developed criteria which are still found as standard models to date are the pressure independent formulations after von Mises and Tresca, respectively. The von Mises criterion[14] [280] bases on the further assumption that besides the independence on the hydrostatic pressure (3.139) also the third invariant has no influence:

$$F(J_2) = 0 \tag{3.140}$$

Translated into Haigh-Westergaard coordinates, equation (3.140) means that the resulting yield surface has equal radial loci over the angle θ and is therefore a circle in the deviatoric plane and a circular cylinder in the principal stress space. The resulting *octahedral shear stress* criterion is therefore:

$$J_2 - \kappa^2 = 0 \tag{3.141}$$

where the radius of the cylinder is

$$r = \sqrt{2}\kappa \tag{3.142}$$

and its projection into the two-dimensional plane stress surface σ_1, σ_2 forms an ellipse:

$$\sigma_1^2 - \sigma_1\,\sigma_2 + \sigma_2^2 = 3\;\kappa^2 \tag{3.143}$$

Under one-dimensional stress conditions as present in a standard uniaxial tension test, the limit stress σ_Y is found to be

$$\sigma_Y = \sqrt{3}\kappa \tag{3.144}$$

whereas in a pure shear stress test, the parameter κ is identified directly in the yield stress

$$\tau_Y = \kappa \tag{3.145}$$

A combined axial tension and torsion test with a cylindrical material sample delivers critical yield values

$$\frac{1}{3}\sigma^2 + \tau^2 = \kappa^2 \tag{3.146}$$

The other very famous criterion was proposed already 1864 by Tresca [395] for granular and soil material. It differs from the von Mises criterion with respect to the dependency on the third invariant. Formulated not as octahedral shear stress criterion but as maximum shear stress model:

$$\tau_{max} = \kappa \tag{3.147}$$

[14] As Khan [223] points out, the so called von Mises criterion "in fact was first published by Huber [199], and there is evidence that even earlier, in 1856, Maxwell came up with this criterion in a letter to Kelvin (Bell 1973)" [41].

it shapes a hexagon in the deviator plane (compare illustration in Figures 3.8 and 3.12) as well as in the plane stress state:

$$\sigma_1 - \sigma_2 = \pm 2\ \kappa \tag{3.148}$$

From a uniaxial stress test κ is determined through

$$\kappa = \frac{\sigma_Y}{2} \tag{3.149}$$

and from a shear test by

$$\kappa = \tau_Y \tag{3.150}$$

In case of the Tresca criterion, a combined axial tension and torsion test shows critical yield values at

$$\frac{1}{4}\sigma^2 + \tau^2 = \kappa^2 \tag{3.151}$$

Pressure Dependent Isotropic Formulations

Though for metals the influence of hydrostatic pressure on the yield locus can be neglected, there are many materials , e.g. concrete, rock or other geological materials, for which the yield criterion or the failure model must describe an influence of that kind. Specifically some simple formulations for the simulation of dynamic loading situations in concrete often use a so called *tensile cutoff* which is a combination of either the von Mises or the Tresca criterion with the *Rankine maximum principal stress* criterion. The latter forms three planes perpendicular to the principal stress axes:

$$\sigma_1 = \sigma_0, \qquad \sigma_2 = \sigma_0, \qquad \sigma_3 = \sigma_0 \tag{3.152}$$

represented by a triangle in the deviatoric plane (compare Figure 3.12) leading to the surface

$$\sqrt{2}\ \rho\cos\theta + \xi - \sqrt{3}\sigma_0 = 0 \tag{3.153}$$

where σ_0 is the critical stress to failure in any of the principal stress directions.

A generalized version of the Tresca criterion can be found in the *Mohr-Coulomb* formulation which assumes, like the Tresca model, that the maximum shear stress is the decisive parameter for yielding or failure. It is based on the Coulomb rule which considers a minimum amount of shear stress necessary to allow for slip in the material, e.g. in soil material, that depends upon the cohesion C and on the normal pressure σ_n:

$$\tau = C - \sigma_n\ \tan\phi \tag{3.154}$$

where ϕ represents the angle of internal friction[15]. A common depiction of the cohesion and the friction angle uses the plane formed by an equivalent stress,

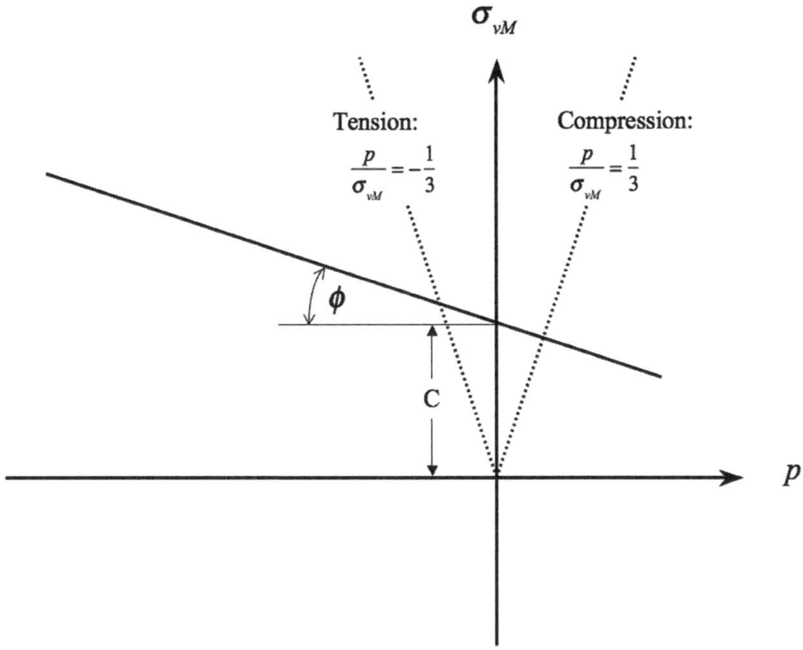

Fig. 3.10. Coulomb type yield criterion depicted in the von Mises equivalent stress - hydrostatic pressure plane with the cohesion parameter C and the friction angle ϕ.

usually the von Mises stresses, and the hydrostatic pressure p. A schematic example is given in Figure 3.10.

The mathematical description of the resulting hexagonal Mohr-Coulomb surface in the principal stress space can be expressed through:

$$\frac{1}{3}I_1 \sin\phi + \sqrt{J_2}\sin\left(\theta + \frac{\pi}{3}\right) + \sqrt{\frac{J_2}{3}}\cos\left(\theta + \frac{\pi}{3}\right)\sin\phi - c\cos\phi = 0 \quad (3.155)$$

or alternatively:

$$\sqrt{2}\,\xi\sin\phi + \sqrt{3}\,\rho\sin\left(\theta + \frac{\pi}{3}\right) + \rho\cos\left(\theta + \frac{\pi}{3}\right)\sin\phi - \sqrt{6}c\cos\phi = 0 \quad (3.156)$$

Knowing the yield stresses σ_T and σ_C from uniaxial tension and compression tests, respectively, the parameters c and ϕ can be derived as follows:

$$\phi = \sin^{-1}\frac{\sigma_C - \sigma_T}{\sigma_C + \sigma_T} \quad (3.157)$$

[15] Detailed derivations and graphical explanations through Mohr's circle are found in textbooks on plasticity, e.g. Khan [223]

$$c = \frac{\sigma_T(1+\sin\phi)}{2\,\cos\phi} \tag{3.158}$$

Another straight forward way to model the pressure influence can be achieved by adding a term to the von Mises criterion that depends on the first stress tensor invariant. This leads to the *Drucker-Prager* model that can be written as:

$$\alpha\, I_1 + \sqrt{J_2} - \kappa = 0 \tag{3.159}$$

Formulated with the Haigh-Westergaard coordinates the criterion is written as:

$$\sqrt{6}\alpha\,\xi + \rho - \sqrt{2}\,\kappa = 0 \tag{3.160}$$

With that modification the yield surface takes the shape of a circular cone as illustrated in Figure 3.11. In a plane stress state $\sigma_3 = 0$ the Drucker-Prager surface forms an ellipse described by:

$$\alpha\,(\sigma_1+\sigma_2) + \sqrt{\frac{1}{3}\,(\sigma_1^2 - \sigma_1\sigma_2 + \sigma_2^2)} - \kappa = 0 \tag{3.161}$$

Again, two experiments are needed to derive the parameters α and κ. The relation to the Mohr-Coulomb parameters is as follows:

$$\alpha = \frac{2\,\sin\phi}{\sqrt{3}\,(3 \mp \sin\phi)} \tag{3.162}$$

$$\kappa = \frac{6\,c\,\cos\phi}{\sqrt{3}\,(3 \mp \sin\phi)} \tag{3.163}$$

where, in the denominator, the minus signs account for Drucker-Prager circular cone coincident with the outer apexes of the Mohr-Coulomb hexagon and the plus signs represent coincidence with the inner apexes.

An often applied isotropic yield surface model that is quadratic in the stress is:

$$\sigma^T\,\mathbf{F}\,\sigma + \mathbf{B}\,\sigma + F_0 \leq 0 \tag{3.164}$$

where σ, $\mathbf{F}$ and $\mathbf{B}$ using a Voigt notation (3.35) can be written as:

$$\sigma = \begin{bmatrix} \sigma_{xx} \\ \sigma_{yy} \\ \sigma_{zz} \\ \sigma_{xy} \\ \sigma_{yz} \\ \sigma_{zx} \end{bmatrix} \quad \mathbf{F} = \begin{bmatrix} F_{11} & F_{12} & F_{12} & & & \\ F_{12} & F_{11} & F_{12} & & & \\ F_{12} & F_{12} & F_{11} & & & \\ & & & F_{44} & & \\ & & & & F_{44} & \\ & & & & & F_{44} \end{bmatrix} \quad \mathbf{B} = \begin{bmatrix} F_1 & & & & & \\ & F_1 & & & & \\ & & F_1 & & & \\ & & & 0 & & \\ & & & & 0 & \\ & & & & & 0 \end{bmatrix} \tag{3.165}$$

To guarantee an equivalence of pure shear with a biaxial tension-compression state,

$$F_{44} = 2\,(F_{11} - F_{12}) \tag{3.166}$$

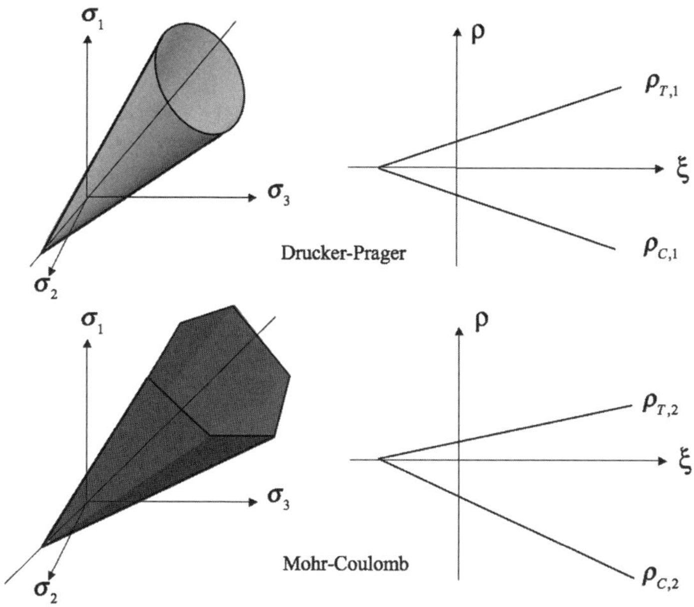

Fig. 3.11. Pressure dependent yield surfaces after *Drucker-Prager* and *Mohr-Coulomb* in the principal stress space with their corresponding tensile and compressive meridians plotted in the $\rho - \xi$-planes.

must be required to the parameters. Using the invariants I_1 and J_2 (3.164) can also be formulated as:

$$J_2^2 + A_0 + A_1\, I_1 - A_2\, I_1^2 \leq 0 \tag{3.167}$$

where the coefficients A_i are identified by:

$$A_0 = F_0 \quad A_1 = 3\, F_1 \quad A_2 = 9\,(1 - F_{11}) \tag{3.168}$$

To determine these parameters, three material tests under uniaxial tension, uniaxial compression and simple shear are needed, respectively. The correlation between the the yield stress σ_t, σ_c and σ_s derived in the tests and the coefficients is found in:

$$A_0 = 3\,\sigma_s^2 \quad A_1 = 9\,\sigma_s^2\, \frac{\sigma_c - \sigma_t}{\sigma_c \sigma_t} \quad A_2 = 9\, \frac{\sigma_c \sigma_t - 3\,\sigma_s^2}{\sigma_c \sigma_t} \tag{3.169}$$

Formulations to Model the Influence of Stress Triaxiality

The dependency of the surface of critical stresses on the third invariant and thus on the angle θ can be an important aspect for materials with strongly different yield or failure thresholds under tensile and compressive stress states, respectively. That is another reason why the Tresca criterion is often applied to soil materials. A disadvantage of the Tresca as well as of the Mohr-Coulomb models is, however, the non-continuous characteristic at angles which are multiples of 60°, the edges of compressive meridians. For implementation in codes a smooth continuous shape of the yield or failure surface avoids difficulties with discontinuities in the return algorithms discussed later.

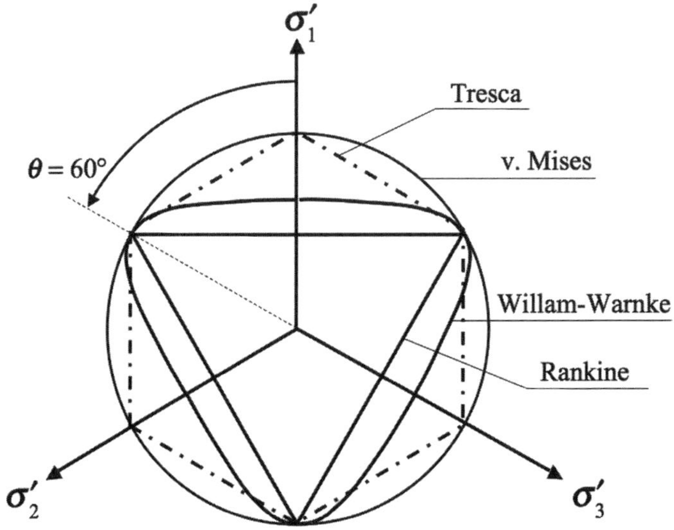

Fig. 3.12. Dependence of yield criteria on the third invariant illustrated in the deviator plane.

An example three invariant criterion with a continuous surface is the *Ottosen four parameter model*:

$$a\, J_2 + \lambda\, \sqrt{J_2} + b\, I_1 - 1 = 0 \tag{3.170}$$

where the third invariant is expressed through the function λ:

$$\lambda = \begin{cases} k_1 \cos\left[\frac{1}{3}\cos^{-1}\,(k_2 \cos 3\,\theta)\right] & \text{for} \quad \cos 3\theta \geq 0 \\ k_1 \cos\left[\frac{\pi}{3} - \frac{1}{3}\cos^{-1}\,(k_2 \cos 3\,\theta)\right] & \text{for} \quad \cos 3\theta < 0 \end{cases} \tag{3.171}$$

with the material constants a, b, k_1 and k_2.

Chen and Han [86] discuss the Ottosen model, find good correlation with experimental data but the *lambda* expression somewhat "quite involved". Besides other alternatives, they also point out the quality of the *Willam-Warnke five parameter model*:

$$a + a_1\rho_T + a_2\rho_T^2 - \frac{I_1}{3} = 0 \qquad (3.172)$$

$$a + b_1\rho_C + b_2\rho_C^2 - \frac{I_1}{3} = 0 \qquad (3.173)$$

with the parameters a, a_1,a_2, b_1 and b_2.

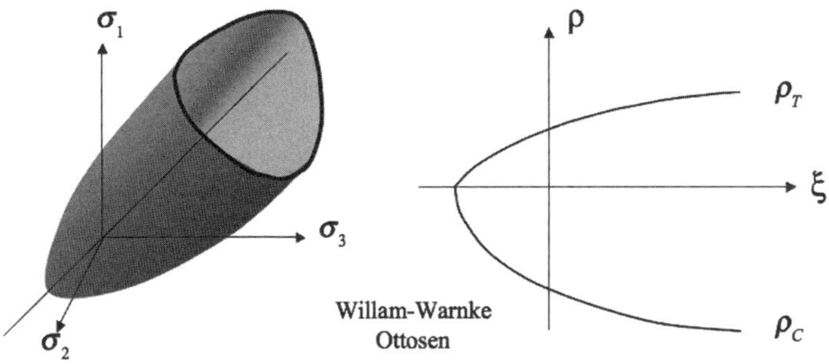

Fig. 3.13. General shape of the pressure dependent yield surfaces as described with the *Willam-Warnke* and *Ottosen* models, respectively, sketched in the principal stress space with their corresponding tensile and compressive meridians plotted in the $\rho - \xi$-planes.

Criteria for Anisotropic Behaviour

So far we discussed isotropic criteria for plastic flow or fracture. For materials with anisotropic strength behaviour corresponding formulations of yield surfaces $F(\sigma_{xx}, \sigma_{yy}, \sigma_{zz}, \sigma_{xy}, \sigma_{xz}, \sigma_{yz},) = 0$ exist. A quadratic formulation for orthotropic materials was identified by Hill [183] as

$$F\left(\sigma_{yy} - \sigma_{zz}\right)^2 + G\left(\sigma_{zz} - \sigma_{xx}\right)^2 + H\left(\sigma_{xx} - \sigma_{yy}\right)^2 + L\sigma_{yz}^2 + M\sigma_{zx}^2 + N\sigma_{xy}^2 = 1 \qquad (3.174)$$

where the material constants F, G, H, L, M and N are to be characterized by three uniaxial tension tests along the three principal directions of orthotropy together with three pure shear tests on the orthogonal planes of anisotropy. The correlation between the test results and Hill's material parameters is explicated in (6.56) where the model is further discussed to describe failure mechanisms in fibre reinforced plastics. In the same chapter 6.3.1 additional anisotropic formulations are presented.

3.3.3 Flow Rules

Once a criterion for the initiation of plastic flow is fulfilled, the subsequent paths of stress and strain in terms of inelastic deformation and the resulting changes in shape of the yield surface must be described. Only under the idealization of assumed perfectly plastic behaviour (see Figure 3.6) the yield surface does not change upon loading and unloading and the yield stress stays constant. In reality either strain hardening (metals) or strain softening (geological, or soil materials) or combinations of both forms (thermoplastic polymers) are observed.

Isotropic Hardening

Isotropic hardening expands the initial yield surface $f(\sigma) = 0$ isotropically and is described by

$$f(\sigma) - \kappa(\alpha) = 0 \tag{3.175}$$

where κ is a monotonically increasing function of the hardening parameter α. That hardening parameter can either be described through the effective plastic strain $\alpha = \bar{\varepsilon}^p$ or via the total plastic work $\alpha = W_p$. The first form is called the strain-hardening hypothesis, the second one is the work-hardening hypothesis (see e.g. Khan and Huang [223] for details).

Kinematic Hardening

Under kinematic hardening a translation of the yield surface in the stress space is understood where the shape and size of the surface remain constant. The change from an initial yield surface $f(\sigma) = 0$ is described by the so called *back stress tensor* σ_B that leads to subsequent yield surfaces after loading and unloading:

$$f(\sigma - \sigma_B) = 0 \tag{3.176}$$

A linear hardening rule, e.g. proposed by Prager [306], leads to an evolution formulation for the back stress tensor through a constant parameter c:

$$\mathrm{d}\,\sigma_{B\,ij} = c\,\mathrm{d}\,\varepsilon^p_{ij} \tag{3.177}$$

A nonlinear kinematic hardening rule has been suggested by Armstrong and Frederick using the material parameters c_1 and c_2 and the equivalent plastic strain $\bar{\varepsilon}_{ij}$:

$$\mathrm{d}\,\sigma_{B\,ij} = \frac{2}{3}\,c_1\,\mathrm{d}\,\varepsilon^p_{ij} - c_2\,\sigma_{B\,ij}\,\mathrm{d}\,\bar{\varepsilon}_{ij} \tag{3.178}$$

Linear kinematic hardening shifts the center of the yield surface up and down on the same line for loading and unloading, respectively. Nonlinear kinematic hardening of the type (3.178) shifts the surface on a nonlinear path for loading and on a different, linear one, for unloading[16].

[16] See the derivation and graphical illustration in Wu [429], pp. 299 f

Associated and Non-Associated Flow Rules

An important law that classifies materials, or better material behavior, for the treatment by methods of classic plasticity theory is *Drucker's postulate.* It is not a physical or thermodynamical law but a condition for describing plasticity with the related theory. The postulate, in its original formulation involves quasi-cycles of load and demands for positive changes in plastic work for any quasi-cycle. A more practicable formulation of the postulate demands for a *convex yield surface* and *normality* of the flow rule, i.e. plastic strain increments normal to the yield surface. Mathematically these two conditions for *stable material* can be formulated by:

$$\mathrm{d}\,\sigma_{ij}\,\mathrm{d}\,\varepsilon^{p}_{ij} > 0 \tag{3.179}$$

and

$$\mathrm{d}\,\varepsilon^{p} = \mathrm{d}\,\lambda\,\frac{\partial\, f}{\partial\,\sigma} \tag{3.180}$$

with a scalar function of proportionality $\mathrm{d}\,\lambda$. If the yield surface is described by a yield function $f(\sigma,\, q_i) = 0$ with hardening parameters q_i to distinguish between elastic and plastic stress states then the derivative $\frac{\partial f}{\partial \sigma}$ describes the surface normal on the yield surface. Thus, equation (3.180) demands that plastic strain increments are of coinciding direction with the plastic stress increments.

Plastic behavior that is described with the flow rule (3.180) is called *associated flow.* If a function $g(\sigma)$ is defined as *plastic potential,* then a flow rule describing increments in plastic strain by:

$$\mathrm{d}\,\varepsilon^{p} = \mathrm{d}\,\lambda\,\frac{\partial\, g}{\partial\,\sigma} \tag{3.181}$$

is called *non-associated flow rule.*

Consistency Condition

Basically, the yield function f is negative for elastic states. Furthermore, the change of the Lagrange multiplier $\mathrm{d}\,\lambda$ is zero in the elastic regime since, only then the yield surface remains constant. On the other hand, given a current stress state to be plastic ($f = 0$), the further loading or unloading, and accordingly the hardening effects, are described through the conditions:

$$f = 0 \quad \mathrm{d}\,f = f(\sigma + \mathrm{d}\sigma) = \frac{\partial\, f}{\partial\,\sigma}\mathrm{d}\,\sigma = 0 \tag{3.182}$$

for loading, and

$$f = 0 \quad \mathrm{d}\,f = f(\sigma + \mathrm{d}\sigma) = \frac{\partial\, f}{\partial\,\sigma}\mathrm{d}\,\sigma < 0 \tag{3.183}$$

for unloading. Both, loading and unloading, can be described through the so called *Kuhn-Tucker* conditions:

$$\mathrm{d}\,\lambda \geq 0, \quad f(\sigma) \leq 0, \quad \mathrm{d}\,\lambda\, f(\sigma) = 0 \tag{3.184}$$

The last condition in (3.184) is called *consistency condition*. It means that

- whenever plastic flow is happening ($\mathrm{d}\,\lambda > 0$), the stress state must be located on the yield surface ($f(\sigma) \equiv 0$) and,
- on the other hand, when the stress state is elastic ($f(\sigma) < 0$), then no plastic flow can occur ($\mathrm{d}\,\lambda = 0$)

and is used to determine the value of $\mathrm{d}\,\lambda$. For an arbitrary flow rule and hardening behavior described by $f(\sigma,\, \varepsilon^p,\, q_i) = 0$, the increment in the yield function is zero $\mathrm{d}\,f = 0$ for an infinitesimal change from one plastic state to another plastic state:

$$\mathrm{d}\,f = \frac{\partial\,f}{\partial\,\sigma_{ij}}\,\mathrm{d}\,\sigma_{ij} + \frac{\partial\,f}{\partial\,\varepsilon^p_{ij}}\,\mathrm{d}\,\varepsilon^p_{ij} + \frac{\partial\,f}{\partial\,q_i}\,\mathrm{d}\,q_i = 0 \tag{3.185}$$

Using (3.180) we find

$$\mathrm{d}\,f = \frac{\partial\,f}{\partial\,\sigma_{ij}}\,\mathrm{d}\,\sigma_{ij} + \mathrm{d}\,\lambda\,\frac{\partial\,f}{\partial\,\varepsilon^p_{ij}}\,\frac{\partial\,f}{\partial\,\sigma_{ij}} + \frac{\partial\,f}{\partial\,q_i}\,\mathrm{d}\,q_i = 0 \tag{3.186}$$

that leads to

$$\mathrm{d}\,\lambda = \frac{\frac{\partial\,f}{\partial\,q_i}\,\mathrm{d}\,q_i + \frac{\partial\,f}{\partial\,\sigma_{ij}}\,\mathrm{d}\,\sigma_{ij}}{-\frac{\partial\,f}{\partial\,\sigma_{kl}}\,\frac{\partial\,f}{\partial\,\varepsilon^p_{kl}}} \tag{3.187}$$

3.3.4 Strain Rate Dependent Yield Criteria

Inelastic deformation behavior of many materials depends on the deformation rate. For application to dynamic processes, e.g. impact or deep drawing of metals, strain rate dependent formulations of yield criteria, as well as of other mechanical properties, are of interest. Various formulations are implemented in commercial codes with explicit time integration. An overview and critical discussion of some typical models for strain rate dependent plasticity models is given in the following.

Phenomenological Plasticity Models

Johnson-Cook Model

A typical phenomenological model is the Johnson-Cook [214] strain, strain rate and temperature dependent yield model:

$$\sigma_Y = \left(\sigma_0 + B\,\bar{\varepsilon}_p^n\right)\,\left(1 + C\,\ln \dot{\bar{\varepsilon}}_p^*\right)\,\left(1 - T^{*m}\right) \tag{3.188}$$

with the equivalent plastic strain $\bar{\varepsilon}_p$, the equivalent plastic strain rate

$$\dot{\bar{\varepsilon}}_p^* = \frac{\dot{\bar{\varepsilon}}_p}{\dot{\varepsilon}_0} \tag{3.189}$$

normalized to $\dot{\varepsilon}_0 = 1\,s^{-1}$ and with the homologous temperate

$$T^* = \frac{T - T_A}{T_M - T_A} \tag{3.190}$$

where T_A is the ambient or reference temperature and T_M the melting temperature. The five parameters σ_0, B, n, C, m allow for a description of strain hardening as well as strain rate and temperature dependency. The first three of them can be derived from quasi-static tension tests at constant strain rate. The strain rate parameter C is identified through dynamic tension tests at varying strain rates, e.g. using servo-hydraulic facilities, drop towers, Hopkinson-Bars or Taylor tests.

Since temperature influences in the course of dynamic characterization tests are hard to separate from the rate effects, the temperature influence represented through the third term in the Jonson-Cook model is often omitted. Measured stress-strain curves are accordingly fitted by the first two terms and the influences of strain rate and temperature modeled without discrimination.

A critical point of discussion is the range of strain rates over which the Johnson-Cook model is applicable[17]. The formulation (3.188) was originally derived to account for experimental results from quasi-static to moderate dynamic strain rates measured with servo-hydraulic facilities, Hopkinson bars and Taylor tests. Clearly the spectrum of available test types and results defines the quality of the analytical fit and flyer plate tests results for strain rates up to $10^6\,s^{-1}$ were not used in the course of the model's derivation. Pronounced nonlinearities in the yield stress for strain rates of $10^5\,s^{-1}$ and beyond are, thus, always a problem for that model due to the logarithmic formulation of the strain rate sensitivity. This becomes apparent with the data given in Figure 3.14 where the Johnson-Cook model is apparently limited to strain rates up to $10^4\,s^{-1}$.

Based on these findings, Rohr [331] proposed a modified Johnson-Cook model with an initial yield stress $\sigma_0(\dot{\bar{\varepsilon}}_{el})$ depending on the equivalent elastic, not plastic, strain rate. In addition, the modified model allows for an improved representation of measured temperature dependence of the yield stresses, observed in the case of steel (35NiCrMoV109):

[17] This topic is for example discussed in Meyers [276], Zukas et al. [436] and Rohr [331]

$$\sigma_Y = \sigma_0 \left(1 + A \ln \dot{\bar{\varepsilon}}_{el}\right) \left(1 - T^{*m_1}\right) + B \bar{\varepsilon}_{el}^{n} \left(1 - T^{*m_2(T)}\right) \tag{3.191}$$

where the initial yield stress for the investigated steel is found to be

$$\sigma_0 = \sigma_0(\dot{\bar{\varepsilon}}_{el}) = \begin{cases} 755[\text{MPa}] & for\ \dot{\bar{\varepsilon}}_{el} \leq 2 \cdot 10^3 s^{-1} \\ 280[\text{MPa}] & for\ \dot{\bar{\varepsilon}}_{el} > 2 \cdot 10^3 s^{-1} \end{cases} \tag{3.192}$$

with a parameter A also depending on $\dot{\bar{\varepsilon}}_{el}$ according to

$$A(\dot{\bar{\varepsilon}}_{el}) = \begin{cases} 0.015 & for\ \dot{\bar{\varepsilon}}_{el} \leq 2 \cdot 10^3 s^{-1} \\ 0.27 & for\ \dot{\bar{\varepsilon}}_{el} > 2 \cdot 10^3 s^{-1} \end{cases} \tag{3.193}$$

Constants B and n were identified to be:

$$B = 570\,[\text{MPa}] \qquad n = 0.39 \tag{3.194}$$

The temperature dependency is described by parameters $m_1 = 1.15$ and:

$$m_2(T) = \begin{cases} 0.8 & for\ \ T \leq 623\,[\text{K}] \\ -0.0028\,T + 2.52 & for\ \ T > 623\,[\text{K}] \end{cases} \tag{3.195}$$

, respectively.

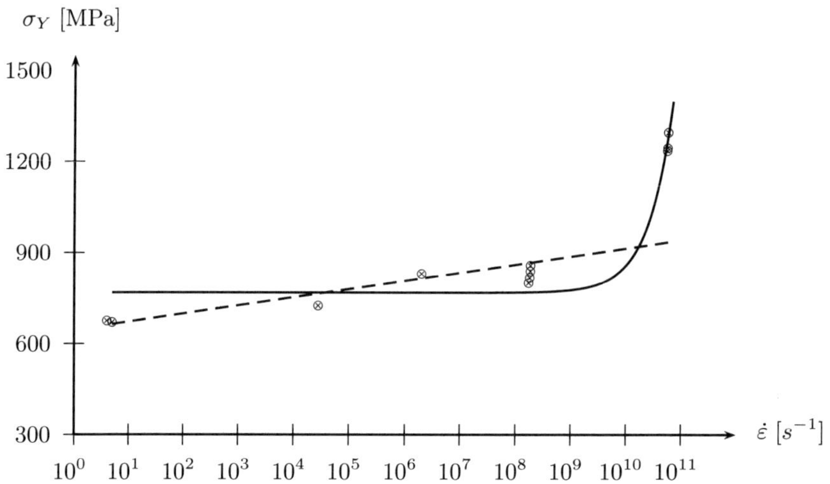

Fig. 3.14. Strain rate sensitivity in the yield stress of a 35NiCrMoV109 steel modelled with Johnson-Cook (dashed line) and Zerilli-Armstrong (solid line) models (Data courtesy of Rohr [331]).

Steinberg-Cochran-Guinan Model

Another empirical yield model was introduced by Steinberg, Cochran and Guinan [365] . It does not contain any strain rate dependency in its original version. Instead it is formulated explicitly for metals at the high strain rate regime of $\dot{\varepsilon} \geq 10^5\,[s^{-1}]$:

$$\sigma_Y = \sigma_{Y,0}(\varepsilon_p)\,\frac{G(p,T)}{G_0} \tag{3.196}$$

using a pressure and temperature dependent shear modulus $G(p,T)$:

$$G(p,T) = G_0\left\{1 + A\frac{p}{\eta^{1/3}} - B(T-300)\right\} \tag{3.197}$$

$$A = \frac{1}{G_0}\frac{\partial G}{\partial p} \tag{3.198}$$

$$B = \frac{1}{G_0}\frac{\partial G}{\partial T} \tag{3.199}$$

$$\eta = \frac{V_0}{V} \tag{3.200}$$

Later, a modified version was presented by Steinberg and Lund [366] containing a strain rate dependency according to:

$$\sigma_Y = \{\sigma_{Y,T}(\dot{\varepsilon}_p) + \} \tag{3.201}$$

with the thermally activated yield stress $\sigma_{Y,T}$ implicitly defined through

$$\dot{\varepsilon}_p = \frac{1}{\frac{1}{C_1}\exp\left[\frac{2\,U_k}{k\,T}\left(1-\frac{\sigma_{Y,T}}{\sigma_p}\right)^2\right] + \frac{C_2}{\sigma_{Y,T}}} \tag{3.202}$$

with parameters C_1 and C_2 depending on the dislocation density, the Burgers vector and the Debey frequency.

Physical Based Yield Models

Zerilli-Armstrong Formulation

Zerilli and Armstrong [434] continued the partly physical path of Steinberg and Lund to propose a physically based model for Armco iron that considers the specific differences in its plastic behavior depending on whether it is body-centered cubic (bcc) or face centered cubic (fcc). Their model bases on the dislocation motion and describes the yield stresses for bcc and fcc Armco iron, respectively, by:

$$\sigma_{Y,bcc} = \sigma_G + C_1 \exp\left(-C_3 T + C_4 T \ln \dot{\varepsilon}\right) + C_5 \varepsilon^n + k\, d^{-\frac{1}{2}} \tag{3.203}$$

and

$$\sigma_{Y,fcc} = \sigma_G + C_2 \sqrt{\varepsilon} \exp\left(-C_3 T + C_4 T \ln \dot{\varepsilon}\right) + k\, d^{-\frac{1}{2}} \tag{3.204}$$

with athermal, strain rate and temperature independent initial yield stress σ_G and a set of parameters C_i describing the strain rate and temperature dependent behavior.

3.3.5 Plasticity Effects at Shock Compression States

Dynamic compressive behavior of materials at strain rates in the regime of 10^6 $[s^{-1}]$ is typically testes with shock loading devices, e.g. the flyer plate test to be discussed later. Results from that kind of material tests show an elastic-plastic characteristic which is at least for metals surprising at first glance. These tests are performed to characterize materials with respect to their equation of state properties and, thus, describe hydrostatic compressive behavior. Having that in mind, effects of plasticity appear strange at first, since we are used to assume plastic incompressibility (3.139) for many materials.

The mentioned effect is, however, not in contradiction to our traditional understanding. The reason is that the shock loading tests are performed in a way that a uniaxial strain state is initiated in the specimen, say in the $xx-$direction:

$$\varepsilon_{xx} \neq 0 \quad \varepsilon_{yy} = \varepsilon_{zz} = 0 \tag{3.205}$$

On the other hand, the stress state in the sample is described by

$$\sigma_{xx} = (\lambda + 2\,\mu)\ \varepsilon_{xx} \quad \sigma_{yy} = \sigma_{zz} = \lambda\, \varepsilon_{xx} \tag{3.206}$$

Due to the uniaxial strain state, the yield stress σ_Y can be described with both the von Mises (3.140) and the Tresca model (3.147) through the related strain state as:

$$\sigma_Y = \sigma_{HEL} = 2\,\mu\, \varepsilon_{xx} \tag{3.207}$$

Its affiliation with shock loading conditions lead to the name *Hugoniot elastic limit* for that yield stress.

The stress state (3.206), in turn, defines a hydrostatic pressure at the position of the shock wave of:

$$p = \frac{1}{3}\left(\sigma_{xx} + \sigma_{yy} + \sigma_{zz}\right) = \left(\lambda + \frac{2}{3}\,\mu\right)\, \varepsilon_{xx} \tag{3.208}$$

With equations (3.207) and (3.208) the stress component in $xx-$, or shock-direction, is defined as:

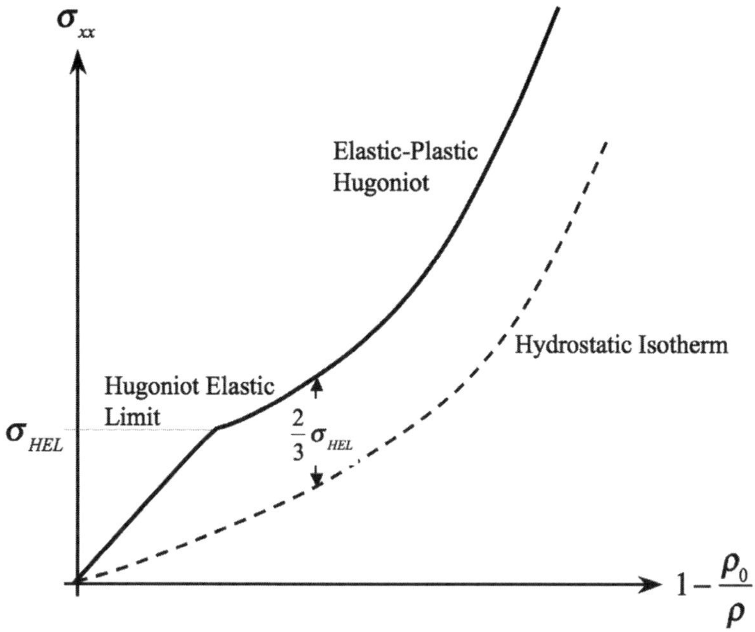

Fig. 3.15. Schematic illustration of plasticity effects under shock loading situations with a characteristic Hugoniot elastic limit.

$$\sigma_{xx} = p + \frac{4}{3}\,\mu\,\varepsilon_{xx} = p + \frac{2}{3}\,\sigma_{HEL} \tag{3.209}$$

Figure 3.15 illustrates the relations schematically. Further detailed introduction to the nature of shock wave conditions and the meaning of Hugoniot lines will be provided in chapter 4. Figure 3.16 gives an impression on how precise shock load application can be uses to derive dynamic yield strength data provided a related fast instrumentation is applied. The graphs show free surface velocities measured by Rohr [331] on the rear side of 3 $[mm]$ thick 35NiCrMoV109 steel samples. They were impacted by 1.5, 3 and 8 $[mm]$ thick flyer plates at velocities between 588 $[\mathrm{m\,s^{-1}}]$ and 1027 $[\mathrm{m\,s^{-1}}]$.

The Hugoniot elastic limit is characterized by the first velocity step in the curves, denoted as Hugoniot elastic limit speed v_{HEL}. The relation between σ_{HEL} and v_{HEL} is described with the longitudinal sound speed c_L as:

$$\sigma_{HEL} = \frac{1}{2}\,\rho\,c_L\,v_{HEL} \tag{3.210}$$

Since the Hugoniot elastic limit is a property related to a three-dimensional stress state (3.208), another relation is needed to compare it to yield stresses

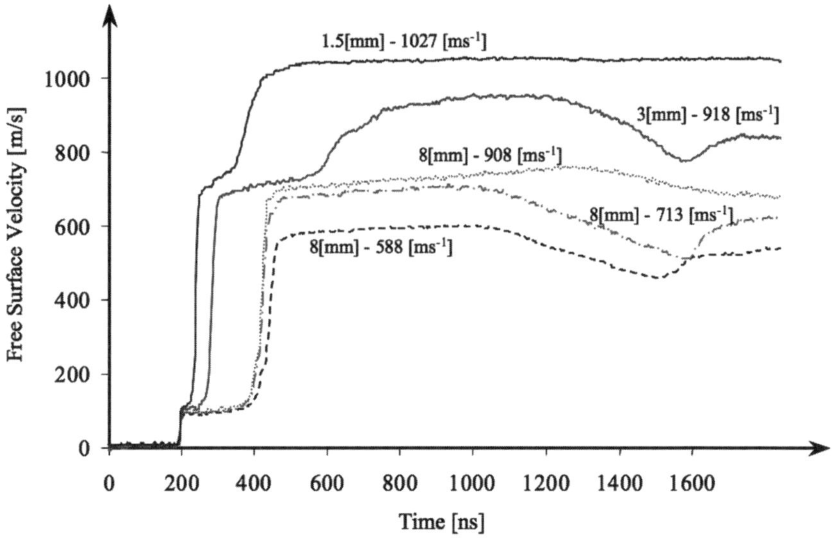

Fig. 3.16. Measured free surface velocities in flyer plate tests with 35NiCrMoV109 steel samples of 3 $[mm]$ thickness. Projectile plates were 1.5, 3 and 8 $[mm]$ thick, respectively. Impact velocities ranged from 588 $[ms^{-1}]$ to 1027 $[ms^{-1}]$. (Reprinted with permission from Rohr [331])

σ_{Y-1D} under uniaxial stress states:

$$\sigma_{Y-1D} = \frac{1-2\,\mu}{1-\mu}\,\sigma_{HEL} \tag{3.211}$$

Table 3.2 reflects the strain rate dependency in the Hugoniot elastic limit and the uniaxial yield stress derived according to (3.211) for the steel investigated by Rohr.

Measured data of Hugoniot elastic limits for many materials can be found in literature. Excellent data are for example provided for Tungsten and Tantalum along with comparing data under quasi-isentropic loading conditions by Chhabildas and Barker [89] and in Chhabildas and Asay [88].

3.3.6 Meso-Mechanical Calculation of Yield Loci

Experimental methods to characterize material behavior and to identify the parameters needed in the mathematical formulations of constitutive laws are sometimes limited. An example for such limitations is the measurement of yield loci under complex loading conditions, e.g. hydrostatic tension or combinations of hydrostatic and arbitrary deviatoric stresses. Numerical simulation can support in such situations, provided a sufficient basic knowledge of

Table 3.2. Strain rate dependency of the Hugoniot elastic limit σ_{HEL} and the related uniaxial yield stress σ_Y in 35NiCrMoV109 steel measured by Rohr [331].

Impact velocity $[ms^{-1}]$	588	713	908	918	1027
σ_{HEL} $[MPa]$	2089	2106	2193	2296	2406
σ_Y $[MPa]$	1235	1245	1297	1358	1423
$\dot{\varepsilon}$ 10^5 $[s^{-1}]$	5.53	5.65	5.80	13.1	17.8

the investigated material is at hand. If for example the inelastic behavior of a composite material is of interest and the individual material properties of the components, e.g. matrix and fiber or aggregate, are sufficiently characterized, then a meso-mechanical simulation of the composite material can be performed to numerically identify composite properties.

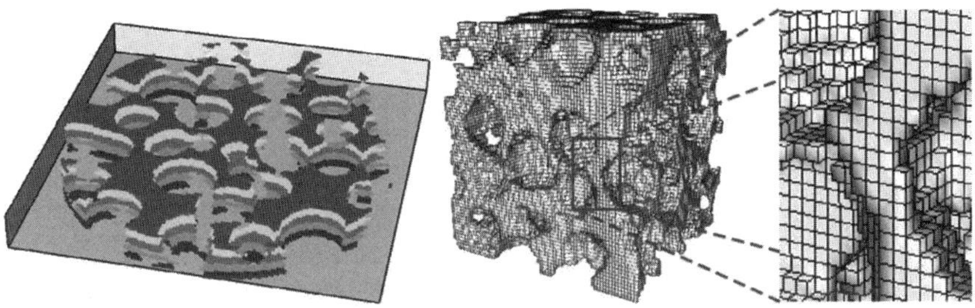

(A) (B) (C)

Fig. 3.17. Derivation of a meso-mechanical discretization of aluminum foam samples utilizing results from X-ray tomography. (A): Slice data from X-Ray scans, (B): Mapping of the scanned data into a 3D grid, (C): Local grid resolution in the chosen model. (Reprinted with permission from Wicklein [420]).

An exemplary procedure of such a meso-mechanical simulation for the purpose of characterizing the yield stress locus of aluminum foam under complex stress states was performed by Wicklein [420]. The basic matrix material aluminum was characterized in terms of its elastic plastic and failure behavior. Samples of the foam were investigated by X-ray tomography providing full three dimensional information on the distribution and size of pores in foam samples. These samples were then tested under not so complex loading conditions, e.g. uniaxial compression, uniaxial tension or confined compression. The same test conditions were applied as boundary conditions in meso-mechanical simulations with the information from the X-ray scans for the discretization of the samples. Figure 3.17 illustrates the derivation of a meso-mechanical discretization of foam samples utilizing results from X-ray scans. The results of both the tests and the simulation were then used to compare and validate the material model for the matrix material. With that validated model in turn complex loading conditions were applied to further characterize the foam material.

With the results from various meso-mechanical simulations a complex analytical yield locus was derived. The relation between numerical results and analytical fit is shown in Figure 3.18. Its shape in the principal stress space along with the deformed shapes after strain hardening is illustrated in Figure 3.19.

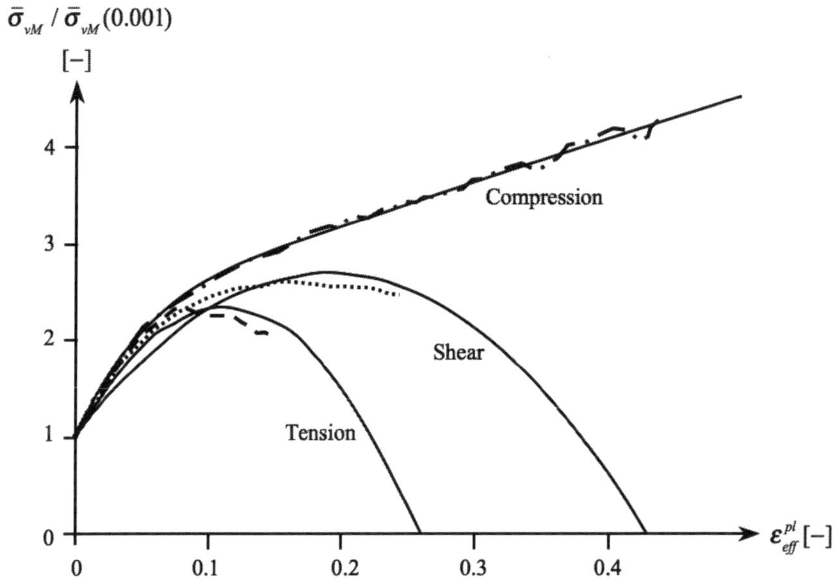

Fig. 3.18. Results of the meso-mechanical simulation under tension, compression and shear along with analytical fits to describe the yield locus. (Reprinted with permission from Wicklein [420])

An application of the Wicklein model for simulations of foamed aluminum under dynamic loading conditions is illustrated in Figure 3.20. A cylindrical sample of the characterized aluminum foam material was impacted at 15 $m\,s^{-1}$. Clearly, the continuum model assuming homogeneity of the material cannot reflect localized processes originating from inhomogeneities in the real

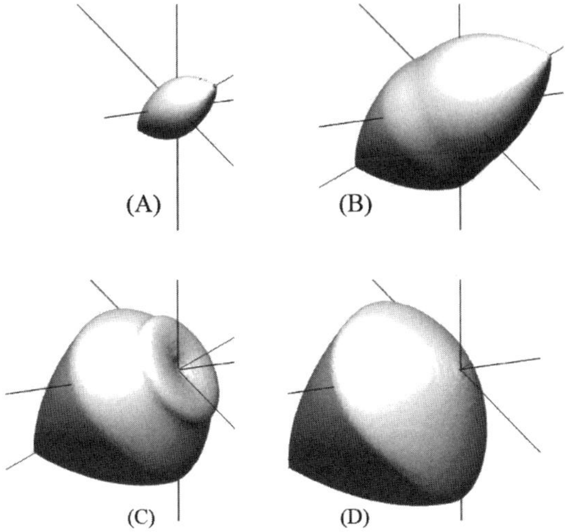

Fig. 3.19. Yield locus in its initial shape (A), after strain hardening (B), and after failure with respect to tensile (C) and shear (D) loads. (Reprinted with permission from Wicklein [420])

structure. However, the simulation using the numerically derived material model is able to describe the crushing behavior of the material, as illustrated by the force time graphs in Figure 3.21. Clearly, the overall behaviour in terms the structure's energy absorption as well as the force-time history at which it takes place is well predicted.

3.3.7 Polymers - Nonlinear Elasticity, Initial Plastic Softening, Visco-Plastic Hardening

There is a big variety of polymeric materials which are macromolecular configurations of long, randomly twisted, possibly entangled and interconnected chains. A material specific temperature, the so called *glass transition temperature*, separates the so called glassy state, i.e. a solid phase, from the rubbery which can be described as a viscous fluid type phase. With respect to the molecular structure, three basic groups of polymers are distinguished[18]:

- *Thermoplastics*, consisting of long non-linked molecular chains arranged in an amorphous, i.e randomly oriented, or partly directed manner. Depending on whether such distinct local zones of preferred molecular orientations exist, the thermoplastics are called *semi-crystalline* or amorphous, respectively.

[18] For detailed introductions to he physics of polymers see for example Sperling [362] or Strobl [377]

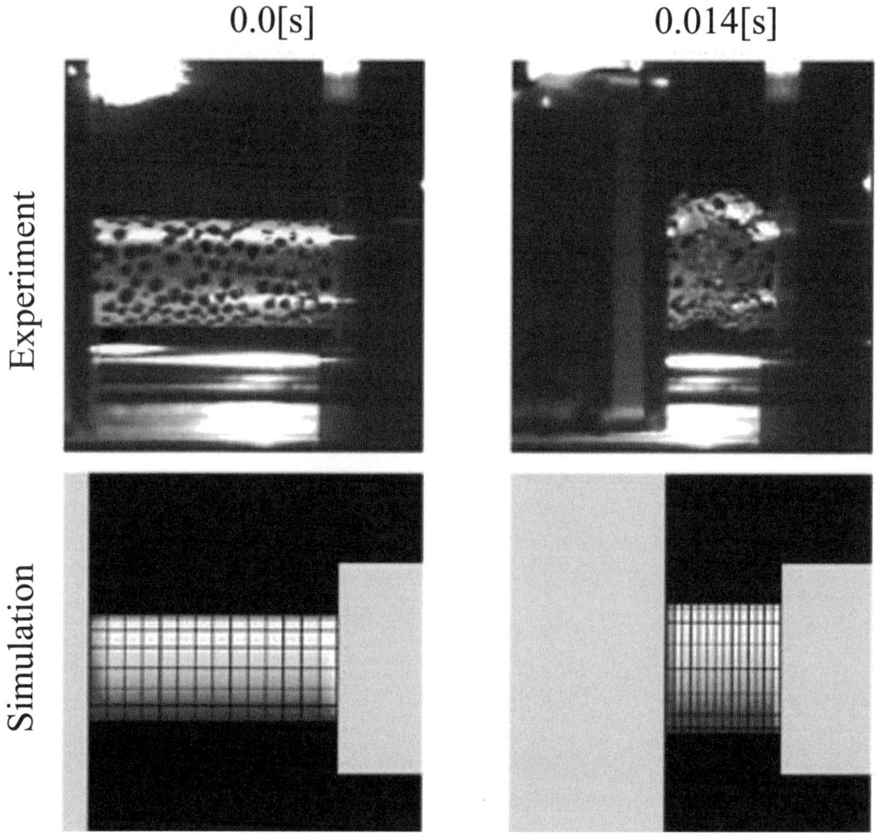

Fig. 3.20. Comparison of experimental and numerical results for the crushing of a cylindrical specimen made from the aluminum foam at 15 $m\,s^{-1}$. As a consequence of the homogenization in the model, loacal effects cannot be simulated. (Reprinted with permission from Wicklein [420])

- *Elastomers* are rubber materials always showing amorphous structure. Their typical wide meshed intermolecular covalent cross-links of the enable a large nonlinear elasticity.
- *Duromers* display a even closer volumetric network of cross-links between the molecular chains. That leads to higher stiffnesses and a much more brittle behavior compared to thermoplastics and elastomers.

Depending on the type of their molecular structure, the mechanical properties of polymers range from rubber elasticity of elastomers in temperature regimes above the material dependent glass transition temperature T_G to hyperelastic ductile as well as brittle behavior of thermoplastics in the glassy regime below T_G. In the context of automobile applications of polymers under crash-relevant loading conditions, amorphous and semi-crystalline

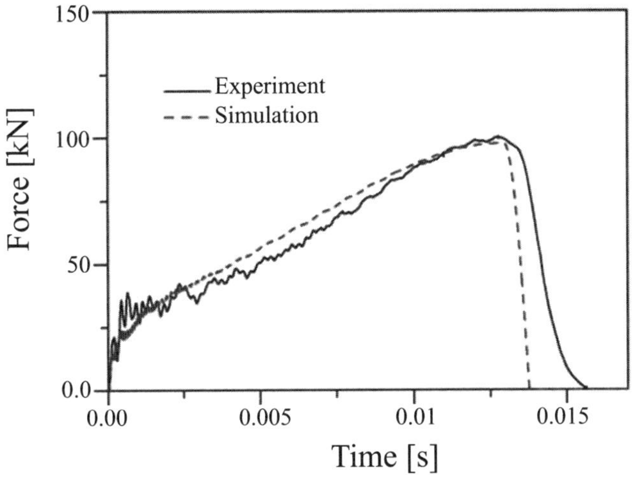

Fig. 3.21. Comparison of the load histories measured and calculated for the crushing process shown in Figure 3.20. (Reprinted with permission from Wicklein [420])

thermoplastics are the most common and important representatives of polymers.

Some of the most often found representatives applied in automotive interieur and exterieur are:

- Polycarbonate (PC)
- Acrylonitrile Butadiene Styrene (ABS)
- Polyamide (PA)
- Polypropylene (PP)
- Polyethylene (PE)
- Polyacetal (POM)

Whereas plastics in older cars until the late eighties were primarily used for optical design purposes, the motivation for their nowadays adoption results from passive safety regulations, e.g. the US Federal Motor Vehicle Safety Standard (FMVSS). In order to fulfil these standards, complex layered structures of polymer materials were developed to guarantee acceptable decelerations during occupants' impact against interior structures like pillars, headers, roof, doors or panels (see Figure 3.22 for example definitions of some impact zones). Certification experiments include impact tests with instrumented head or knee dummies under well defined impact velocities and angles. Criterion for the certification of structural components in the defined locations are characteristic values derived from the measured deceleration-time curves.

In order to distinguish polymer behaviour from what is observed with metals, the different sections of typical stress-strain-curves need new types of

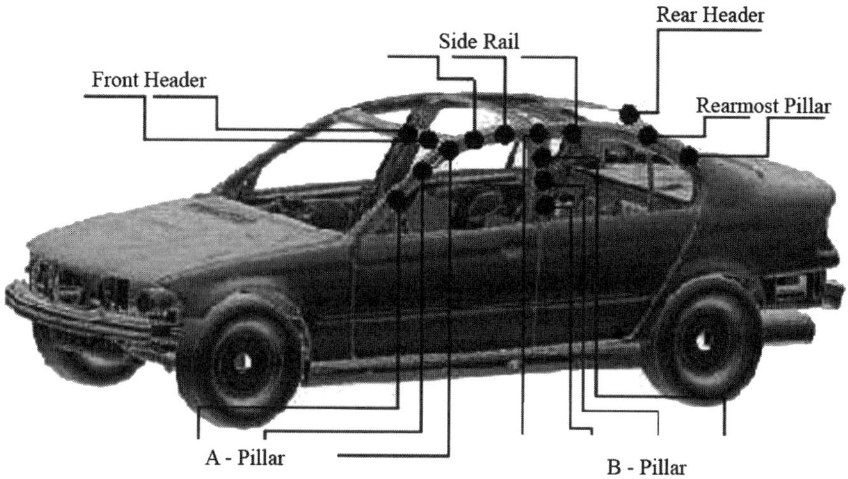

Fig. 3.22. Definition of impact zones for head impact tests (Reprint with permission from Junginger [220]).

mathematical formulations. Since the micro-physics are so very different from crystalline materials, a specific understanding of the processes at that scale and its influence on the macro-phenomena is necessary.

As indicated before in the subsection on statistical mechanics based elastic potentials 3.2.5, micro-mechanical models for polymers have a long tradition. Significant progress in linking the statistical mechanics to continuum mechanics towards a more physical, thermodynamically consistent description of the elastic, inelastic and failure behavior of polymers was achieved over the last twenty years as demonstrated with the publications of Arruda and Boyce [18], Anand [4], Anand and Gurtin [6] and Gearing [133]. Still, these models have barely found their way into application when crash and impact processes are simulated. Specifically in the automobile crash environment, where ever new kinds of polymers are used, a consistent description of the fabrication of plastic parts, of their process induced mechanical properties and of the resulting structural behavior demands for physical based models.

In the course of this section, a brief overview on the typical inelastic behavior of thermoplastic materials and on phenomenological as well as physical based constitutive formulations will be given.

Phenomenology of Inelastic Polymer Deformation

Strain Localization

Amorphous networks of molecular chains are the basic structure of polymers which, in the case of semi-crystalline materials, also contain crystalline phases. The network structure of polymers leads to the characteristic deformation behavior described in many textbooks, e.g. showing pronounced *localization* of strain and distinct differences under tensile, compressive or shear loading.

The strain localization leads to very different results in stress-strain curves depending on the size of the measurement zone. Figure 3.23 illustrates that phenomenon observed in a PC-ABS material by measured true stress true strain curves of a quasi-static uniaxial tension test comparing global deformation based strain with the results of local optical strain measurement. The global and local strain measurements involved the total free specimen length of 35 $[mm]$ and a small frame observed optically of length 1.25 $[mm]$, respectively. Whereas the global technique delivers a dedicated softening phase during the initial inelastic deformation and a maximum strain to failure of 36 %, the local measurement shows almost no softening and a maximum strain of 36 %.

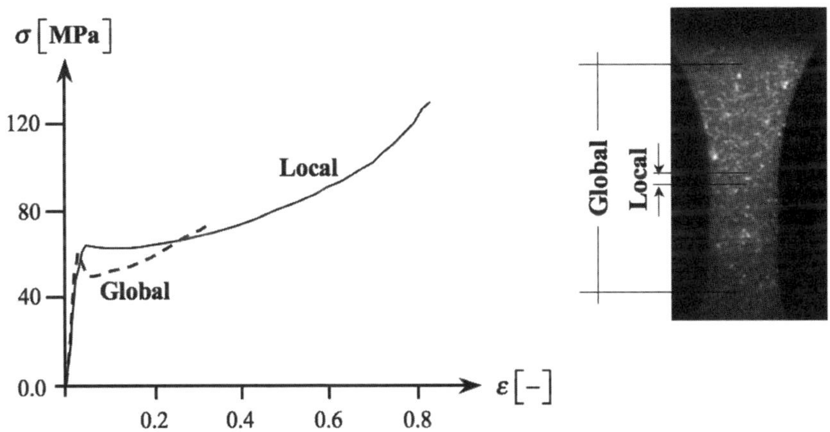

Fig. 3.23. Comparison of global (gauge length 35 $[mm]$) and local (gauge length 1.25 $[mm]$) true stress - true strain curves for a PC-ABS under quasi-static uniaxial tension (Data from Junginger [220]).

Strain data as measured from the global and the local technique, respectively, are illustrated in the graphical plots of Figure 3.24. Clearly, both the time-dependent evolution and the achieved total strains are remarkably different for the two methods. The total deformation process observed in that test shows three distinct phases. In a first homogeneous elastic deformation phase the global and local measurement cannot be distinguished from each other. With the beginning strain localization, observable by the necking process forming shoulders in the specimen, both local principal strains change in slope whereas the global strain remains constant in slope. That second phase of necking continues until the shoulders are drawn out o f the local measurement window. From that moment on the slope of the local and global strain in tensile direction are identical again.

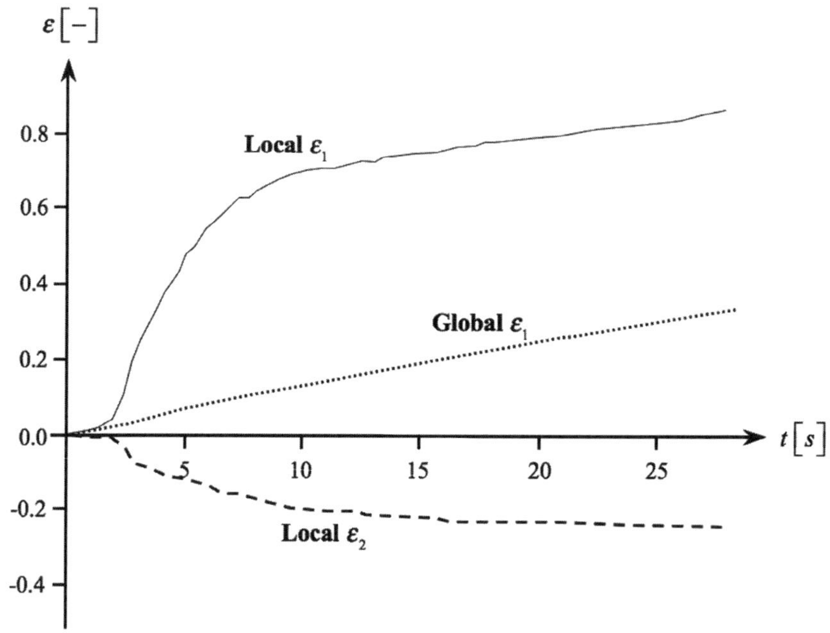

Fig. 3.24. Influence of the gauge length on the measured true strain in uniaxial tensile tests. Compared are local and global strains in tension direction (ε_1). Additional information is given on locally measured strain in perpendicular direction (ε_2). (Data from Junginger [220]).

Comparison of the strain rates observed in global or local strain measurement techniques used for the tension test of Figure 3.23 shows an almost constant global strain rate in the tensile direction $\dot{\varepsilon}_1$. The local strain rate, both in longitudinal direction ($\dot{\varepsilon}_1$) and perpendicular to it ($\dot{\varepsilon}_2$), is varying over the test duration. Moreover, the local tensile strain rate is bigger by an order of magnitude compared to the global one (see Figure 3.25).

Specifically these dependencies of the measured stress-strain and strain rate curves on the gauge sizes point out the importance of either associating a measurement technique to the scale of discretizing elements used in simulation, or of an understanding and modelling of the micro-mechanics of polymers. Both, the phenomenological and the physical approach have to reflect the specific characteristics of that type of material.

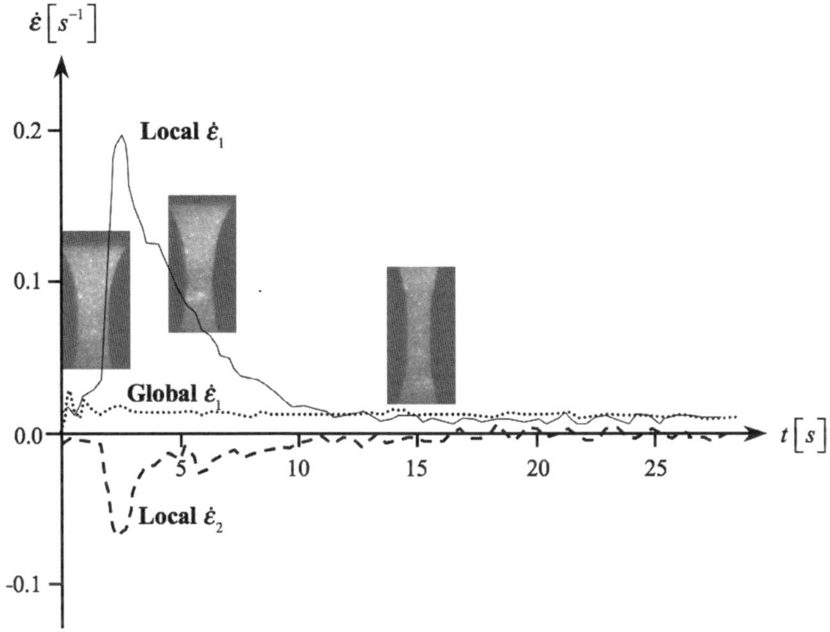

Fig. 3.25. Influence of the gauge length on the measured strain rates in tensile direction ($\dot{\varepsilon}_1$) and in perpendicular direction ($\dot{\varepsilon}_2$). (Data from Junginger [220]).

As can be seen from Figure 3.23, on the *macroscopic level* a sequence of three, partly overlapping, phases in the stress-strain relation is taking place:

- In the first phase of uni-axial tension and compression tests, both amorphous and semi-crystalline polymers show a nonlinear reversible deformation behavior which in addition can also be rate dependent. Here, linear elastic models are either only a course approximation or must be restricted to a minimum of strain.

- In a second phase, initial irreversible processes are taking place accompanied by global strain softening macroscopically observed as strain localization and related necking.
- A third and final phase is a visco-plastic irreversible deformation characterized by strain hardening until finally failure occurs.

Strain Rate Sensitivity

In addition to the localization processes, a pronounced *strain rate dependency* of the inelastic stress-strain behavior can be observed. Figure 3.26 illustrates the strain rate dependent behavior under uniaxial tension measured for a PC-ABS material.

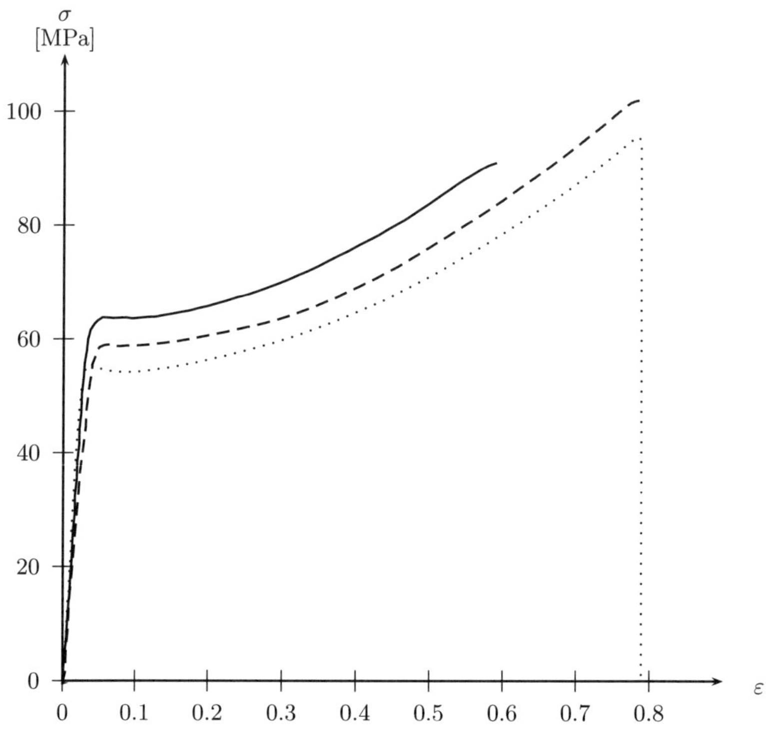

Fig. 3.26. Stress-strain curves for an amorphous polymer (PC-ABS) loaded at cross-head speeds of 5 $[mm/s]$ (dotted), 50 $[mm/s]$ (dashed) and 500 $[mm/s]$ (solid).

De Facto Strain Rates During Polymer Tensile Test

The strain rate sensitivity of many polymers combined with its sometimes large strains to failure poses a particular challenge to the strain rate dependent

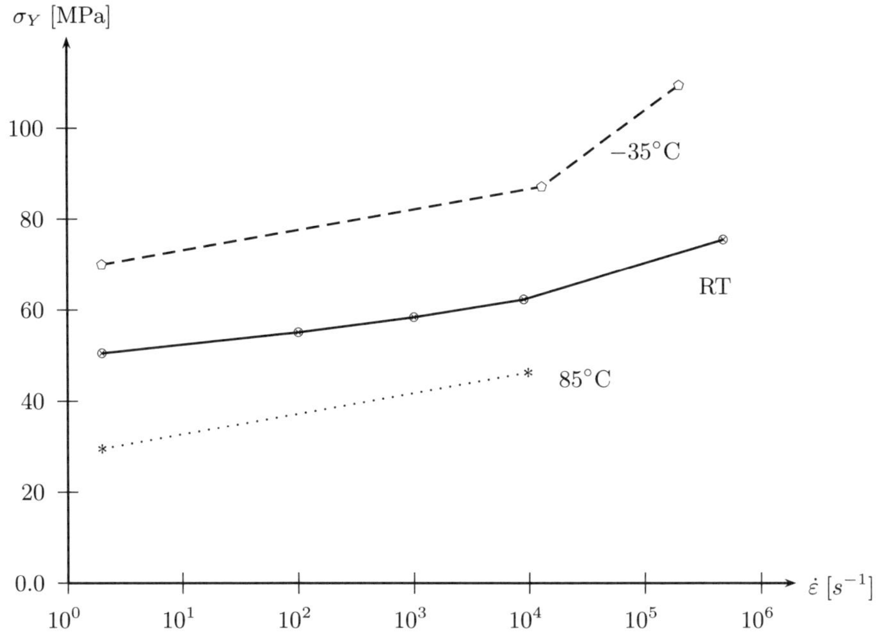

Fig. 3.27. Strain rate dependent yield stresses under uniaxial tension for Bayblend T65 at room temperature (solid line), −35°C (dashed line) and 85°C, respectively.

experimental characterization of polymers. For example, tensile tests at varying loading speeds up to failure strains around 100 percent cannot guarantee a constant strain rate during the whole test duration. This arises from the fact that the sample size changes by an order of magnitude during the loading time which leads to a related change of the reference length for the strain measure. Figure 3.28 gives an impression of the change in strain rates during tests at constant deformation speed. The variations result partly from the mentioned change in the specimen length. In addition the measured rates are influenced by oscillations which are typical and not totally avoidable in dynamic material tests.

In order to arrive at experimental data for a stress-strain relation at constant strain rate $\sigma = \sigma(\varepsilon, \dot{\varepsilon} = const.)$, both the current strain and strain rate during the test needs to be measured. After tests at various loading speeds, the measured data need to be mapped into a three dimensional stress, strain and strain rate table. Connecting these stress-strain-strainrate lines by an interpolation scheme provides a surface in that space which can then be used to extract stress-strain curves at constant strain rate. The procedure is illustrated

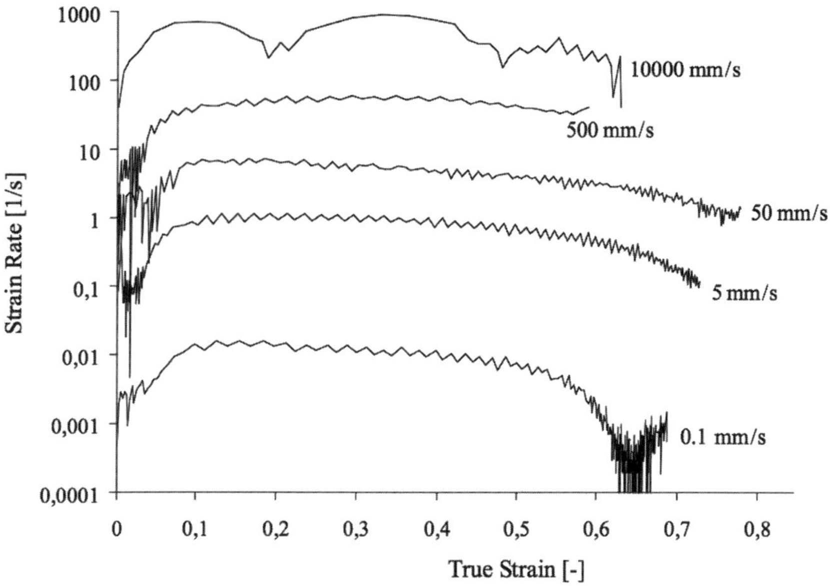

Fig. 3.28. De facto strain rates in tensile tests of an amorphous polymer with constant deformation (crosshead) speed.

in Figure 3.29 showing the results from various tests and the construction of related data by interpolation between the actually measured curves.

Anisotropy with Respect to Tension, Compression and Shear

Another direct consequence of the basic network structure on the molecular scale is the typical difference in the stress-strain behavior of polymers under tension, compression and shear loads. The characterization tests performed in Junginger [220] reflect this dependency as illustrated in Figure 3.30 and point out the need for polymer specific material models and adequate characterization methods. To improve the strain measurement, an optical instrumentation allowing for high-speed camera-based observation of deformations on the specimen and derivation of strain measures was developed by Junginger.

Plastic Volume Dilatation

Craze formation, a polymer typical failure mechanism (see Figure 3.32), and the related development of micro voids in amorphous polymers lead to a volume dilatation under irreversible deformation. That specific property of some thermoplastics requires that the increase in volume is precisely measured and that the related plasticity model reflects the phenomenon. Huberth et al. applied an optical strain measurement system to the top and side surfaces of a

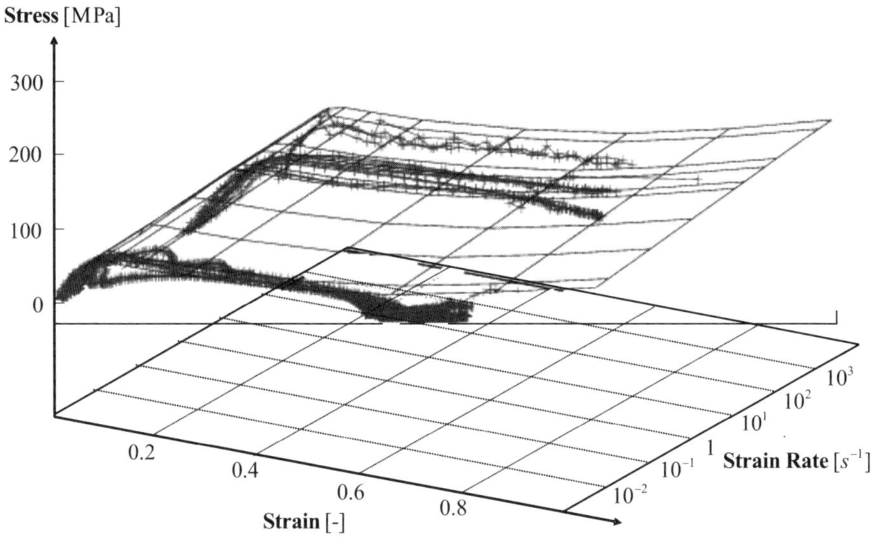

Fig. 3.29. Construction of stress-strain relations at constant strain rates through interpolation between experimental results containing stress-strain data at non-constant strain rates.

tension specimen as illustrated in Figure 3.31. With the double-view optical instrumentation volume elements of the specimen can be observed during a material test. Tracking of the volume elements over time enables the measurement of volume increase as shown in the lower part of Figure 3.31. The results for two individual tests of a PA-ABS material as published by Huberth et al. [200] illustrate the reproducible character of the test set-up and of the effect in the material. With an increase of about 60 % in volume for a 100 % longitudial strain in the tensile test, the significance of plastic volume dilatation for the chosen material is evident.

Phenomenological Models

To account for the empirical experiences with mechanical properties of polymers, some of which are shown in the last paragraph, various modifications to existing material models have been developed. Motivated by the distinct anisotropy regarding the elastic limit under tension and compression, existing plasticity formulations were modified. Common to most of these modified plasticity models is that the yield surfaces is shifted towards the compressive quadrant in a plane stress projection.

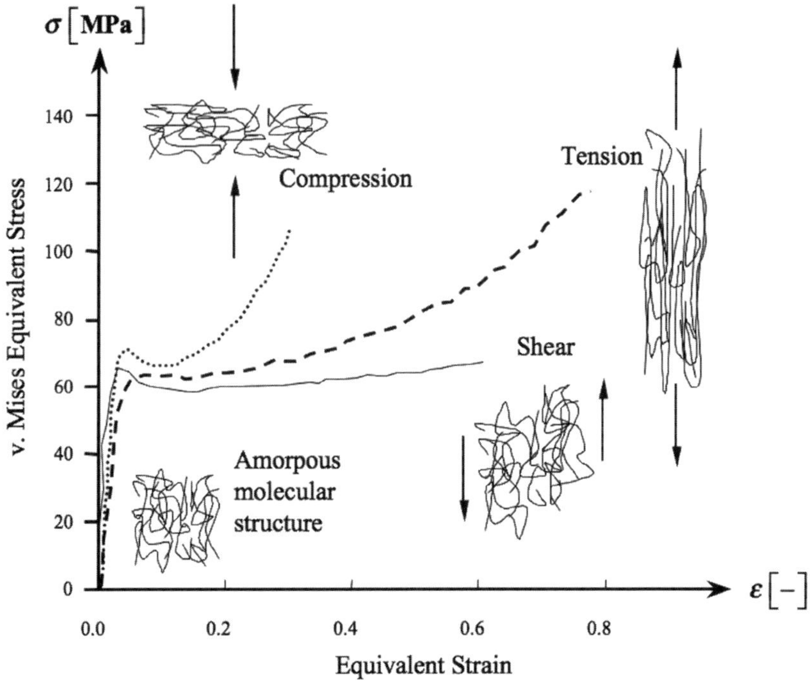

Fig. 3.30. Influence of the amorphous structure on the macroscopic mechanical behavior in terms of stress-strain curves. Measured von Mises equivalent stress against equivalent strain under uniaxial tension, uniaxial compression and pure shear loads for an amorphous polymer (PC-ABS). (Reprint with permission from Junginger [220]).

Pressure Modified von Mises Models

One example for a shift of the yield surface is the so called *pressure-modified von Mises model*, originally proposed by Schleicher [346], and later formulated by Raghava et al. [310] as yield surface depending on the first invariant of the stress tensor I_1, the second deviator invariant J_2, the yield stress under uniaxial compression C and the yield stress under uniaxial tension T:

$$f(I_1, J_2, C, T) = J_2 + \frac{1}{3}(C - T) I_1 - \frac{1}{3} CT = 0 \qquad (3.212)$$

For identical yield stresses under uniaxial tension and compression $C = T$, the model converges to the von Mises criterion.

A formulation that is implemented in mot commercial codes and that is often used as basic model for polymer simulations is the Drucker-Prager

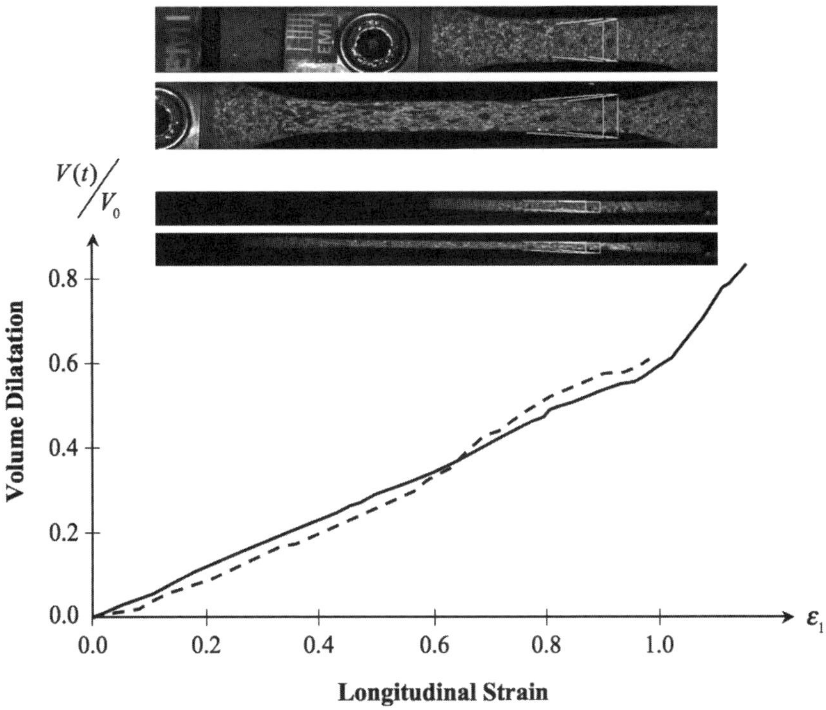

Fig. 3.31. Top and side view on PA-ABS specimen instrumented with optical strain measurement techniques to derive the plastic volume dilatation illustrated in the graphs below (See Huberth et al. [200] for details). The solid and dashed lines represent two individual tests on the same material type and, thus, the good reproducibility of the phenomenon.

pressure dependent yield model (3.159):

$$\alpha I_1 + \sqrt{J_2} - \kappa = 0 \tag{3.213}$$

Further examples for similar modes with yield surfaces shifted to the compressive quadrant are found in Bardenheier [34], Troost and Schlimmer [396] and in Caddell, Raghava and Atkins [75]. Models for polymer yielding under uniaxial tension or compression with superimposed hydrostatic pressure were proposed and discussed by Spitzig and Richmond [363] , Maers et al. [261], Rabinowitz, Ward and Parry [308] and Sauer et al. [342].

Caddell and Woodliff [74] investigated plastic deformation induced formation of anisotropic yielding. Uniaxial tensile deformations to different states of increasing plastic pre-deformation displayed a rotation of the yield locus in the plane stress projection along with a shift to the tensile quadrant. These

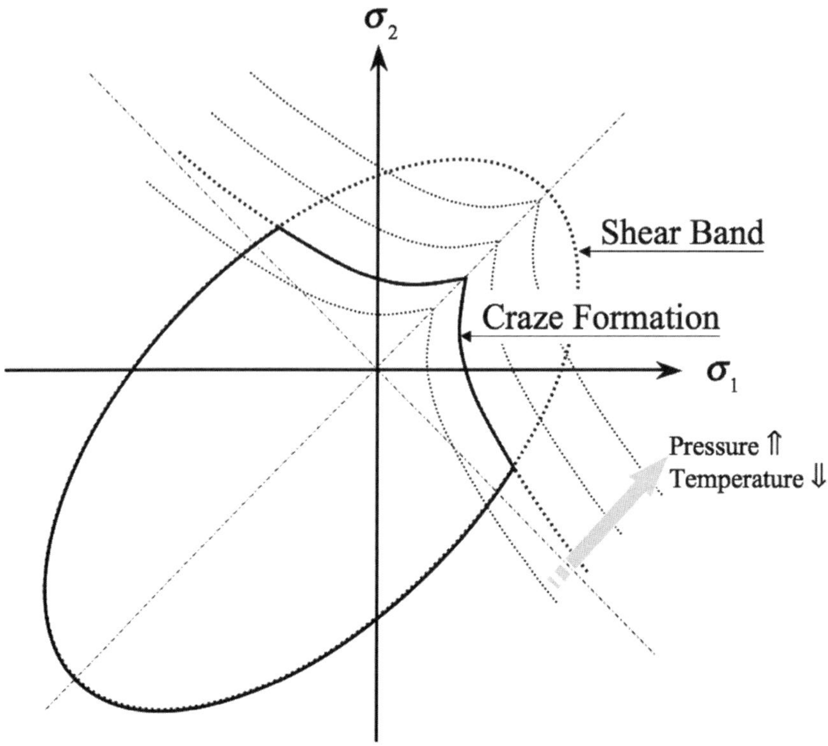

Fig. 3.32. Schematic yield loci in thermoplastics resulting from shear band and craze formation. The yield locus representing craze formation is found to shift towards the first quadrant for increasing pressure and decreasing temperature, respectively.

observations lead to modifications of the Hill criterion (3.174). The same approach was taken by Stassi-D'Alia [364].

Stress Bias Criterion

Based on different processes at the molecular scale, thermoplastics show two distinct forms of inelastic deformation mechanisms[19] under tensile loads in terms of *shear band* formation and *craze formation*, respectively. Shear bands lead to inhomogeneous orientations under an angle of about 45° with respect to the tension direction. In the case of craze formation, additional development of lens shaped cavities leads to density inhomogeneities and to volume growth.

[19] See Retting [319] or Sperling [362]

Sternstein and Ongchin [372] and Sternstein and Myers [371] formulated a so called *stress bias criterion*indexYield criteria!stress bias allowing the onset of plastic flow through craze formation after

$$\sigma_{bias} = |\sigma_1 - \sigma_2| \geq A + \frac{B}{I_1} \tag{3.214}$$

is fulfilled. The shear band formation, on the other hand, s described through the octahedral shear stresses and the Coulomb rule 3.154

$$\tau_{oct} = \tau_0 - \mu\, p \tag{3.215}$$

The resulting combination of two separate surfaces projected into the two-dimensional stress plane is illustrated in Figure 3.32.

Visco-Elastic Models

A strain rate and temperature dependent visco-elastic model for the stress-strain behavior of thermoplastics was suggested by G'Sell and Jonas[152] formulating the yield stress by:

$$\sigma_Y(\varepsilon, \dot{\varepsilon}, T) = K\, e^{\frac{-a}{T}} \left(1 - e^{-w\,\varepsilon}\right) e^{\left(\frac{\gamma}{2}\,\varepsilon^2\right)} \dot{\varepsilon}^m \tag{3.216}$$

An application of the G'Sell model implemented with an associated flow rule was presented by Schang et al. [344].

Junginger Model

Junginger performed a complex test program on the PC-ABS Bayblend T65 including tensile, compressive and shear tests with optical strain measurement. Positions on the stress strain curve under uniaxial tension, uniaxial compression and shear at equal plastic strain values make up coulomb type yield surfaces. In contrast to a standard Mohr-Coulomb type, the developed model consists, of two individual branches in the $\sigma_{vM} - p$ plane. The two branches can be described by plastic strain dependent friction angles $\phi_T(\varepsilon_{pl})$ and $\phi_C(\varepsilon_{pl})$ for the tensile and compressive direction, respectively. The cohesion $C(\varepsilon_{pl})$ is derived from the shear test results, represented by squares □ on the $p = 0$ axis in Figure 3.33. With these data, the following formulation for the yield loci was found:

$$f(\sigma_{vM}, p, \phi_T, C, \bar{\varepsilon}_{pl}) = \sigma_{vM} - p\, \tan\phi_T(\bar{\varepsilon}_{pl}) - C(\bar{\varepsilon}_{pl}) = 0 \quad \text{for} \quad p \leq 0 \tag{3.217}$$

$$f(\sigma_{vM}, p, \phi_C, C, \bar{\varepsilon}_{pl}) = \sigma_{vM} - p\, \tan\phi_C(\bar{\varepsilon}_{pl}) - C(\bar{\varepsilon}_{pl}) = 0 \quad \text{for} \quad p > 0 \tag{3.218}$$

The proposed yield locus is not convex as can also be seen from the plane stress illustration in Figure 3.34. With the tensile, compressive and shear tests performed under varying loading velocities, Junginger established the yield loci for the related strain rates according to the strain dependency illustrated in Figure 3.33.

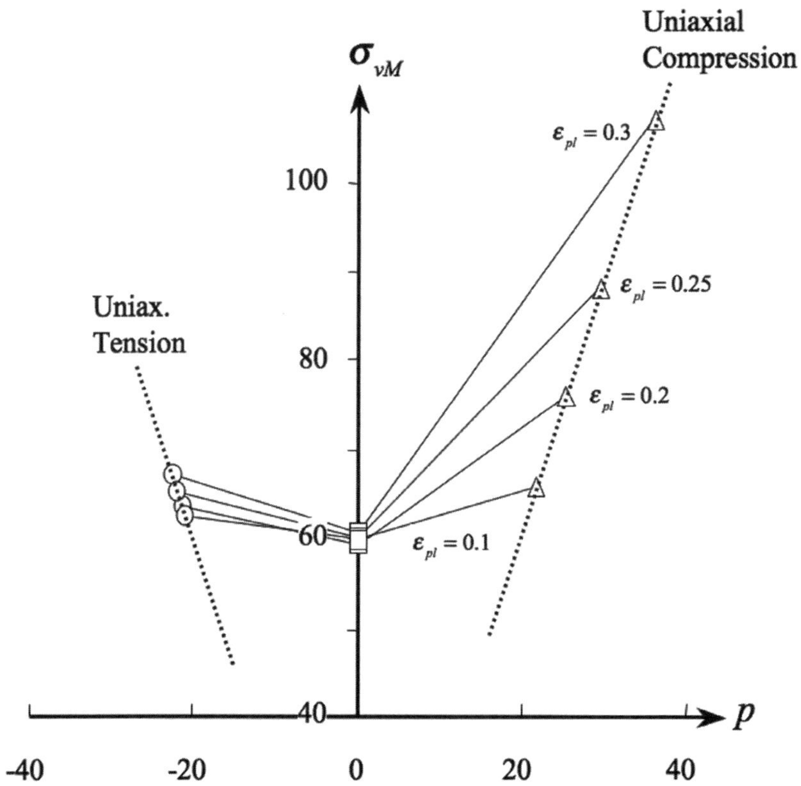

Fig. 3.33. Equivalent stresses over hydrostatic pressures at varying plastic strain under uniaxial tension (○), uniaxial compression (△) and shear (□). Formation of related strain dependent yield loci with non-convex shape. (Reprint with permission from Junginger [220])

SAMP Model

A so called semi-analytical model for the simulation of polymers (SAMP) was recently proposed by Kolling et al. [229]. The SAMP formulation reflects the major complexities of polymers mentioned above providing a new standard for the simulation of thermoplastics under crash loads. In particular the anisotropic yield behaviour under tensile, compressive and shear loads is modelled using the quadratic yield surface in invariant form (3.167) :

$$J_2^2 + A_0 + A_1\, I_1 - A_2\, I_1^2 \leq 0 \tag{3.219}$$

or by the alternative formulation which is linear in the second invariant:

$$J_2 + A_0 + A_1\, I_1 - A_2\, I_1^2 \leq 0 \tag{3.220}$$

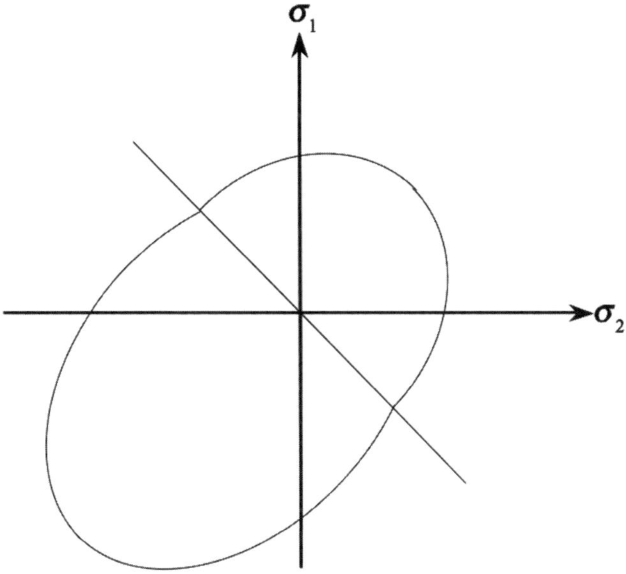

Fig. 3.34. Illustration of the Junginger yield locus under plane stress conditions. (Reprint with permission from Junginger [220])

Whereas the correlation between the parameters for the quadratic model is given in (3.169), the parameters in the linear formulation (3.220) are obtained from uniaxial tension, uniaxial compression and simple shear tests according to:

$$A_0 = \sqrt{3}\,\sigma_s \quad A_1 = 3\left(\frac{\sigma_t - \sigma_c}{\sigma_t + \sigma_c} - \sqrt{3}\,\sigma_s\,\frac{\sigma_t - \sigma_c}{\sigma_t\,\sigma_c}\right)$$

$$A_2 = 18\left(\frac{1}{\sigma_t + \sigma_c} - \frac{\sqrt{3}\,\sigma_s}{2\,\sigma_t\,\sigma_c}\right) \tag{3.221}$$

where σ_t, σ_c and σ_s represent the experimentally measured yield stresses in the three types of tests.

The pressure dependent plastic potential, its strain rate dependency and the damage controlled failure use tabulated input providing an interface to directly use experimentally measured data.

Micro-Physics of Inelastic Polymer Deformation

Microscopic Deformation Behavior

As found in the first paragraph of this section, inelastic deformation of polymers can be totally different from the plasticity effects commonly known from crystalline materials. Since polymer glasses often show volume change during inelastic deformations, a plasticity model assuming plastic incompressibility, as for example the von Mises type criteria, is not appropriate. The corresponding mechanisms taking place on the*microscopic* level during the individual phases can be understood through investigations of the *intra-* and *inter-molecular* processes and the behaviour of the neighbouring crystals.

On a microscopic scale, two types of physically distinct barriers are identified that need to be overcome in a glassy polymer before large strain inelastic deformation can occur[20]:

- First, in a deformation phase prior to initial inelastic processes, *intermolecular resistance* against segment rotation must be exceeded. The intermolecular resistance represents a free energy barrier to the mobility of the molecules.
- As a second phase, *entropic resistance* impedes deformation when molecular alignment takes place changing the configurational entropy of the material.

Boyce [63] picked up earlier work including Argon's "double-kink" model [14] on the intermolecular resistance, to model the pressure dependent strain softening after the onset of initial inelastic deformation using an initial value of the athermal shear resistance and an evolution equation

Under isothermal conditions, the change in Helmholtz free energy with respect to the two resistances can be formulated as:

$$\Delta\psi = \Delta e_1 + \Delta e_2 + T\,\Delta s \tag{3.222}$$

where e_1 represents the intermolecular barrier and e_2 the intra-molecular one, e.g. due to bulky side-groups and energy barriers between different isomeric states. With the second barrier overcome, an affine orientation is taking place leading to permanent deformations. The fact that these permanent deformations can in turn be fully recovered from by heating above the glass transition temperature T_G, motivated to use statistical mechanics known from rubber elasticity (Boyce et al. [64]). The idea was to model the entropic resistance characterized through the stress state that is needed to avoid the relaxation process to the initial deformation state if heated to a temperature $T > T_G$.

[20] See for example Haward and Thackray [171], Argon [14] or Boyce, Parks and Argon [64]

This stress state defines the components of the backstress tensor σ^B which is, in that theory, uniquely related to the texture, i.e. the locked in plastic stretches expressed through the plastic stretches $\mathbf{V}^p$ (see equation (2.55)) and its principal values λ_i^p, respectively. For temperatures above T_G the inter- and intra-molecular energy terms in equation (3.222) vanish and the components of the back stress tensor are calculated from:

$$\sigma_i^B = \lambda_i^p \frac{\partial \psi}{\partial \lambda_i^p} \approx -T \lambda_i^p \frac{\partial \Delta s}{\partial \lambda_i^p} \tag{3.223}$$

Expression of the entropy change through the Langevin chain statistics (3.108) allowing for descriptions of Δs up to the locking state, the principal components of the backstress tensor are formulated in [64] as:

$$\sigma_i^B = G \frac{\sqrt{N}}{3} \left[\lambda_i^p L^{-1} \left(\frac{\lambda_i^p}{\sqrt{N}} \right) - \frac{1}{3} \sum_{j=1}^{3} \lambda_j^p L^{-1} \left(\frac{\lambda_j^p}{\sqrt{N}} \right) \right] \tag{3.224}$$

Anand-Gurtin Elasto-Viscoplastic Model

Based on physical ideas developed in earlier works on the modeling of rate dependent elastic-plastic deformations including deformation-induced anisotropies in amorphous polymeric glasses[21], Anand and Gurtin [6] provided a thermo-mechanically consistent extension to the framework of that theory.

Fundamental characteristics of their model are:

- a Kröner-Lee decomposition (2.52) of the deformation gradient: $\mathbf{F} = \mathbf{F}^e \mathbf{F}^p$ without prescription of a symmetric $\mathbf{F}^e$ leading to frame indifference under all rotations in the reference and (independently) in the relaxed configuration,
- Description of the initial plastic deformation phase through the formulation of a *local free volume* by an internal state variable η,
- Incompressible flow rule formulation on base of the Clausius-Duhem inequality and a $\mathbf{F}^p$-dependent Helmholtz potential leading to the back-stress tensor.

Whereas the elastic part of free energy is described through

$$\psi^e = G \left| \mathbf{E}_0^e \right|^2 + \frac{1}{2} K \left| tr\, \mathbf{E}^e \right|^2 \tag{3.225}$$

with the elastic shear and bulk moduli G and K, the plastic part is discussed by two alternatives. The first one uses a neo-Hookean form

[21] See for example the publications of Haward and Thackray [171], Parks et al. [304], Boyce et al. [64], Arruda and Boyce [18], Arruda and Boyce [19], Wu and van der Giessen [430] and Gurtin [155].

$$\psi^{pl} = \mu \frac{3}{2} \left\{ (\lambda^p)^2 - 1 \right\} \tag{3.226}$$

leading to the constant back stress modulus μ. This formulation is assumed to be valid for moderate plastic stretches $\lambda^p < 0.35$. For larger stretches, an alternative model using the Langevin function L is proposed:

$$\psi^{pl} = \mu_R \lambda_L^2 \left[\left(\frac{\lambda^p}{\lambda_L} \right) x + \ln \left(\frac{x}{\sinh x} \right) - \left(\frac{1}{\lambda_L} \right) y - \ln \left(\frac{y}{\sinh y} \right) \right] \tag{3.227}$$

where $x = L^{-1} \left(\frac{\lambda^p}{\lambda_L} \right)$ and $y = L^{-1} \left(\frac{1}{\lambda_L} \right)$, using the inverse Langevin function L^{-1}, the so-called rubbery modulus μ_R and the network locking stretch λ_L. This leads to a non-constant backstress modulus of:

$$\mu = \mu_R \left(\frac{\lambda_L}{3\, \lambda^p} \right) L^{-1} \left(\frac{\lambda_p}{\lambda^L} \right) \tag{3.228}$$

The flow rule is formulated through an evolution equation for the plastic deformation gradient:

$$\dot{\mathbf{F}}^p = \mathbf{D}^p \, \mathbf{F}^p \tag{3.229}$$

with the plastic stretch tensor

$$\mathbf{D}^p = \nu^p \left(\frac{\mathbf{T}_0^e - \mu \, \mathbf{B}_0^p}{2 \, \bar{\tau}} \right) \tag{3.230}$$

an equivalent plastic shear strain

$$\nu^p = \nu_0 \left(\frac{\bar{\tau}}{s + \alpha \, \pi} \right)^{1/m} \tag{3.231}$$

applying the parameters ν_0, m, and α, and the equivalent plastic shear stress

$$\bar{\tau} = \frac{|\mathbf{T}_0^e - \mu \, \mathbf{B}_0^p|}{\sqrt{2}} \quad , \tag{3.232}$$

as well as the deviatoric component of the left Cauchy-Green tensor $\mathbf{B}_0^p$ as described before in equation (3.122).

With the backstress modulus (3.228), the evolution equation (3.229) and the deviatoric plastic left Cauchy-Green tensor (3.122), the backstress tensor is described by

$$\sigma^B = \mu \, \mathbf{B}_0^p \tag{3.233}$$

The entropy term in equation (3.231) is accumulated via the evolution equation

$$\dot{s} = h_0 \left(1 - \frac{s}{\tilde{s}(\eta)} \right) \nu^p \tag{3.234}$$

where the *saturation value* of s is described via the local free volume η by

$$\tilde{s}(\eta) = s_{cv}\left[1 - b\left(\eta_{cv} - \eta\right)\right] \tag{3.235}$$

and the local free volume evolves through

$$\dot{\eta} = g_0 \left(\frac{s}{s_{cv}} - 1 \right) \nu^p \tag{3.236}$$

with the additional parameters h_0, g_0, s_{cv}, b and η_{cv}.

Implementation of this model in ABAQUS/Explicit provided a platform to adjust the various parameters and allowed for a validation of the model through comparison with experimental results on quasi-static tensile and compressive tests of polycarbonate at atmospheric pressure and room temperature. Detailed derivations of the model and the mentioned comparisons with tensile and compressive experimental data are given in Anand and Gurtin [6]. An application of the model to micro-indentation problems was published by Anand and Ames [5] showing reasonable agreement with the experimental results on PMMA.

4

Shock Waves and Related Equations of State

Information about loads applied to a structure is propagated through the medium by waves and accordingly at the local speed of sound. In the case of quasi-static loads, wave effects are not investigated since the loading duration is long compared to the duration of multiple reflections throughout the structure. In addition, the resulting structural deformation and material state is not influenced in a comparable way by single wave transitions. However, if the induced waves take the shape and amplitude of so called shock waves or the load speed is in the order of magnitude of the local sound speed, then wave effects and their propagation through the structure needs to be resolved in time and space.

In the course of this chapter the conditions for initiation and stable propagation of shock waves will be deerived after a short introduction to general wave propagation basics. Dispersion and the related concave nonlinearity of equations of state wil be indentified as preconditions for shocks. Therefore, nonlinear equations of state, their derivation from basically fluid dynamic considerations and a collection of the most often used formulations will be given. A critical discussion of the limitations of equations of state and in particular their application to anisotropic materials concludes the chapter.

4.1 Elastic Wave Propagation in Solids

Waves in solids are basically perturbations in the velocity field propagating through the continuum in different forms and at related different velocities. The propagating perturbation leads to wave form specific motion of the particles. The most important wave forms in solids are:

- *Longitudinal waves* of compressive or tensile type which cause particle deflections along the propagation direction. Since longitudinal waves are the fastest wave forms in solids they are also called *primary waves*. Its wave propagation speed is usually denoted by c_L

- The next fastest waves propagating at c_S are the *shear* or *secondary waves* causing particle motion perpendicular to the wave propagation.
- Along the surfaces of solids propagate so called *Rayleigh waves* setting surface particles into elliptic motion.
- In structures of finite bending stiffness *flexural waves* propagate upon dynamic loading.

A mathematical description of wave propagation in solids can be established via the momentum equation or equilibrium considerations. The latter method will be applied next to find a relation that describes the motion of waves involving the sound speed of the medium as well as displacements and their spatial and time derivatives.

4.1.1 Wave Equation and Sound Speeds

To derive the one-dimensional wave equation, a planar wave situation as illustrated in Figure 4.1 is assumed. Equally, a longitudinal wave in a slender rod could be observed. Equilibrium on the front and rear side of a differential element with dimensions $\mathrm{d}x$ an $\mathrm{d}y$ requires

$$\left(\sigma_x + \frac{\partial \sigma_x}{\partial x}\,\mathrm{d}x\right)\mathrm{d}y - \sigma_x\,\mathrm{d}y = \rho\,\mathrm{d}x\,\mathrm{d}y\,\frac{\mathrm{d}^2 u_x}{\mathrm{d}t^2} \tag{4.1}$$

or

$$\frac{\partial \sigma_x}{\partial x} = \rho\,\frac{\mathrm{d}^2 u_x}{\mathrm{d}t^2} \tag{4.2}$$

Application of a constitutive relation, e.g. Hooke's law $\sigma_x = E\,\varepsilon_x$ for the chosen linear elastic solid, and a linear strain-displacement relation $\varepsilon_x = \partial u_x/\partial x$, describes a one-dimensional wave by:

$$\frac{\partial^2 u_x}{\partial x^2} = \frac{\rho}{E}\,\frac{\mathrm{d}^2 u_x}{\mathrm{d}t^2} \tag{4.3}$$

The one-dimensional wave equation (4.3) can equally be expressed using the *longitudinal wave's sound speed*

$$c_L = \sqrt{\frac{E}{\rho}} \tag{4.4}$$

as

$$\frac{\partial^2 u_x}{\partial x^2} = \frac{1}{c_L^2}\,\frac{\mathrm{d}^2 u_x}{\mathrm{d}t^2} \tag{4.5}$$

In the more general three-dimensional case, the additional components in the stress strain relation lead to a *primary sound speed* of:

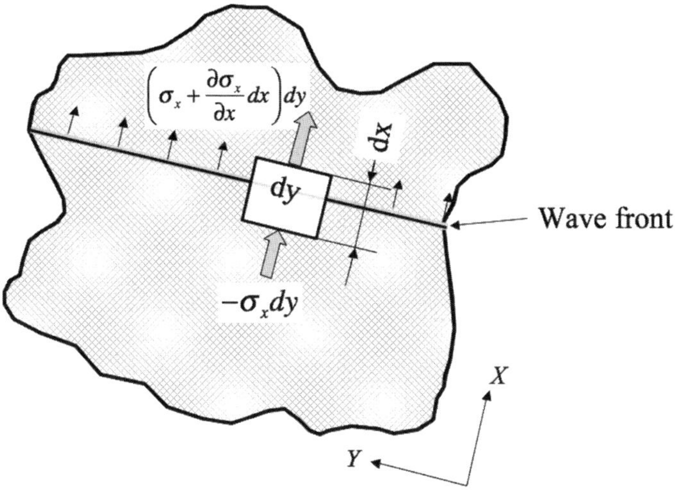

Fig. 4.1. Differential element in a one-dimensional wave and related force terms.

$$c_p = \sqrt{\frac{K + 4/3\,G}{\rho}} \tag{4.6}$$

and, thus, to the three-dimensional wave equation

$$\nabla \mathbf{u} = \frac{1}{c_p^2} \frac{\mathrm{d}^2 u_x}{\mathrm{d}\, t^2} \tag{4.7}$$

Sound speeds for the remaining wave types can be identified accordingly for shear waves

$$c_S = \sqrt{\frac{G}{\rho}} \tag{4.8}$$

and as so called *bulk sound speed*

$$c_B = \sqrt{\frac{K}{\rho}} \tag{4.9}$$

for hydrostatic pressure waves in a perfect fluid with $G = 0$.

Anticipating a generalization to non-linear material behaviour, it should be noted that then the constant elastic modulus E is replaced by the tangent stiffness $\partial\,\sigma/\partial\,\varepsilon$ of the material at a specific loading state $(\sigma,\,\varepsilon)$. For the associated sound speed of a one-dimensional longitudinal wave the non-linear case yields:

$$c_L = \sqrt{\frac{1}{\rho}\frac{\partial \sigma_x}{\partial \varepsilon_x}} \tag{4.10}$$

which, in the case of fluids and the adiabatic relations for the change in internal energy e with volume V at pressure p

$$\mathrm{d}e = -p\,\mathrm{d}V \tag{4.11}$$

$$\mathrm{d}s = 0 \tag{4.12}$$

turns into

$$c^2 = \left.\frac{\partial p}{\partial \rho}\right|_s = \left.\frac{\partial p}{\partial \rho}\right|_e + \frac{p}{\rho^2}\left.\frac{\partial p}{\partial e}\right|_\rho \tag{4.13}$$

4.1.2 Solution to the One-Dimensional Wave Equation

For most technically relevant applications there are no closed form analytical solutions of the related wave equation due to the multi-dimensional space, complex boundaries or nonlinearities. That is why numerical methods are used to solve wave propagation related problems as we will see in chapter 5.

However, the comparably simple and analytically solvable case of a one-dimensional wave in a linear-elastic medium is still of significant relevancy. In the context of specimen definition and stress analyses for dynamic material testing solutions these specific solutions are very useful as we will see in section 7.4.1. It will allow us to evaluate velocities, strains and ultimately stresses providing well defined experimental set-ups.

An analytical solution for the one-dimensional wave equation (4.5) in terms of displacements $u(x,\,t)$ was provided by d'Alembert. The so called *d'Alembert solution* starts with a coordinate transformation from x, t to ζ, η in the form:

$$\zeta = x - c_L t \qquad \eta = x + c_L t \tag{4.14}$$

Application of the chain rule yields for the first derivatives $\frac{\partial u(\zeta,\,\eta)}{\partial x}$ and $\frac{\partial u(\zeta,\,\eta)}{\partial t}$:

$$\frac{\partial u(\zeta,\,\eta)}{\partial x} = \frac{\partial u}{\partial \zeta}\frac{\partial \zeta}{\partial x} + \frac{\partial u}{\partial \eta}\frac{\partial \eta}{\partial x} \tag{4.15}$$

and

$$\frac{\partial u(\zeta,\,\eta)}{\partial t} = \frac{\partial u}{\partial \zeta}\frac{\partial \zeta}{\partial t} + \frac{\partial u}{\partial \eta}\frac{\partial \eta}{\partial t} = -c\,\frac{\partial u}{\partial \zeta} + c\,\frac{\partial u}{\partial \eta} \tag{4.16}$$

The second derivatives turn out to be:

$$\frac{\partial^2 u(\zeta,\,\eta)}{\partial x^2} = \frac{\partial^2 u}{\partial \zeta^2} + 2\,\frac{\partial^2 u}{\partial \zeta\,\partial \eta} + \frac{\partial^2 u}{\partial \eta^2} \tag{4.17}$$

and

$$\frac{\partial^2 u(\zeta, \eta)}{\partial t^2} = c_L^2 \frac{\partial^2 u}{\partial \zeta^2} - 2 c_L^2, \frac{\partial^2 u}{\partial \zeta \, \partial \eta} + c_L^2 \frac{\partial^2 u}{\partial \eta^2} \tag{4.18}$$

Implementation in the wave equation (4.5) leads to its formulation in terms of ζ and η:

$$\frac{\partial^2 u}{\partial \zeta^2} + 2 \frac{\partial^2 u}{\partial \zeta \, \partial \eta} + \frac{\partial^2 u}{\partial \eta^2} = \frac{\partial^2 u}{\partial \zeta^2} - 2 \frac{\partial^2 u}{\partial \zeta \, \partial \eta} + \frac{\partial^2 u}{\partial \eta^2} \tag{4.19}$$

which is equivalent to

$$\frac{\partial^2 u}{\partial \zeta \, \partial \eta} = 0 \tag{4.20}$$

With the coordinate transformation (4.14) a form of the wave equation results that can be solved much easier. Integration of (4.20) with respect to ζ delivers:

$$\frac{\partial u}{\partial \eta} = g(\eta) \tag{4.21}$$

with an arbitrary function $g(\eta)$. Secondly, integration with respect to η yields:

$$u(\zeta, \eta) = F(\zeta) + G(\eta) = F(x - c_L t) + G(x + c_L t) \tag{4.22}$$

where $G(\eta) = \int g(\eta) \, \mathrm{d}\eta$. To complete the solution, equation (4.22) must be specified by initial conditions

$$u(x, 0) = f(x) \qquad \frac{\partial u(x, 0)}{\partial x} = h(x) \tag{4.23}$$

Thus, the solution function is a combination of two arbitrary functions $F(\zeta)$ and $G(\eta)$, for example $F = \sin(x - c_L t)$ and $F = \sin(x + c_L t)$. Physically, the two functions represent a left and a right running wave in the one-dimensional bar.

With the d'Alembert solution (4.22) the resulting spatial derivative for a strain measure $\partial u / \partial x$ as well as the velocity $\partial u / \partial t$ are:

$$\frac{\partial u}{\partial x} = \frac{\partial F(\zeta)}{\partial \zeta} \frac{\partial \zeta}{\partial x} + \frac{\partial G(\eta)}{\partial \eta} \frac{\partial \eta}{\partial x} = \frac{\partial F(\zeta)}{\partial \zeta} + \frac{\partial G(\eta)}{\partial \eta} \tag{4.24}$$

and

$$\frac{\partial u}{\partial t} = \frac{\partial F(\zeta)}{\partial \zeta} \frac{\partial \zeta}{\partial t} + \frac{\partial G(\eta)}{\partial \eta} \frac{\partial \eta}{\partial t} = -c_L \frac{\partial F(\zeta)}{\partial \zeta} + c_L \frac{\partial G(\eta)}{\partial \eta} \tag{4.25}$$

If the left running wave $F(\zeta)$ is ignored, we find

$$\frac{\partial u}{\partial x} = \frac{\partial G(\eta)}{\partial \eta} \tag{4.26}$$

and

$$\frac{\partial u}{\partial t} = c_L \frac{\partial G(\eta)}{\partial \eta} = c_L \frac{\partial u}{\partial x} \tag{4.27}$$

Equation (4.27) is the basic relation between spatial and time derivative often used in the context of wave propagation analyses in bars. Specifically, the stress at the wave position in a linear elastic bar is then found using equation (4.27) and $c_L = \sqrt{\frac{E}{\rho}}$:

$$\sigma = E\,\varepsilon = E\,\frac{\partial u}{\partial x} = \frac{E}{c_L}\frac{\partial u}{\partial t} = \rho\, c_L \frac{\partial u}{\partial t} \tag{4.28}$$

4.2 Shock Wave Formation

An important result of the last section is the dependency of wave speeds on the adiabatic compression modulus $\left.\frac{\partial p}{\partial \rho}\right|_s$. In other words, equation (4.13) describes a pressure dependent propagation speed for the case of nonlinear constitutive laws. First this resulting variable wave speed enables the formation of so called *shock waves* which are the topic of this section.

Characteristic properties of all shock waves are extremely short rise times as well as high pressure, density and temperature amplitudes. Basically, shock waves can arise as a consequence of both wave superposition and dispersion effects:

- If the source of a pressure disturbance is moving at the speed of sound of the surrounding medium or faster, superposition of the propagated disturbances and thus pressure waves leads to increased amplitudes and pressure gradients.
- In case of nonlinear pressure-density relations the corresponding dispersion effects lead to the formation of shock waves if faster wave components overtake earlier induced waves of lower propagation speed.

Already in the late nineteenth century scientists like Christian Doppler and Ernst Mach began to explain the physical nature of shock waves. Huygens' principle of wave front construction lead the basics to Doppler's[1] theoretical study on the formation of head waves generated by moving disturbances. *Schlieren-optical* techniques allowed for experimental visualization of shock waves in air. It is the involved density gradient along the wave front that invokes changes in the refraction index of the gas and, hence, enables the visualization of shocks. Therefore, sharp white lines mark the location of shock

[1] Doppler C. A.: Über den Einfluß der Bewegung des Fortpflanzungsmittels auf die Erscheinung der Äther-, Luft- und Wasserwellen, Abhandl. Königl. Böhm. Gesellsch. Wiss. Prag 5 [V]: 293-306, (1848)

waves in schlieren images. As shown in Fig. 4.2, one of the first schlieren pictures of supersonic projectiles taken by Mach and Salcher [259], the *head wave* forms a cone of approximately 80° originating from the tip of the projectile which was accelerated to a velocity of about 530 [m/s]. The vertical white lines are wires used to trigger the spark light source (compare Krehl [232] for details).

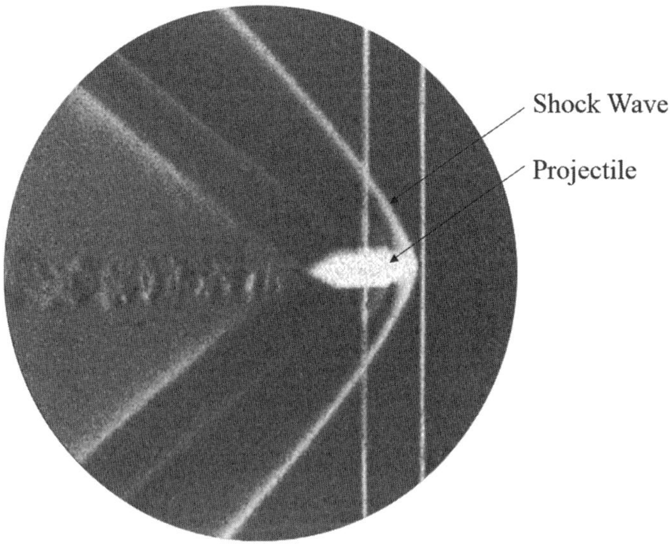

Fig. 4.2. Schlieren image of a supersonic projectile and its corresponding head wave. Ernst Mach archive, EMI, Freiburg

As mentioned before, the formation of the head wave cone results from wave superposition, predicted first by Doppler [112], as envelope of all elementary disturbances. The moving projectile is the source of these disturbances. If it moves at the speed of sound or faster, the superposition of wave fronts leads to a *shock cone*[2] with a characteristic angle α that is defined by

$$sin\,\alpha = \frac{c_0}{v} = \frac{1}{\mathrm{Ma}} \quad , \tag{4.29}$$

[2] The shock cone is often called *Mach cone* though Doppler predicted it 40 years before Mach and Salcher took the first pictures. According to Krehl [232] Mach even missed to cite Doppler though using the same explanation for the evolution of the shock wave as Doppler did so many years earlier.

where v is the speed of the projectile. The speed of sound in the surrounding un-shocked medium is denoted by c_0. *Ma* is the so called *Mach number* representing the ratio of the two velocities.

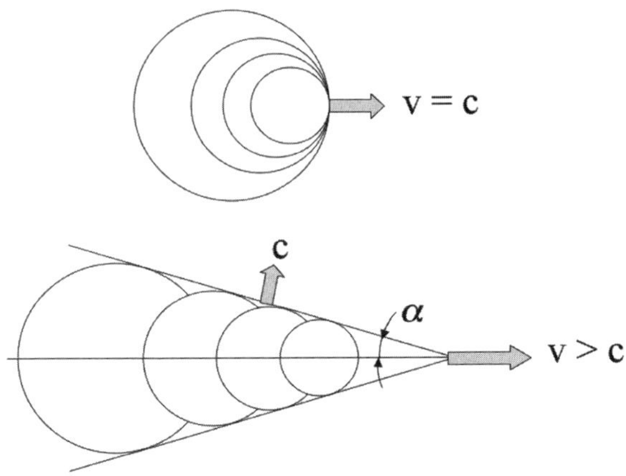

Fig. 4.3. Wave formations arising from moving disturbances as originally explained by Doppler [112].

Shock wave research has developed a wide spectrum of experimental and numerical methodologies to better understand and to predict shock wave effects. Figure 4.4 illustrates results of an experimental setup developed by Neuwald [296] applying colour schlieren techniques to visualize not only the position of shock waves in air but also the orientation of the density gradient and, thus, the propagation direction.

A recent result of a combined experimental and numerical investigation of shock waves and their influence on the trajectory of a spin stabilized projectile is illustrated in Figure 4.5. Topic of the investigation was the effect of a small charge detonated right on the surface of the projectile in order to shift its trajectory in a controlled way. The complexity of that manoeuvre arises from the spin, the necessity positioning the charge in order to guarantee a momentum free transverse force and the interaction of stationary and in-stationary shock waves. Where the first two aspects pose noteworthy problems to an experimental investigation of that process, the latter sorts out many existing numerical codes which are designated to solve either one of the wave types. With the combined investigations, Klomfass and Warken [228] were able to quantify the achievable transverse thrust as well as the flight instability which arises from the interaction of the detonation driven shock wave with the head wave and the related disturbance of the head wave.

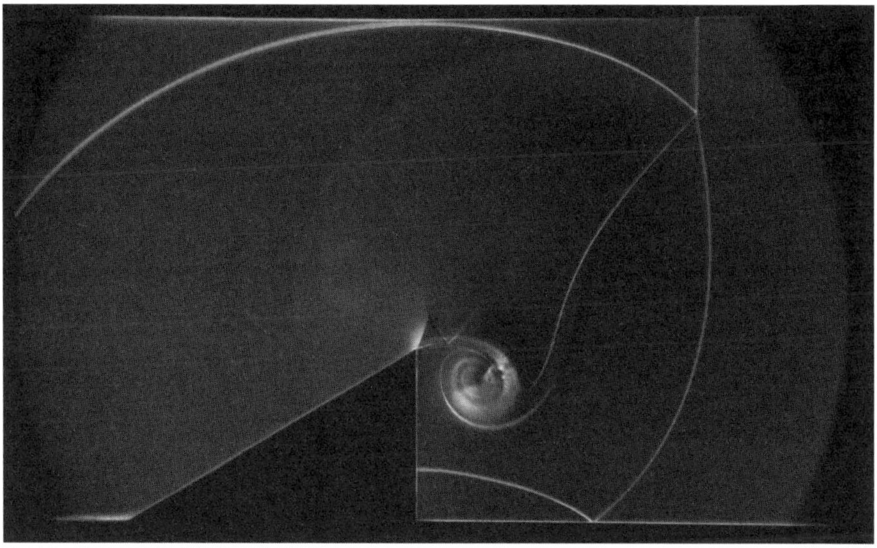

Fig. 4.4. Colour schlieren optical visualization of shock waves in a shock tube reflected on a wedge. The related experimental setup was developed by Neuwald, Ernst-Mach-Institute. See Neuwald and Reichenbach [296].

4.3 Shock Wave Propagation in Solids

4.3.1 Conditions for Shock Waves - Phenomenological Aspects

While in the framework of fluid-dynamics, a source moving inside a gaseous medium was observed in the last section to cause a shock wave, obviously another physical reason must be present in case of shock waves in condensed matter. Experience shows that external dynamic compressive loads, initiated e.g. by impact or detonation, can possibly cause very strong waves with extremely short rise times inside structures. As in the gas-dynamics example discussed before, superposition of different wave components is responsible for the steepening of the wave front. To be more precise, the corresponding wave superposition takes place as a consequence of *dispersion*, an effect that arises in media with nonlinear compressive behavior according to equation (4.10). The consequences of nonlinearities in the constitutive laws with respect to wave propagation will be discussed in some more details now.

Figure 4.6 gives a schematic example for a nonlinear elastic-plastic compression curve. Under a given compression state $p(V)$ the corresponding material has a specific sound speed that is proportional to the adiabatic compression modulus $(\partial p \,/\, \partial V)$. In Figure 4.6 compression moduli are represented by tangents at different locations in the graph. If we follow the pressure-volume path for a rapid compression of that material from $(p_0\,, V_0\,)$ to $(p_3\,, V_3\,)$ we

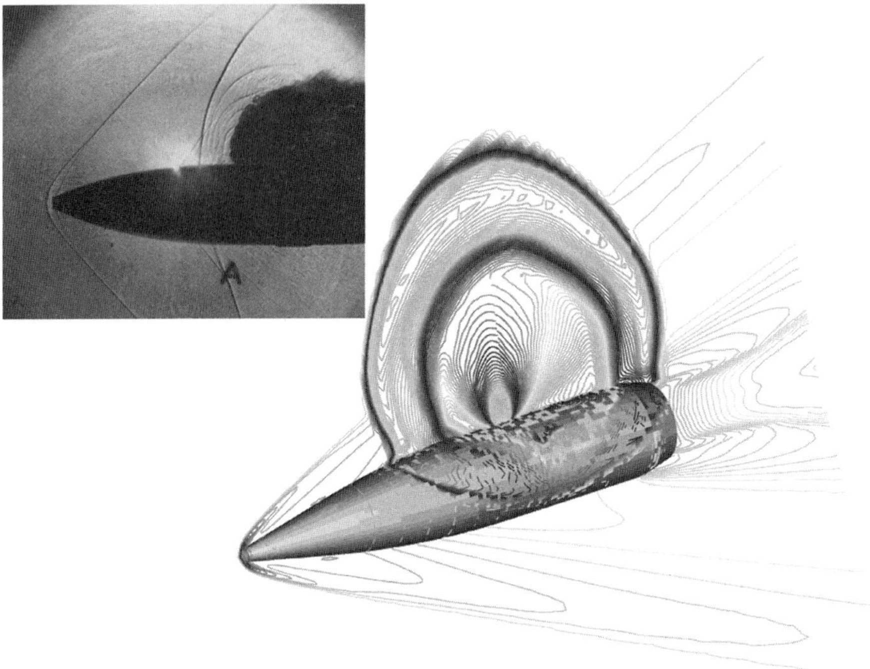

Fig. 4.5. Experimental and numerical investigation of maneuvering a spin stabilized projectile by initiation of a charge located underneath the projectile surface (Courtesy of Klomfass and Warken [228], Ernst-Mach-Institute).

find a smooth change in the propagation speed of the initiated pressure waves. In other words, *dispersion* is enabled. In the initial elastic regime compressive waves are propagated at the elastic wave speed

$$c_{elastic} = c_0 = -\frac{1}{V^2}\sqrt{\left.\frac{\partial p}{\partial V}\right|_0} \tag{4.30}$$

As the load rises to higher pressures beyond the plastic threshold, the gradient and thus the propagation speed decreases drastically. Enhancement of pressure beyond the state of $(p_1\,, V_1\,)$ leads to a gradual increase of the modulus. From that turn around point onwards, pressure waves are initiated that propagate faster than others before. Consequently, a superposition of slower wave packages by faster ones with higher amplitude occurs. Ultimately, a ever steeper wave profile and related shorter rise time evolves, as illustrated in Figure 4.7.

In the light of these observations and with a mathematical description for the slopes in the $p - V$ diagram of Figure 4.6, conditions for the formation of shock waves can be formulated as:

$$\frac{\partial p}{\partial V} < 0 \qquad \textit{and} \tag{4.31}$$

$$\frac{\partial^2 p}{\partial V^2} > 0 \tag{4.32}$$

In materials with an elastic-plastic compressive behavior according Figure 4.6, only loading conditions achieving pressures of p_1 or more can lead to shock waves. In gases and fluids, however, compression shocks can arise from ambient pressures onwards since no regions with $\left(\partial^2 p \,/\, \partial V^2\right) \leq 0$ exist.

Another necessary precondition for the shock formation is the mentioned rapid loading. Imagine a quasi-static load application to a pressure level indicated by p_1 in Figure 4.6. Still, information about the applied load would be transported by waves at the sound speed defined by dispersion effects, i.e. depending on $\frac{\partial p}{\partial V}$. But time delay for each pressure increment along a certain equilibrium path would avoid the formation of a shock wave.

Only if the load application is fast enough, the wave fronts of the faster packages keep up with the earlier wave fronts. The result is a steepened wave front and shorter rise times to higher pressures. Often, the wave components from the elastic regime are fast enough that a so called *elastic precursor* is formed. It is, however, also possible that even the elastic precursor is overtaken by very fast plastic waves. Whether or not this happens is only a matter of the load application speed and the achieved maximum pressure level.

Under slow or quasi-static loading conditions, low pressure wave components propagate into the loaded structure before faster waves are able to keep up. However, in case of dynamic loading situations, as sketched in the left graph of Figure 4.7, steep wave fronts arise. The right graph gives the pressure contour at three different locations away from the load initiation point. As the contour travels into the medium, the faster components overtake slower ones according to their individual sound speed indicated in the middle graph. At the time instant marked as t_3 a shock wave is formed right behind the still leading elastic precursor.

In a spatially one dimensional *Lagrange diagram*, the effect of shock formation through wave interaction becomes evident. Individual wave fronts are represented by lines. The higher the velocity of the wave the smaller is the slope of the corresponding line. In the case of fast waves overtaking slower ones, the location and time of a shock formation is found in the intersection point of the respective lines. Fig. 4.8 gives an example where at a specific location x_0 waves of different propagation velocity are initiated. At t_0 an elastic wave starts to propagate and over time it is followed by waves of continuously increasing speeds. At $(x_s\,,t_s)$ those waves form a shock and - only typical for strong shocks - an *expansion fan*. A shock wave resulting from two or more superimposed waves propagates on with a velocity that is between the speeds

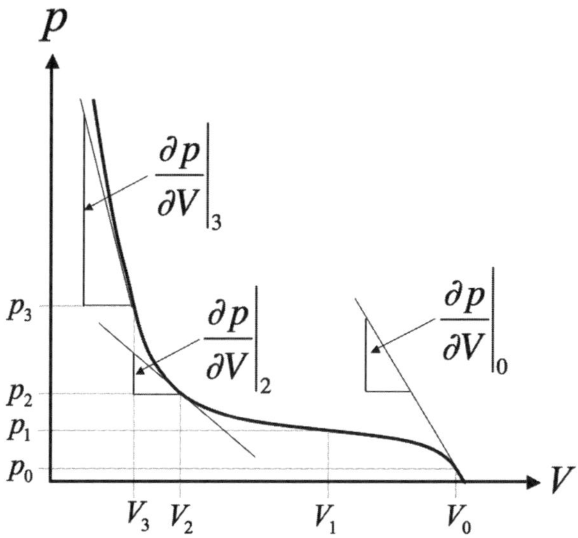

Fig. 4.6. Nonlinear compression curve of a solid elastic-plastic material allowing for dispersion driven shock waves.

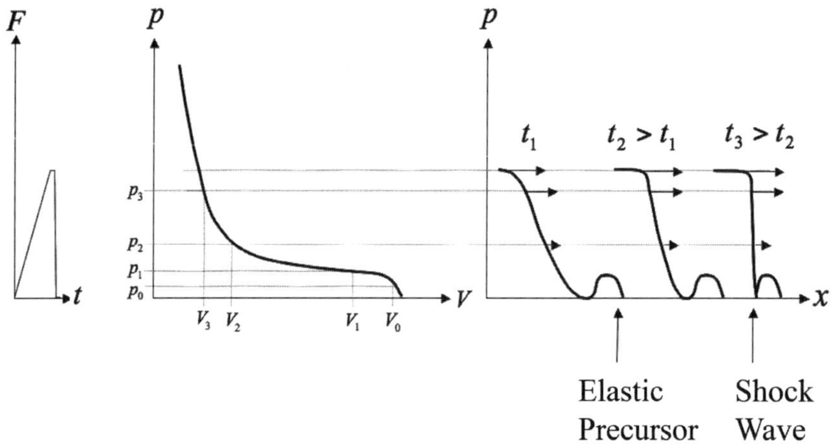

Fig. 4.7. Shock wave formation due to dispersion.

of the two individual waves and typically measured by the so called *shock Mach number* as a multiple of the local sound speed. In gases, there is no elastic-plastic transition in the compression curve. Therefore, the equation of state does not allow for an elastic precursor. Thus, any shock wave in gases propagates at a shock Mach number larger than one. In solid media this is usually different as can be seen from the Fig. 4.6. The elastic precursor in the example given in Fig. 4.7 will never be reached by the following shock wave. A higher propagation speed of the shock and thus an increased amplitude of the pressure pulse would be necessary to achieve this.

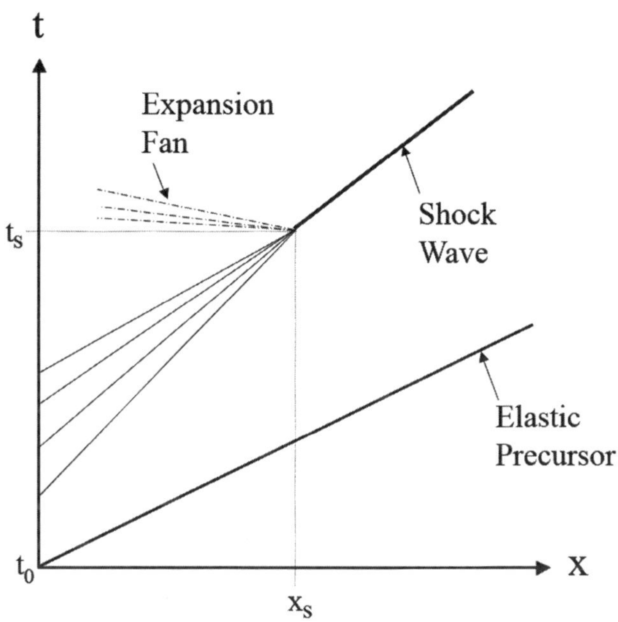

Fig. 4.8. Lagrange diagram representing the formation of a shock wave and a related expansion fan.

4.3.2 Shock Front Dimensions

Often, shock waves are idealized as singularities with instantaneous pressure rises and mathematical descriptions as Dirac pulses. In physical reality, however, also a shock front has a finite rise time, albeit a very short one. Already 1906 Prandtl [307] made an early attempt to quantify the thickness of shock fronts in single atom gases. His ideas were picked up by Oertel [303] who found dependencies between the dimensions of the shock front thickness δx and the mean free path $\bar{s}$ of the gas atoms ahead of the shock wave. For small shock Mach numbers he found thicknesses of about 10 $\bar{s}$, and less than $2\,\bar{s}$ for high shock Mach numbers. Thus, in gases typical shock front dimensions are found between 10^{-7} [m] and 10^{-6} [m].

Shock rise times in various metals were investigated by Swegle and Grady [380]. They related the rise times to the achieved Hugoniot pressure and found dimensions of $0.01[\mu s]$ for shocks of about $10[GPa]$ and up to $0.1[\mu s]$ for weaker shocks of around $2\,[GPa]$. Thus for metals, the material dependent properties of shock front dimensions are in the order of magnitude of 10^{-5} [m] to 10^{-4} [m].

4.4 Thermo-Mechanics of Shock Waves

So far, conditions for the formation of shock waves have been explained in a more intuitive way. Impact and detonation driven shock wave propagation has motivated very sophisticated research work on the thermo-mechanics of shock waves. Picking up earlier works by Riemann [327], Rankine [315], Hugoniot [204] and Rayleigh [318], basic theories for arbitrary media and equations of state have been formulated by von Neumann [294], Bethe [58] and Weyl [419]. These theories shall be picked up in this section to set a thermo-mechanical basis for later models that will be used to predict material behavior under shock loading.

4.4.1 Dispersion - Precondition for Shock Wave Evolution and Stability

With the compression curves in the $p-\rho$ or $p-V$ plane we were able to identify an intuitive criterion for shock wave formation. That criterion is the convex nonlinearity in the compression curve causing dispersion and thus allowing for the steepening in the wave shapes.

Obviously, the compression curve $p(V)$ alone can only cover an isothermal part of the thermodynamic conditions that a solid material may undergo under dynamic deformations. Therefore, criteria for the driving dispersion effect need to be found with respect to the shape of *state surfaces* $p(V,\,e)$. Whether, at a given thermodynamic state, dispersion occurs or not, depends on the shape of the state surfaces and thus on the equation of state describing it. A first generalized thermodynamic description of conditions for a stable shock wave propagation in media with an arbitrary equation of state was established in 1942 by Hans Bethe [58]. Investigating the behavior of pressure, internal energy, volume and entropy for various classes of solid, liquid and gaseous matter, Bethe established two conditions for the existence and one for the stability of shock waves. The two sufficient criteria for its existence are:

$$\left.\frac{\partial^2 p}{\partial V^2}\right|_S > 0 \tag{4.33}$$

and

$$\Gamma = V \left.\frac{\partial p}{\partial e}\right|_V > -2 \tag{4.34}$$

where Γ denotes the Grüneisen parameter that will discussed later.

The criterion for a shock wave to be stable against break-up, Bethe used the inequality

$$\left.\frac{\partial p}{\partial V}\right|_e < 0 \tag{4.35}$$

With respect to an isothermal case reflected in a compression curve, the intuitively derived equations (4.31) and (4.32), are equivalent with conditions (4.33) and (4.35). Bethe's investigations were partly based on an earlier publication by Duhem [118], where the first condition (4.33) was already found though there was no link formulated to equations of state by Duhem. Discussions between von Neumann and Bethe lead to the fundamental theoretical work of Bethe.

How do the conditions (4.33) - (4.35) relate to dispersion? There is a single thermodynamic parameter $\mathcal{G}$ called *fundamental derivative*[3] that describes the link uniquely:

$$\mathcal{G} = -\frac{1}{2} V \frac{\left.\frac{\partial^3 e}{\partial V^3}\right|_S}{\left.\frac{\partial^2 e}{\partial V^2}\right|_S} = -\frac{1}{2} V \frac{\left.\frac{\partial^2 p}{\partial V^2}\right|_S}{\left.\frac{\partial p}{\partial V}\right|_S} \tag{4.36}$$

Using the sound speed c in the medium ahead of a shock wave defined by:

$$c^2 = -V^2 \left.\frac{\partial p}{\partial V}\right|_S \tag{4.37}$$

the fundamental derivative

$$\mathcal{G} = \frac{V^3}{2\,c^2} \left.\frac{\partial^2 p}{\partial V^2}\right|_S = 1 + \frac{\rho}{c} \left.\frac{\partial c}{\partial \rho}\right|_S \tag{4.38}$$

quantifies more obviously the rate of change in sound speed with density as well as the curvature of insentropes in the $p-V$ plane and, thus, links the dispersion to the convex shape of the equation of state. The fundamental meaning of the derivative (4.38) will become more apparent after the introduction

[3] The fundamental derivative, also known as *Duhem parameter* was introduced by Thompson [389]

to some more basic thermodynamics of shock waves including the Rankine-Hugoniot equations and Hugoniot curves in the following sub-sections and non-linear equations of state that will be discussed in section 4.5. After these introductions, the fundamental derivative will be picked up once more in section 4.6 to investigate some characteristics of the equations derived before.

4.4.2 Thermodynamic Conditions upon Shock Wave Transit

Moving disturbances at supersonic velocities or dynamic compression of media with nonlinear compressive behavior according to (4.31) and (4.32) have above been identified as origins of shock waves. The next important questions are how a shock wave propagates and changes its shape in different media and how it influences the state variables on its way through these media.

A well known experimental set up in fluid dynamics to investigate these questions for shock waves in air is the so called *shock tube*. It consists of a gas filled tube in which a high pressure region is separated from a low pressure region by a diaphragm. Sudden opening of the diaphragm allows for the formation of different waves and discontinuities:

- a pressure wave travelling into the low pressure volume. If the equation of state of the gas satisfies the condition (4.32) a shock wave is formed by the pressure wave.
- a release wave moving backwards into the high pressure zone
- and a contact front representing the boundary layer separating the gas volumes originating out of the high pressure zone from the gas that originally was in the low pressure zone.

The conditions at a certain time after opening the diaphragm are demonstrated in Fig. 4.9 together with the evolution of the propagating discontinuities represented by a *Lagrange diagram*.

Shock tubes are used to observe shock waves and their interaction with structures of various shapes. Reflections of the waves at surfaces and specifically the so called *irregular* or *Mach reflections* became a branch of science of their own but will not be discussed in this book. Interested readers will find detailed information on shock tubes and related investigations in Oertel [303].

One may wander why an experiment used in fluid dynamics is investigated in the course of deriving formulations for shock wave formation and propagation in solids, which is the topic of this section. The reason for this approach is found in the fundamental sense of the process, called *Riemann problem*, investigated in shock tubes: a moving discontinuity producing entropy in a well defined initial-boundary-condition system. Its mathematical description by hyperbolic conservation equations for mass, momentum and energy with an additional equation of state provides the basic understanding of such processes

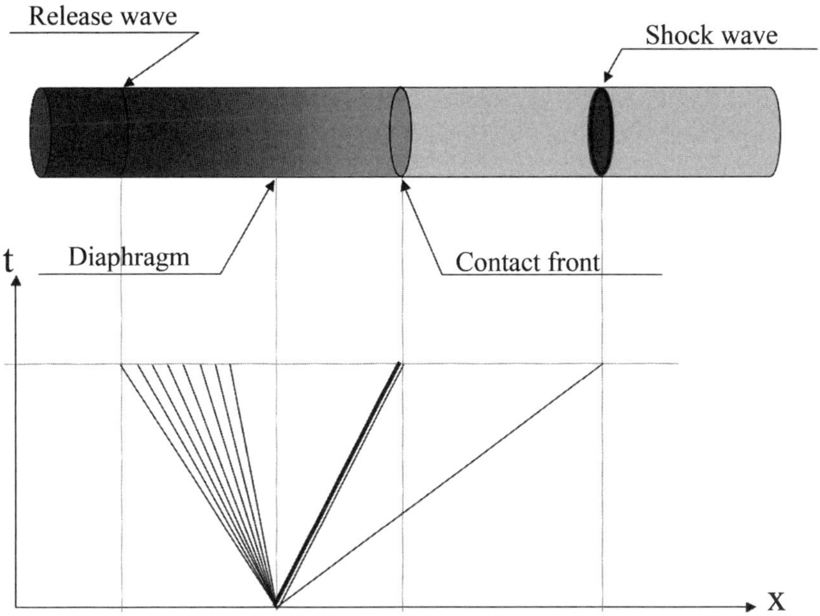

Fig. 4.9. Discontinuities forming in a shock tube and related Lagrange diagram.

in any kind of matter, be it gaseous, liquid or solid.

Any investigation of shock wave effects, which is a key component of dynamic loading conditions, demands for a precise understanding and formulation of its continuum and thermo-mechanics. Since that is exactly what is investigated in the Riemann problem, we will next focus on the fundamental components of this, basically fluid-dynamical, approach to find the needed mathematical framework describing shock waves in solids.

4.4.3 Riemann Problem and Rankine-Hugoniot Equations

In order to quantify the change of state variables after the passage of a shock wave, these variables are defined and evaluated before and behind the shock front. The conservation equations, i.e. in their integral conservative version (2.179), (2.187) and (2.206), are used to balance the conditions at both sides of the shock and to derive a set of equations describing the change. That approach of investigating a plane shock wave as it arises in a shock tube is called formulation of the *shock tube problem* or the *Riemann problem*. In Fig. 4.10 the Riemann problem is shown as a plane shock wave propagating through a cross sectional area of the size A. The interesting variables of state ahead and behind the shock denote the pressures by p_0, p_1, the densities by ρ_0, ρ_1 and specific integral energies by e_0, e_1. Also the sound speeds are denoted as c_0 and c_1 for the conditions ahead and behind the shock front, respectively. The shock propagation velocity is denominated by v_S. With these quantities and according to (4.29) we have the following additional parameters to describe the shock front:

Shock Mach number: $Ma = v_S/c_0$
Shock compression: $\eta = \rho_1/\rho_0$
Shock strength: $y = p_1/p_0$

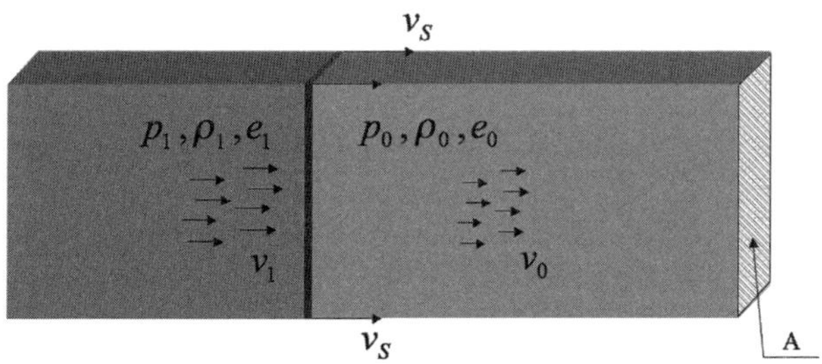

Fig. 4.10. State variables defining the conditions ahead and behind a shock wave.

Basically, the relative velocity between the un-shocked medium and the shock wave is equal to $v_S - v_0$. A common assumption for the Riemann problem is that the medium ahead of the shock is at rest ($v_0 = 0$). We will take use of this assumption later together with the postulates that the density and pressure in that region are constant ($p_0 = const$, $\rho_0 = const$). For now, a shock wave is regarded as an infinitesimally thin discontinuity or transition zone leading to a sudden change in the state variables.

The conserved variables in a certain volume that is passed by a transient shock wave during a time interval *dt* are expressed by the thermo-mechanic state variables. To balance the conserved variables mass, momentum and energy over the transition zone, the respective conditions ahead of the shock and behind it are set into the following relations:

Mass Conservation

During the time interval *dt* the volume covered by the transient shock can be described by $A\, v_S\, \mathrm{d}t$ and $A\,(v_S - v_1)\, dt$. Accordingly, the mass flowing into the transition zone is expressed by

$$\rho_0 \, A \, (\, v_S - v_0) \, \mathrm{d}t$$

and the same amount of mass, now described in terms of the post-shock condition is

$$\rho_1 \, A \, (\, v_S - v_1) \, \mathrm{d}t$$

Therefore, *conservation of mass* demands that

$$\rho_0 \, A \, (\, v_S - v_0) \, \mathrm{d}t = \rho_1 \, A \, (\, v_S - v_1) \, \mathrm{d}t \tag{4.39}$$

and under the assumption that $v_0 = 0$, we find

$$\rho_0 \, v_S = \rho_1 \, (\, v_S - v_1) \tag{4.40}$$

Momentum Balance

During the same time interval $\mathrm{d}t$, the mobilized mass accelerated to the velocity v_1 represents a change in the momentum of the covered volume. The resulting rate of change of momentum is

$$\rho_1 \, A \, (\, v_S - v_1) \, \mathrm{d}t \, v_1 - \rho_0 \, A \, (\, v_S - v_0) \, \mathrm{d}t \, v_0$$

where the second term equals zero if $v_0 = 0$. This leads to a formulation

$$\rho_1 \, A \, (\, v_S - v_1) \, \mathrm{d}t \, v_1$$

which must be balanced by the impulse due the change of pressure forces:

$$(\, p_1 - p_0) \, A \, \mathrm{d}t$$

Thus, to guarantee *momentum conservation* the two expressions must be equal:

$$\rho_1 \, (\, v_S - v_1) \, v_1 = p_1 - p_0 \tag{4.41}$$

Using the mass conserving equation (4.40) to replace ρ_1 yields:

$$\rho_0 \, v_S \, v_1 = p_1 - p_0 \tag{4.42}$$

Equation (4.42) can be understood as relation between the pressure jump $\Delta p = p_1 - p_0$ and the particle velocity v_1 that are to be expected in a material initially at rest and of density ρ_0 upon a shock transition with velocity v_S. In other words, if ρ_0 is known and the shock and particle speeds v_S and v_1 are measured, the change in pressure can easily be derived. Accordingly, the product $\rho_0 \, v_S$ describes the slope of the pressure jump upon shock transit in a $p - v_1$ diagram. This relation is used in the so called *impedance-match* technique for experimental equation of state derivations (see section 7.4.2).

Energy Conservation

The increase in specific internal and kinetic energy of the volume ahead of the shock that passes through the transition zone can be written as

$$e_1\,\rho_1\,A\,(\,v_S - v_1)\,\mathrm{d}t - e_0\,\rho_0\,A\,(\,v_S - v_0)\,\mathrm{d}t \\ + \frac{1}{2}\,\rho_1\,A\,(\,v_S - v_1)\,\mathrm{d}t\,v_1^2 - \frac{1}{2}\,\rho_0\,A\,(\,v_S - v_0)\,\mathrm{d}t\,v_0^2$$

and the corresponding work done by the pressure change is

$$p_1 A\, v_1\,\mathrm{d}t - p_0 A\, v_0\,\mathrm{d}t$$

leading to the *conservation of energy* across the shock front for a volume initially at rest ($v_0 = 0$):

$$p_1\,v_1 = e_1\,\rho_1\,(\,v_S - v_1) - e_0\,\rho_0\,v_S + \frac{1}{2}\,\rho_1\,(\,v_S - v_1)\,v_1^2 \tag{4.43}$$

Using the mass conservation once again we find:

$$p_1\,v_1 = (\,e_1 - e_0)\,\rho_0\,v_S + \frac{1}{2}\,\rho_0\,v_S\,v_1^2 \tag{4.44}$$

Equations (4.39),(4.41) and (4.43) are called *Rankine Hugoniot Relations*, named after the two mathematicians who independently applied them first[4].

In the derived form, the Rankine Hugoniot relations are given in an Eulerian description with respect to the shock front. If they are transformed into a frame of reference that moves with the shock front[5], direct relations are established between variables in the un-shocked and the post shock regime without shock velocities. For that purpose, a velocity transformation needs to be performed:

$$v_0 \Rightarrow -v_S \qquad \textit{and} \qquad v_1 \Rightarrow v_1 - v_S$$

to achieve the following set of equations, called the *jump conditions* of the shock front:

$$\rho_1\,v_1 = \rho_0\,v_0 \tag{4.45}$$

$$p_1 + \rho_1\,v_1^2 = p_0 + \rho_0\,v_0^2 \tag{4.46}$$

$$e_1 + \frac{p_1}{\rho_1} + \frac{v_1^2}{2} = e_0 + \frac{p_0}{\rho_0} + \frac{v_0^2}{2} \tag{4.47}$$

---

4 Both original publications of Rankine [315] and Hugoniot[203] are available in Johnson J. and Cheret R. (eds): *Classic Papers in Shock Compression Science*, Springer, New York, 1998

5 The resulting moving coordinates are also called *shock fixed coordinates*.

Furthermore, a substitution of (4.45) and (4.46) into (4.47) delivers a version of the energy equation that is free of any velocity terms. Hence, the resulting *Hugoniot equation* is a relation between thermodynamic quantities only:

$$e_1 - e_0 = \frac{1}{2}(p_1 + p_0)\left(\frac{1}{\rho_0} - \frac{1}{\rho_1}\right) \tag{4.48}$$

For the more general application of the Rankine Hugoniot relations in numerical codes and for the further derivations in this book we will, however, stay with the formulations found in (4.40), (4.42) and (4.44).

4.4.4 Hugoniot Curves and v_S-v_1 Relations

The Hugoniot equation (4.48) describes one specific curve in the state surface of a material and represents the sum of all possible peak conditions achieved when shock waves arise. That curve is called *Hugoniot curve*, *Shock Hugoniot* or *Shock adiabatic*. It is not an equation of state, neither is it a path along which shock waves arise. Figures 4.11 and 4.12 are plotted to illustrate the differences: The sketched generic state surfaces shows solid, liquid, gaseous and transition phases. The steep bold line represents the Hugoniot line of the assumed material. Starting at a position A an isentrope leads to a compressed condition B. From there, two differently strong transient shock waves are assumed leading to the positions C and C, respectively. That process is not following a line of consecutive states of thermodynamic equilibrium, as an isentrope would. A shock compressed condition is assumed to be achieved along a straight line in the $p-V-e-$space, called *Rayleigh line*. The relative positions of a Rayleigh line, shock adiabatic and some equilibrium paths is plotted in Fig. 4.13. The equilibrium path along the compressive isentrope and the Rayleigh line are both leading to position (2) on the Hugoniot line.

Hence, for any possible shock transition, two characteristics concerning the related change in the thermodynamic state are found:

- it always leads to a condition specified by a point on the material specific *Hugoniot Curve*,
- the shock compression state is reached along a non equilibrium path, assumed to be a straight line in the $p-V$ plane called *Rayleigh Line*.

Thus, if the equation of state of a material is known, the Hugoniot curve can be identified using the Rankine-Hugoniot equations (4.39) - (4.43).

On the other hand, if the equation of state shall be determined, the Rankine-Hugoniot equations plus one additional relation between two of the involved variables can be utilized to identify the EOS. The needed additional equation to solve the Riemann problem can be found in a relation between the velocity behind the shock wave v_1 and the shock velocity v_S. This $v_S - v_1$

relation is of great importance for the shock wave characterization in solid matter and, due to its contribution to formulate an equation of state, often itself called EOS which is not correct. It should rather be called a Hugoniot curve in the $v_S - v_1$ plane.

An experimental approach to characterize a $v_S - v_1$ relation can be performed as follows:

- A flat sample of the investigated material is dynamically loaded by a planar impact of another plate with known impact velocity. The experiment is called *flyer-plate-test* (see chapter 7).
- A shock wave is initiated in both plates reaching the rear side of the impacted plate after a certain amount of time.
- The shock velocity v_S is measured from the time needed for the transition and the initiated particle velocity v_1 (also indicated as v_P) from the velocity of the impacted plate at the end of the acceleration process.
- Repeated experiments with varying impact velocities lead to a number of v_S-v_1 couples.
- The measured data are used to define a relation

$$v_S = v_S\left(v_1\right) \tag{4.49}$$

  between the shock velocity and particle velocity.
- Application of the Rankine-Hugoniot equations (4.39) - (4.43) for each measured v_S-v_1 couple delivers one single point on the Shock Hugoniot, each.

In section 4.5 we will see how the found way to identify a Shock Hugoniot leads to the so called Shock Equation of State and thus how the Shock Hugoniot can be used as a baseline to formulate a specific nonlinear equation of state of a material.

Basically, arbitrary formulations can be chosen to describe v_S-v_1 relations (4.49). Often polynomials are used with

$$v_S = c_0 + \sum_{i=1}^{n} S_i \, v_1^i \tag{4.50}$$

where the number of summands is usually limited to three for gases or porous solids and to only a single one for most other solids, specifically for metals. For the derivation of the parameters c_0 and S in a linear formulation two experiments yielding two v_S-v_1 couples are needed.

For an isotropic homogeneous material, the bulk sound speed c_B is related to the longitudinal and shear sound speeds c_L and c_S through

$$c_B = \sqrt{c_L^2 - \frac{4}{3}\, c_S^2} \tag{4.51}$$

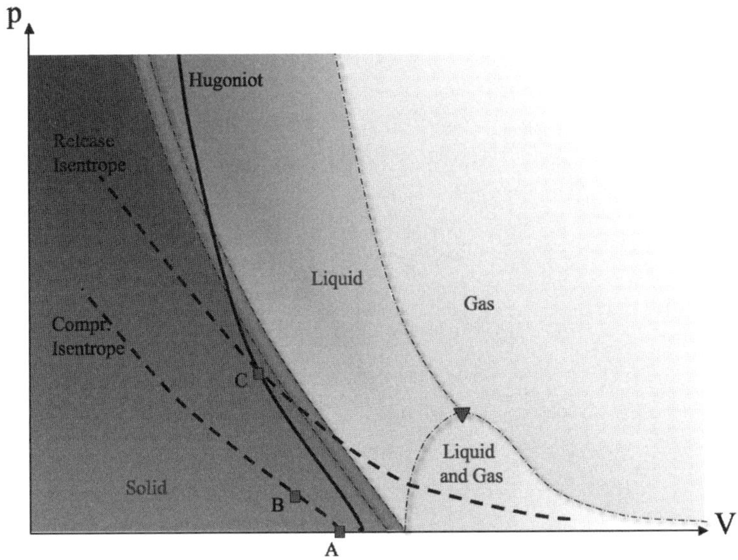

Fig. 4.11. Pressure-volume diagram illustrating a shock compression from an isentropically pre-compressed state (B) to a point (C) in the solid part of the Hugoniot curve. The release isentrope related to (C) expands the material to a mixed liquid-gaseous state.

For some of these isotropic materials, c_B is in good correlation with c_0 and, consequently, a single experiment is sufficient to characterize the slope S in a linear equation of the form

$$v_S = c_0 + S\, v_1 \tag{4.52}$$

However, for most materials the bulk sound speed differs too much from the initial value $v_S(v_1 = 0)$ to fit the residual curve in a linear approach[6].

Metals are a typical representative of linear v_S-v_1 behavior. In chapter 7.4.2 experimental results are given for steel and copper as well as for porous materials like concrete. The latter representing a highly nonlinear Hugoniot characteristic.

Given a relation between shock and particle velocities exists, the Rankine-Hugoniot equations can be solved. Hence, equations (4.40), (4.42) and (4.44) can be reformulated to directly describe the conditions behind the shock wave.

[6] A comprehensive overview on the v_S-v_1 relations for many elements, alloys, minerals and compounds, rocks, plastics, woods, liquids and high explosives is collected in Marsh [267] and in Trunin [400].

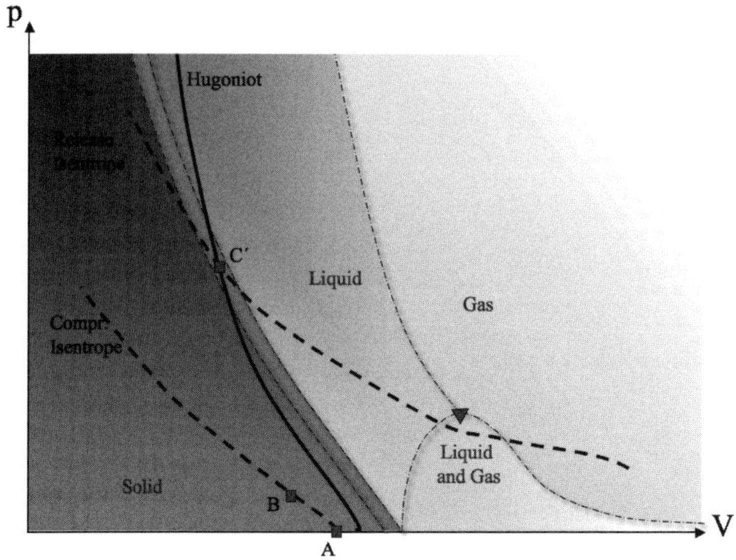

Fig. 4.12. Pressure-volume diagram illustrating a shock compression from an isentropically pre-compressed state (B) to a point (C') in the liquid-solid transition part of the Hugoniot curve. The release isentrope related to (C) expands the material to a gaseous state.

With a linear equation (4.52) and the mass conservation across the shock wave (4.40) :

$$v_1 = v_S \left(1 - \frac{rho_0}{\rho_1}\right) = v_S\, \eta \tag{4.53}$$

the shock velocity can be written in terms of c_0, S and η:

$$v_S = \frac{c_0}{1 - S\,\eta} \tag{4.54}$$

With (4.54), the density, pressure and specific internal energy in the shocked state, which is also called *Hugoniot state*, can now be formulated as:

$$\rho_1 = \rho_0 \, \frac{c_0 + S\, v_1}{c_0 + v_1(S-1)} \tag{4.55}$$

$$p_1 = p_0 + \rho_0\, c_0^2 \, \frac{\eta}{(1 - S\,\eta)^2} \tag{4.56}$$

$$e_1 = e_0 + \frac{\eta}{\rho_0}\, p_1 - \frac{\eta^2}{2}\, \frac{c_0^2}{(1 - S\,\eta)^2} \tag{4.57}$$

In the following chapter 4.5 the Hugoniot curve will be used as baseline in connection with an assumption on equilibrium conditions on the state surface

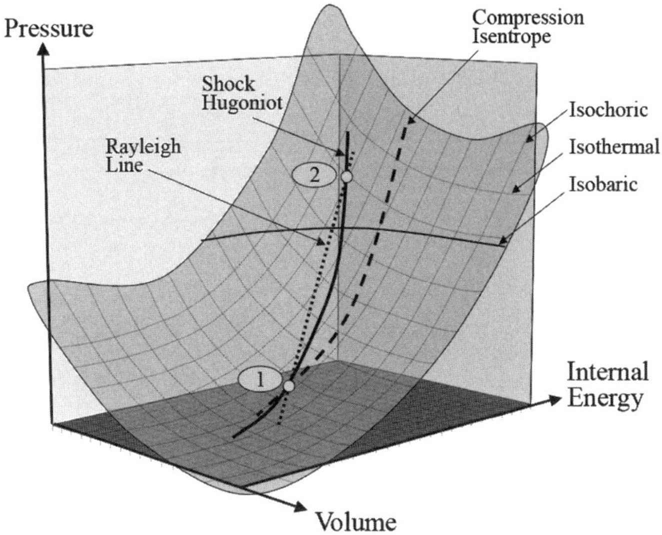

Fig. 4.13. Generic equation of state surface with different kinds of state changes.

off the Hugoniot. This approach will provide the possibility of formulating a nonlinear equation of state capable of describing the thermodynamics of dynamic deformation processes in a material including shock wave formation and propagation. Thus, the found thermodynamic equations in the context of shock wave transition are related to each other as illustrated in the following:

Rankine Hugoniot Equations + EOS $\Rightarrow$ Hugoniot Curve

Rankine Hugoniot Equations + v_S-v_1 relation $\Rightarrow$ Hugoniot Curve

Hugoniot Curve + Assumption (Grüneisen theory) $\Rightarrow$ Nonlin. Shock EOS

4.4.5 Energy Dissipation upon Shock Wave Transition

The irreversible thermodynamic process of an almost instantaneous jump from initial conditions at ambient or other equilibrium conditions to a Hugoniot state enhances the entropy of the material that is passed by the shock wave. The amount of energy dissipation can be calculated from the thermodynamic paths taken by the process. Assuming a shock compression from an ambient state p_0, V_0 to a Hugoniot pressure p_H and a subsequent isentropic release to the initial pressure p_0 at an end volume V_1 the process paths are represented in the $p - V$ plane as illustrated in Figure 4.14.

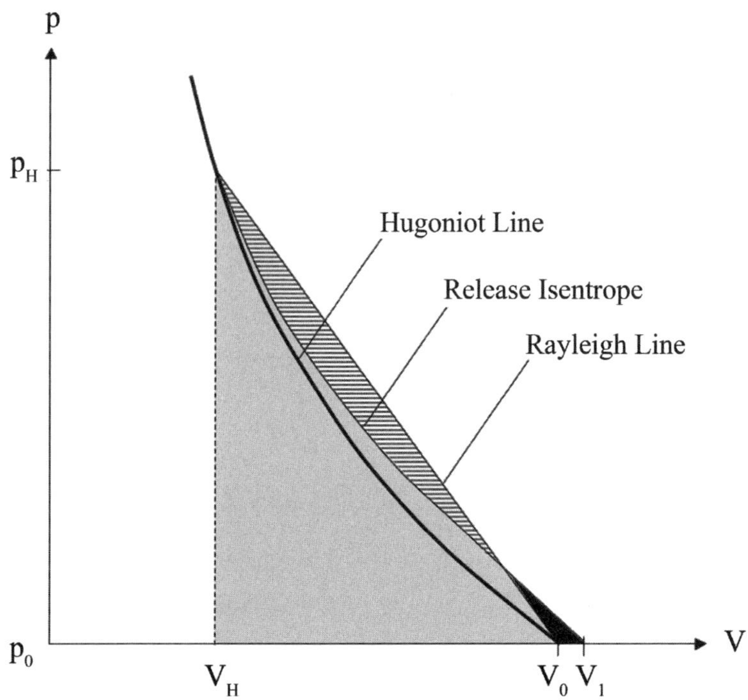

Fig. 4.14. Shock compression along Rayleigh line, isentropic release path and dissipated energy represented in the $p - V$ plane.

Whereas the loading path to the Hugoniot state is described by a Raighley line, i.e. in the $p - V$ plane a straight line of non-equilibrium states, the isentropic release process follows a curved line in the $p - V -$plane. The energy stored and released during shock compression $(p_0, V_0) \Longrightarrow (p_H, V_H)$ and isentropic expansion $(p_H, V_H) \Longrightarrow (p_0, V_1)$, are marked as surfaces in Figure 4.14. A measure for the residual amount of energy dissipated in the irreversible process of a shock wave transition is the difference between the energy stored during shock loading, i.e. the triangular surface, and the recovered energy during isentropic release, represented by the shaded surface. The total dissipated energy can be identified graphically through the surfaces marked in the generic shock and release process of Figure 4.14. In that example, the end state volume after release V_1 is larger than the initial volume V_0. Therefore, the released energy equals the difference between the two surfaces shaded in horizontal lines and in black, respectively.

In hydrocodes, the state variable specific volume V is typically replaced by the density ($\rho = V^{-1}$). Thus, the thermodynamic processes are observed in the $p - \rho - e$ space. This has a consequence for the illustration of shock compression processes along the Rayleigh line in the respective $p - \rho$ plane:

Rayleigh lines are defined in the $p - V$ plane as connecting straight lines between an initial state p_0, V_0 and a related shocked Hugoniot state p_H, V_H as

$$p(V) = p_0 + \frac{p_H - p_0}{V_H - V_0} (V - V_0) \tag{4.58}$$

with a constant negative slope of:

$$\frac{\partial p(V)}{\partial V} = \frac{p_H - p_0}{V_H - V_0} = const. < 0 \tag{4.59}$$

On the other hand, transformed into the $p - \rho$ plane equation 4.58 becomes:

$$p(\rho) = p_0 + \frac{p_H - p_0}{1/\rho_H - 1/\rho_0} \left(\frac{1}{\rho} - \frac{1}{\rho_0}\right) = p_0 + \alpha \left(\frac{1}{\rho} - \frac{1}{\rho_0}\right) \tag{4.60}$$

with α a constant negative term. Thus, the straight line known from the $p-V$ plane is no longer straight since the slope is now:

$$\frac{\partial p(\rho)}{\partial \rho} = \alpha (-1) \frac{1}{\rho^2} > 0 \tag{4.61}$$

Figure 4.16 illustrates a specific Hugoniot state with subsequent release path in the $p - \rho$ plane for a specific aluminium material. The calculation of release isentropes for the material illustrated in the graph will be discussed in more detail in section 4.6.2.

4.5 Nonlinear Equations of State for Shock Waves

In chapter 4.3 the conditions for a stable shock wave formation were found in equations (4.31) and (4.32) demanding for a concave pressure-volume behavior. That means for an equation of state, and thus for any mathematical formulation in a numerical scheme expected to predict shock wave formation and propagation, that it must be non-linear. A polynomial equation of state was already formulated with equation (3.23).

In the following subsection, basic theories and assumptions from statistical mechanics will briefly be collected to formulate a nonlinear EOS. Consideration of the thermo-mechanics formulated for shock wave transition and application of the found Hugoniot relations will then lead to a closed system of equations able to describe the nonlinearity in shock wave transitions.

4.5.1 Grüneisen Theory for Crystalline Oscillators

Theoretical works in solid state physics and statistical mechanics investigated the macroscopic material behavior as a function of distance dependent interatomic forces. To formulate an equation of state, a referential base line of known pressure-energy-density conditions was used. Either theoretical approaches, like the calculation of energy at a 0 K isothermal, or experimentally derived relations, e.g. the above discussed shock Hugoniot, were used to indentify a so called "cold curve" or the Hugoniot curve, respectively, as reference. Conditions off that base line are calculated by the use of assumptions, one of which is the theory described below.

An important parameter with specific meaning in both the aforementioned statistical mechanics and thermodynamics is the so called Grüneisen parameter Γ. Following Grüneisen's theory[7], a crystal is regarded on the atomic level as a set of N quantized harmonic oscillators with a total of $3N$ normal modes of oscillation. Neglecting the thermal excitation of electrons in that system, the total specific internal energy e of the crystal is contributed by

- a volume dependent potential energy $e_{0K}(V)$ of the atoms virtually at rest in their center of oscillation
- and by vibration induced thermal energy e_{th}.

The thermal energy component e_{th} of the 3N oscillators in the crystal can be calculated using the Planck's constant $\hbar$ as:

$$e_{th} = \sum_{i=1}^{3N} \left[\frac{1}{2}\,\hbar\,\nu_i + \frac{\hbar\,\nu_i}{e^{\hbar\,\nu_i/kT} - 1} \right] \tag{4.62}$$

Thus , the total specific internal energy is:

$$e = e_{0K}(V) + e_{th} = e_{0K}(V) + \sum_{i=1}^{3N} \left[\frac{1}{2}\,\hbar\,\nu_i + \frac{\hbar\,\nu_i}{e^{\hbar\,\nu_i/kT} - 1} \right] \tag{4.63}$$

For the same system, the Helmholtz potential Ψ (2.263) is:

$$\Psi = e_{0K}(V) + \sum_{i=1}^{3N} \frac{1}{2}\,\hbar\,\nu_i \; + \; kT \sum_{i=1}^{3N} ln\left(1 - e^{-\hbar\nu i/kT}\right) \tag{4.64}$$

Pressure, expressed as derivative of Ψ with respect to volume at constant temperature T, leads to:

[7] For original and detailed derivations of the theory in German language the reader may be referred to Grüneisen [148] and [149]). For detailed introductions to the basics in statistical mechanics including the Grüneisen theory Slater's "Introduction to Chemical Physics" [359] may be recommended. A more compact derivation of the Mie-Grüneisen equation of state can be found in Rice et al. [320] and similarly in Meyers [276]

$$p(V,e) = -\left.\frac{\partial\,\Psi}{\partial\,V}\right|_T = -\frac{\partial\,e_{0K}(V)}{\partial\,V} + \frac{1}{V}\sum_{i=1}^{3N}\gamma_i\left[\frac{1}{2}\,\hbar\,\nu_i + \frac{\hbar\,\nu_i}{e^{\hbar\,\nu_i/kT}-1}\right] \tag{4.65}$$

The dimensionless variables γ_i represent the relations between volumes V and vibrational frequencies ν_i of oscillation modes i:

$$\gamma_i = -\left.\frac{V}{\nu_i}\,\frac{\partial\,\nu_i}{V}\right|_T = -\left.\frac{\partial\,ln\,\nu_i}{\partial\,ln\,V}\right|_T \tag{4.66}$$

Grüneisen's assumption was to define these terms as equal for all oscillators i, a volume dependent variable that was later called *Grüneisen parameter* $\Gamma(V)$:

$$\Gamma(V) = -\left.\frac{\partial\,ln\,\nu}{ln\,V}\right|_T \tag{4.67}$$

Since $\Gamma(V)$ is equal for all oscillation modes, it can be used as factor outside the summation in (4.65), leading to the following pressure expression:

$$p(V,e) = -\frac{\partial\,e_{0K}(V)}{\partial\,V} + \frac{\Gamma(V)}{V}\sum_{i=1}^{3N}\left[\frac{1}{2}\,\hbar\,\nu_i + \frac{\hbar\,\nu_i}{e^{\hbar\,\nu_i/kT}-1}\right] \tag{4.68}$$

which, with (4.63), can be also expressed as:

$$p(V,e) = -\frac{\partial\,e_{0K}(V)}{\partial\,V} + \frac{\Gamma(V)}{V}\,e_{th} = -\frac{\partial\,e_{0K}(V)}{\partial\,V} + \frac{\Gamma(V)}{V}\,(e - e_{0K}(V)) \tag{4.69}$$

The term $-\frac{\partial\,e_{0K}(V)}{\partial\,V}$ in (4.69) represents the pressure at 0 K and, thus, a first version of an equation of state can be formulated as:

$$p(V,e) = p_{0K} + \frac{\Gamma(V)}{V}\,(e - e_{0K}(V)) \tag{4.70}$$

4.5.2 Equations of State for High-Pressure and High-Energy Regimes

The fundamental meaning of equation (4.70) together with a new formulation for the Grüneisen parameter becomes obvious if we start with the integration of the exact differential of pressure

$$\mathrm{d}\,p = \left(\frac{\partial\,p}{\partial\,V}\right)_e\,\mathrm{d}V\, + \left(\frac{\partial\,p}{\partial\,e}\right)_V\,\mathrm{d}e \tag{4.71}$$

Starting at an arbitrary reference, e.g. a 0 K condition (e_{0K}, V_{0K}), the change in pressure during an isochoric process ($V = V_{0k}$) is:

$$p(V,e) - p(V_{0K}, e_{0K}) = \int_{V_{0K}}^{V}\left(\frac{\partial\,p}{\partial\,V}\right)_e\,\mathrm{d}V + \int_{e_{0K}}^{e}\left(\frac{\partial\,p}{\partial\,e}\right)_V\,\mathrm{d}e \tag{4.72}$$

Integration of the first term in (4.72) yields no contribution for isochoric processes:

$$\int_{V_{0K}}^{V} \left(\frac{\partial p}{\partial V}\right)_e \mathrm{d}V = \left(\frac{\partial p}{\partial V}\right)_e (V - V_{0K}) = 0 \tag{4.73}$$

whereas the second integrates to:

$$\int_{e_{0K}}^{e} \left(\frac{\partial p}{\partial e}\right)_V \mathrm{d}e = \left(\frac{\partial p}{\partial e}\right)_V (e - e_{0K}) \tag{4.74}$$

Equation (4.72) can therefore be written as:

$$p(V,e) - p(V_{0K}, e_{0K}) = \left(\frac{\partial p}{\partial e}\right)_V (e - e_{0K}) \tag{4.75}$$

Comparing equation (4.75) with the equation of state (4.70) shows the identity:

$$\frac{\Gamma(V)}{V} = \left(\frac{\partial p}{\partial e}\right)_V \tag{4.76}$$

Mie-Grüneisen Shock EOS

Thus, with a known reference condition (e_0, V_0) and a known variable $\Gamma(V)$, equation (4.70) can be used as an equation of state. And finally, application of the Hugoniot curve as reference condition leads to the *Mie-Grüneisen shock equation of state*:

$$p(V,e) = p_H - \frac{\Gamma(V)}{V} (e - e_H) \tag{4.77}$$

with p_H and e_H representing the Hugoniot state. Using a linear relation $v_S = c_0 + S\, v_1$ the Hugoniot pressure and energy can be formulated as described before in equations (4.56) and (4.57) :

$$p_H = p_0 + \rho_0\, c_0^2\, \frac{\eta}{(1 - S\,\eta)^2} \tag{4.78}$$

$$e_H = e_0 + \frac{\eta}{\rho_0}\, p_H - \frac{\eta^2}{2}\, \frac{c_0^2}{(1 - S\,\eta)^2} \tag{4.79}$$

with

$$\eta = 1 - \frac{\rho_0}{\rho} \tag{4.80}$$

Hence, besides the Hugoniot condition that can be experimentally characterized using $v_S - v_1$ data of the material, the Grüneisen Γ is needed as final parameter to establish the equation of state. Some further thermodynamical relations derived from (4.76) can be useful to identify data for the Grüneisen parameter:

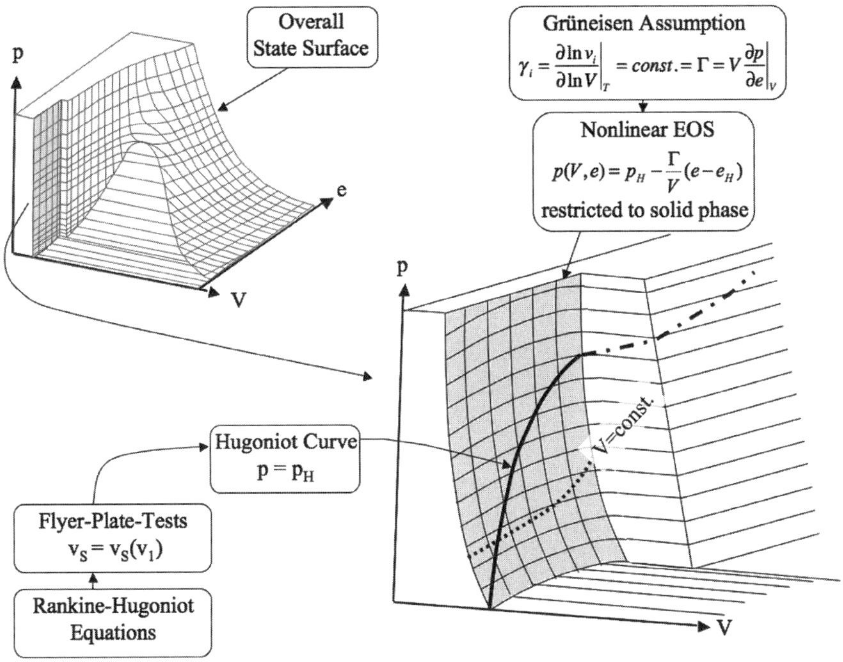

Fig. 4.15. Construction of a nonlinear equation of state for shock propagation processes basing on the Rankine-Hugoniot equations via flyer-plate data and Grüneisen theory.

$$\Gamma = V\,\frac{\partial p}{\partial e}\bigg|_V = \frac{V}{c_V}\,\frac{\partial p}{\partial T}\bigg|_V = -\frac{V}{c_V}\,\frac{\partial p}{\partial V}\bigg|_T\,\frac{\partial V}{\partial T}\bigg|_P = -\frac{V}{c_P}\,\frac{\partial p}{\partial V}\bigg|_S\,\frac{\partial V}{\partial T}\bigg|_P \tag{4.81}$$

with the specific heats at constant volume and pressure c_V and c_P, respectively. In that context, Meyers [276] points out that the term $\frac{1}{V}\,\frac{\partial V}{\partial T}\big|_P$ is equal to the volumetric thermal expansion 3α and that $-\frac{1}{V}\,\frac{\partial V}{\partial P}\big|_T$ represents the isothermal compressibility K. Thus, Γ can be expressed by

$$\Gamma = V\frac{3\,\alpha}{c_V\,K} \tag{4.82}$$

Further ways to calculate the Grüneisen constant were developed by Slater [359] or Dugdale and MacDonald [117]. Detailed discussions on the related theories are to be found in Rice et al. [320] and, more recently, including a new own method in Nagayama et al. [289].

To estimate the accuracy and limitations oft the Mie-Grüneisen EOS (4.77), it can alternatively be understood as isochoric extrapolation off the shock Hugoniot formulated as Taylor series developed around the Hugoniot pressure:

$$p(V,e) = p_H - \left(\frac{\partial p}{\partial e}\right)_V (e - e_H) + \ldots \approx p_H - \frac{\Gamma}{V}(e - e_H) \qquad (4.83)$$

Equation (4.83) explains the assumption needed to establish the Mie-Grüneisen EOS and thus helps to assess the quality and limitations of it. Basically, it is a first order approximation of the state surface in the neighbourhood of the measured Hugoniot curve along an isochoric path. Another important limitation of the Mie-Grüneisen EOS is that it is not capable of phase changes.

Polynomial Equation of State

Polynomial equations, as previously introduced with (3.23), are formulated when nonlinear equations of state are needed. Referring to a line $e = 0$, a polynomial EOS often used for shock wave simulations describes the pressure as function of specific internal energy and density with:

$$p = K_1\mu + K_2\mu^2 + K_3\mu^3 + (\ B_0 + B_1\mu\ )\ \rho_0\ e \qquad (4.84)$$

The cubic formulation in μ of equation (4.84) can be used as an alternative way to describe the Hugoniot pressure p_H. The related parameters K_i must then be correlated with the Hugoniot parameters in order to approximate the formulation of equation (4.78). For the same reference condition as in (4.77), i.e. the Hugoniot curve defined by a linear v_S-v_1-relation as given in (4.52), the parameters K_i of the polynomial EOS are identified as follows:

$$\begin{aligned} K_1 &= \rho_0\, c_0^2 \\ K_2 &= \rho_0\, c_0^2\, [1 + 2\,(S-1)] \\ K_3 &= \rho_0\, c_0^2\, \left[2\,(S-1) + 3\,(S-1)^2\right] \end{aligned}$$

where K_1 equals the bulk compressive modulus of the material.

Using the parameters B_0 and B_1, the second part of (4.84) reflects $p - \rho - e$ conditions off the reference curve. Its two-parameter formulation for the behavior off the reference curve

$$\frac{1}{\rho}\left.\frac{\partial p}{\partial e}\right|_V = \frac{B_0 + B_1\mu}{1+\mu} \qquad (4.85)$$

allows to distinguish between the following cases of:

- Constant $\frac{1}{\rho}\left.\frac{\partial p}{\partial e}\right|_V$, with $B_0 = B_1$.
- Constant $\left.\frac{\partial p}{\partial e}\right|_V$, with $B_0 = \frac{1}{\rho_0}\left.\frac{\partial p}{\partial e}\right|_V$ and $B_1 = 0$

- Linear density dependency in $\frac{1}{\rho}\left.\frac{\partial p}{\partial e}\right|_V$, with $B_0 \neq B_1 \neq 0$

If the polynomial formulation is to replace the Mie-Grüneisen formulation (4.77), the combinations of B_0 and B_1 are accordingly used to specify the density dependency of Γ:

- $\Gamma = const.$ for $B_0 = B_1$.
- $\Gamma\,\rho = const.$ for $B_0 = \frac{1}{\rho_0}\,\Gamma\,\rho$ and $B_1 = 0$
- Linear dependency of Γ in ρ for $B_0 \neq B_1 \neq 0$

Note that though the parameters K_i may be adjusted to fit the Hugoniot pressure p_H, the reference curve in formulation (4.84) is $e = 0$ and not the Hugoniot line with $e = e_H$. If the Hugoniot shall be used as reference, the polynomial formulation needs to be changed to:

$$p = K_1\mu + K_2\mu^2 + K_3\mu^3 + (\ B_0 + B_1\mu\ )\ \rho_0\ (e - e_H) \tag{4.86}$$

Formulations for Expansion States

Expansion from Unloaded States

The whole theory for shock and polynomial type Mie-Grüneisen equations of state discussed so far is related to compression states. Thus, the expansion regime of negative hydrostatic pressures still needs a description. Since application of equations (4.84) or (4.77) to expansion states is not covered by the basic theory and leads to problems, e.g. in producing negative values for the sound speed, other formulations need to be used. An often found expansion formulation is the reduction of the polynomial form (4.84) to a linear or quadratic *expansion equation of state*, i.e.:

$$p = \tilde{K}_1\mu +\ B_0\ \rho_0\ e \tag{4.87}$$

and

$$p = \tilde{K}_1\mu + \tilde{K}_2\mu^2 +\ B_0\ \rho_0\ e \tag{4.88}$$

of which the energy dependent term is sometimes omitted. For the expansion parameter $\tilde{K}_i$ there is no relation to Hugoniot reted parameters as for the compressive ones since shock Hugoniot data only make sense for compression states. Therefore, identification of the expansion parameters is done by experiments with tensile loading states. In the linear case the expansion parameter is often set to the linear compressive value.

Isentropic Release from Hugoniot States

Processes where material is first shock loaded to a Hugoniot state p_H, e_H, ρ_H and then released though an isentropic process were discussed regarding energy release in section 4.4.5. As mentioned in that context, the graphical representation of shock compression and isentropic release is different in a

$p-\rho$ diagram compared to a $p-V$ diagram where the Rayleigh line is known as straight line. Figure 4.16 illustrates the conditions in a hydrocode-typical $p-\rho$ plane with calculated compression and release processes for aluminum with material data according to Table 4.2.

In the following, the isentropic release path shall be calculated for a material with a Mie-Grüneisen equation of state. If we assume an incremental release, the related path for the release isentrope is defined through the isentropic condition:

$$\mathrm{d}e = -p\,\mathrm{d}V = \frac{p}{\rho^2}\,\mathrm{d}\rho \tag{4.89}$$

The start conditions is described by the density at the Hugoniot state $\rho(p_H, e_H)$. The subsequent values of p, ρ and e along the release isentrope shall be calculated stepwise in density according to

$$\rho_{i+1} = \rho_i - \Delta\rho \tag{4.90}$$

where the step-size $\Delta\rho$ is defined through the number of steps n as

$$\Delta\rho = \frac{\rho(p_H, e_H) - \rho_0}{n} \tag{4.91}$$

The changes in energy $\mathrm{d}e$ for each step defined by (4.89) can be approximated by

$$\Delta e_{i+1} \approx \frac{1}{2}\left(\frac{p_{i+1}}{\rho_{i+1}^2} + \frac{p_i}{\rho_i^2}\right)(\rho_{i+1} - \rho_i) \tag{4.92}$$

The values for p_{i+1} and e_{i+1} along the release path need to be calculated iteratively since the pressure

$$p_{i+1} = p(\rho_{i+1}, e_{i+1}) \tag{4.93}$$

and the energy terms

$$e_{i+1} = e_i + \Delta e(p_{i+1}, \rho_{i+1}) \tag{4.94}$$

are implicitly related to each other. An appropriate start value for the change in energy can be chosen by an estimation

$$\langle e_{i+1}\rangle = \lambda\, e_i \tag{4.95}$$

with a factor λ close to 1, e.g. 0.99.

The iteration can be performed in a loop calculating

$$p_{i+1} = p_H + \Gamma\,\rho_{i+1}\left(\langle e_{i+1}\rangle - e_H\right) \tag{4.96}$$

and

$$e_{i+1} = e_i + \Delta e\,(p_i,\, p_{i+1},\, \rho_i,\, \rho_{i+1}) \tag{4.97}$$

to give a normalized residual of

$$\frac{|\delta\, e|}{|e_{i+1}|} = \frac{|\langle e_{i+1}\rangle - e_{i+1}|}{|e_{i+1}|} \tag{4.98}$$

With a selectable stopping criterion $\kappa << 1$, e.g. $\kappa = 0.0001$, the accuracy of the iteration can be controlled if the loop is done until the normalized residual fulfills

$$\frac{|\delta\, e|}{|e_{i+1}|} < \kappa \tag{4.99}$$

The total amount of energy set free along the release path is approximated by summation over the increments $\Delta\, e_i$ in each step:

$$e_{rel} = \sum_{i=1}^{n} \Delta e_i = \sum_{i=1}^{n} \frac{1}{2} \left(\frac{p_i}{\rho_i^2} + \frac{p_{i-1}}{\rho_{i-1}^2} \right) (\rho_i - \rho_{i-1}) \tag{4.100}$$

Since the energy stored during shock compression is equivalent to the related triangular surface in the $p - V$ plane, it can be calculated as

$$e_{shock} = -\frac{1}{2}\,(p_H - p_0)\,(V_H - V_0) = -\frac{1}{2}\,(p_H - p_0)\,(\frac{1}{\rho_H} - \rho_0) \tag{4.101}$$

leading to an energy difference that is stored in the shocked and released material of:

$$e_{stored} = e_{shock} - e_{rel} \tag{4.102}$$

Figure 4.16 illustrates an example calculation of Rayleigh line and release isentrope according to the described procedure. With the material data from Table 4.2 and according to (4.100) and (4.101) the calculated process would heat up the material by $e_{stored} = 0.0328[Terg/g]$.

Treatment of Phase Changes

High pressures and temperatures, occurring for example in hypervelocity impacts, can lead to phase changes of the involved materials. Aluminium, for example, starts to melt at impact velocities around 7 $[km/s]$ and to vaporize at velocities above 11 $[km/s]$. To model the arising fragment clouds and, thus, the impact on subsequent structural components properly, the phase changes must treated by the nonlinear equation of state.

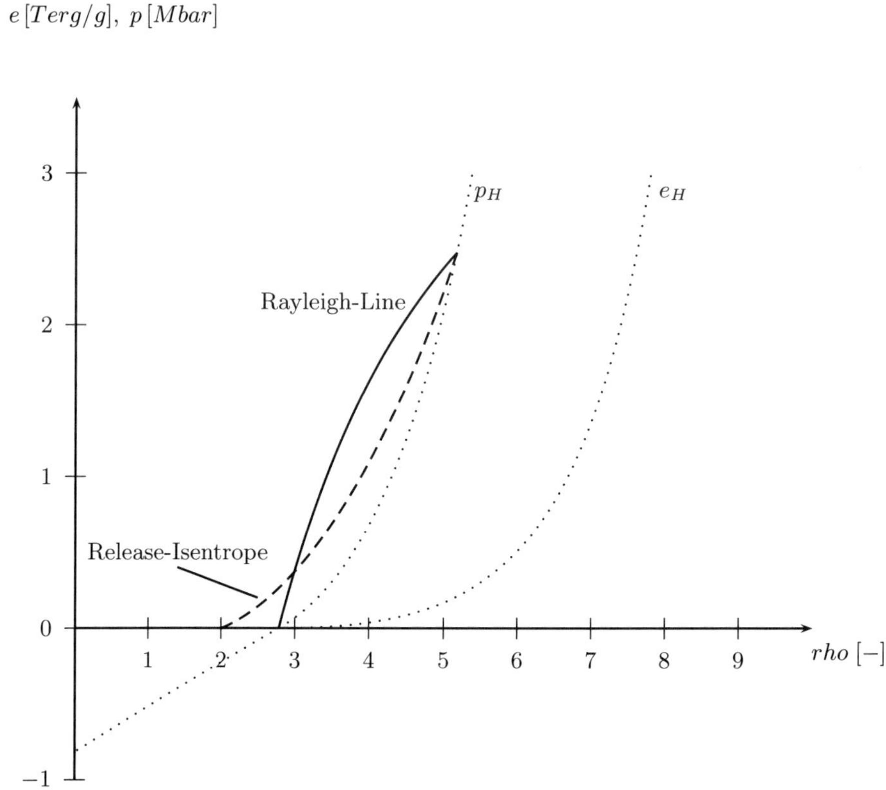

Fig. 4.16. Shock compression along Rayleigh line and calculated isentropic release path represented in the $p-\rho$ plane along with Hugoniot pressure p_H and energy e_H.

SESAME Tabular EOS

One option to describe phase changes in terms of an equation of state is the use of tabular data. The SESAME equation of state (see [351] and Holian [190]) developed at the Los Alamos National Laboratories provides tables for the pressure-density relation and for the energy-density relation of several materials. Since the look-up times of modern computers has become reasonably short, the application of tabular data for material models has lost the drawback of delayed computational times. As a result, tabular relation like the SESAME equation of state are now also implemented in commercial hydrocodes. As for analytical relations, the limitation in its application is of course defined by the range of densities and energies, the data are experimentally derived for. Extrapolations beyond these limitations should be performed carefully, though it is not always clear whether and where a specific code obeys these limits.

Tillotson Equation of State

An analytical alternative to a tabular equation is the Tillotson equation of state (Tillotson [390]) which spans a wide spectrum of validity ranging from comparably low pressures, where it reflects the Mie-Grüneisen equation of state, an to extremely low densities through approximation of the ideal gas equation. Finally, going to extremely high pressures, the Tillotson equation of state has still reasonable validity via convergence to the Thomas-Fermi equation[8].

Specifically derived for hypervelocity impact simulations, the Tillotson equation separates the EOS in to major regions:

- a compressed region where the internal energy e is less than the energy of incipient vaporization
- and one for expanded states, where e exceeds the energy for complete vaporization.

where the expansion states are described by three different equations.

For compressive states, the pressure is calculated from the following equation:

$$p_1 = a\,\rho\,e + b\,\frac{\rho\,e}{1+\frac{e\,\rho_0^2}{e_0\,\rho^2}} + A\left(\frac{\rho}{\rho_0}-1\right) + B\left(\frac{\rho}{\rho_0}-1\right)^2 \tag{4.103}$$

with the parameters a, b and A defined from Hugoniot experiments, B and e_0 adjusted to provide a best fit of the equation of state surface.

Expansion conditions with internal energies below the sublimation energy are described by:

$$p_2 = a\,\rho\,e + b\,\frac{\rho\,e}{1+\frac{e\,\rho_0^2}{e_0\,\rho^2}} + A\left(\frac{\rho}{\rho_0}-1\right) \tag{4.104}$$

The related expansion process do not lead to phase changes.

In cases where the expansion process leads to vaporization, i.e. energies above the sublimation energy are reached, the pressure is calculated from:

[8] See for more details Holian and Holian [192] as well as Anderson and Mullin [8]

$$p_3 = a\,\rho\,e + b\,\frac{\rho\,e\,e^{-\alpha\,\eta}}{1 + \frac{e\,\rho_0^2}{e_0\,\rho^2}} + A\left(\frac{\rho}{\rho_0} - 1\right)e^{\gamma\,\eta}\,e^{-\alpha\,\eta^2} \qquad (4.105)$$

where the material parameters α and γ are chosen to guarantee convergence to the ideal gas equation.

Criterion for the choice of equation (4.105) is, however, not only the energy to be above the sublimation energy e_s but also to above a threshold energy e_2' that is somewhere between the sublimation energy and vaporization energy e_v:

$$e_s' = e_s + \kappa e_v \qquad (4.106)$$

where the parameter κ is usually chosen to be around 0.20[9].

Finally, for expansion states with energies between the sublimation energy e_s and e_s' a fourth formulation is chosen interpolating between p_3 and p_2:

$$p_4 = \frac{p_3\,(e - e_s) + p_2\,(e_s' - e)}{e_s' - e} \qquad (4.107)$$

To clarify which formulation is chosen for which compression and energy state, Table 4.1 summarizes the respective conditions and equations.

Table 4.1. Tillotson EOS: Compression-energy states and related equations.

Compression	Energy	Equation	#
$\frac{\rho}{\rho_0} - 1 > 0$	e	$p_1(\rho,\,e)$	(4.103)
$\frac{\rho}{\rho_0} - 1 < 0$	$e < e_s$	$p_2(\rho,\,e)$	(4.104)
$\frac{\rho}{\rho_0} - 1 < 0$	$e_s < e < e_s'$	$p_4(\rho,\,e)$	(4.105)
$\frac{\rho}{\rho_0} - 1 < 0$	$e_s' < e$	$p_3(\rho,\,e)$	(4.107)

As often stated in literature, the Tillotson equation of state is not able to directly model melting and vaporization since changes in pressure and density in two-phase regions cannot be described. However, the resulting thermodynamic state after phase changes in the course of transient shock waves are well represented. Holian and Burket [191] discuss the Tillotson EOS implemented in a hydrocode. Their investigation of hypervelocity impacts of lead spheres against bumper shield plates compares experimental results with simulations calculated from Tillotson and SESAME equations. The theoretical results are very close to each other in terms of calculated mass per unit area of the fragment cloud. Both also deviate from the experimental results with respect to

9 see, e.g. Anderson and Mullin [8] for material data.

the mass distribution. This was probably due to the discretization techniques available at that time. More recent simulations of aluminium, steel and also of lead hypervelocity impact, show much better results[10].

Also mentioned in Holian and Burket [191] is another deficiency of the Tillotson EOS concerns the calculation of release isentropes. When the release process crosses the sublimation point, an abrupt discontinuous jump is calculated for the related pressure $p(e_s, \rho_s)$. As an improvement of this observation Schonberg [348] proposed an adjustment via the pressure interpolation:

$$p_5 = p_3 - \left(p_2|_{\rho=\rho_s} - p_3|_{\rho=\rho_s} \right) \tag{4.108}$$

for densities $\rho < \rho_s$ in the energy interval $e_s < e < e'_s$. With that improved EOS version Schonberg calculated release isentropes for expanding fragment clouds after aluminum-aluminum impacts at velocities between 4 and 20 km/s. The aim was to calculate the proportionate distribution of solid, liquid and vaporized material in the fragment cloud. The results achieved in [348] are illustrated in Figure 4.17.

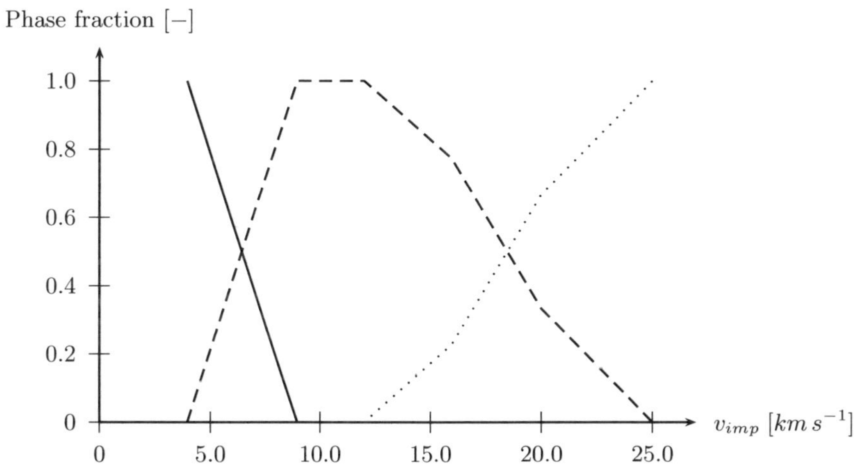

Fig. 4.17. Distribution of solid (solid line), liquid (dashed line) and vaporized material (dotted line) in fragment clouds after hypervelocity impact at various impact velocities calculated with Tillotson EOS. Numerical results from Schonberg [348].

[10] Comparable hypervelocity impact simulations are for example published in Mehra and Chaturvedi [270] or Hiermaier et al. [178], where the latter includes application of the Tillotson EOS.

Equations of State for Porous Materials

There is a wide spectrum of materials with an inherent porosity which may be either intentionally inserted or present to a certain unavoidable minimum due to production reasons. Examples for deliberately produced porosity are cellular materials used to provide thermal, acoustic or shock damping effects with a particular structural stiffness at low specific weight. Others, e.g. polymers, concrete or energetic materials, dispose of some level of porosity that is not always desirable and needs to be reflected in material characterization.

Dynamic deformation behaviour of porous and granular materials is most often investigated regarding two particular aspects:

- The energy dissipation effect of porous materials like polymeric and metallic foams or concrete is one focus. Propagation of acoustic as well as shock waves is damped through irreversible deformation of cells. Kinetic energy of impact and crash processes, ranging from low velocity head impact to hypervelocity impact of micro-meteoroids on space vehicle structures, is transformed into plastic strain and failure. The porosity and the stiffness of the matrix material determine the stress level at which the structure deforms and, thus, for example the deceleration of an impacting body.
- Insensitivity as well as the detonation characteristics of plastic bonded energetic materials is influenced by the porosity. Here, the shock velocities and the strain localization due to the porous nature are of interest.

Energetic materials are not topic of this book. Readers interested in the dynamic deormation behaviour of explosives are for example referred to Davison et al. [105].

Characteristic Deformation Phases in Porous Materials

The complexity in modelling porosity under shock loading conditions arises from the unusual compaction behavior of porous materials including distinct variations in the materials' stiffnesses in three phases:

- Typically, a first deformation phase of porous materials can be described as *elastic compression* of the matrix material. The stiffness is defined by the properties of the matrix material.
- Depending on the type of the matrix material, it is followed by a second phase of *failure and collapse of pores* which can occur as ductile plastic deformation or as brittle fragmentation and includes stability aspects on the scale of the pore wall dimensions. Characteristic for the macroscopic aspect of this deformation stadium is very little stiffness and subsequent deformation at a more or less constant stress level. Hyperelastic matrix materials do not show this phase.
- A third and final deformation stadium, called *compaction phase*, is observed when the pores are all compressed. The stiffness changes again

drastically towards the matrix material's elastic-plastic compressive properties.

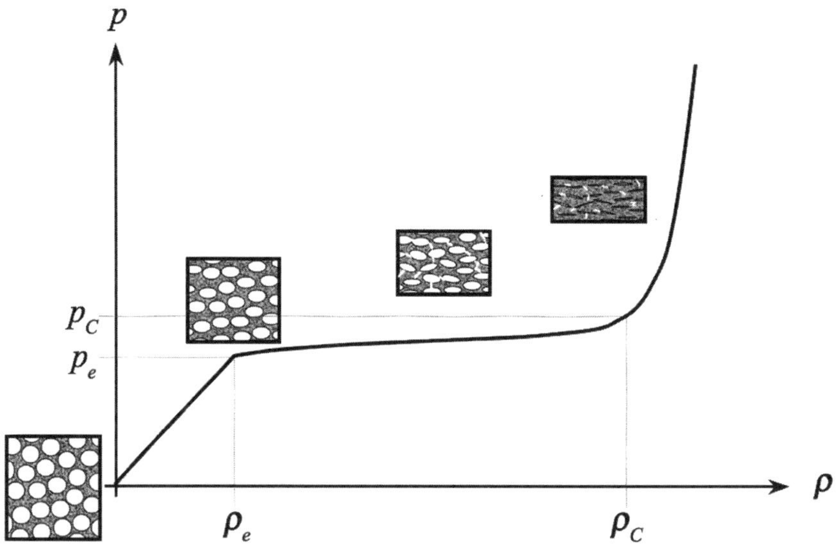

Fig. 4.18. Schematic compression curve for a porous material illustrating initial elastic deformation, pore compression and collapse finally leading to a fully compacted state with the compressive behaviour of the matrix material.

If the sound speed in the matrix material is c_M and c_P is the sound speed of the porous material, the bulk modulus K_P of the porous material in the first elastic compaction phase can be described via the sound speeds:

$$K_P = \rho \frac{\mathrm{d}p}{\mathrm{d}\rho} = \rho\, c_P^2 \tag{4.109}$$

In the third phase, the bulk modulus corresponds to the bulk modulus of the matrix material.

The transition between phase one and three can be described in terms of the material properties of the matrix material using a porosity parameter α:

$$\alpha = \frac{\rho_M}{\rho} = \frac{V}{V_M} \tag{4.110}$$

that relates the matrix and, thus, fully compacted density ρ_M to the current density ρ of the porous material. Often, the inverse of α is used to relate the specific volume of the matrix material V_M at zero pressure to the current specific volume of the porous material. In the context of the following section we will denote that alternative measure of porosity by ϕ:

$$\phi = \frac{V_M}{V} = \frac{\rho}{\rho_M} = \alpha^{-1} \tag{4.111}$$

The choice of ϕ over α was motivated through the use of specific volumes and the preferred variation of ϕ between zero for a non-existing matrix, i.e. a vacuum, and unity for a fully dense material. A preference that seems indeed more convenient than the variation from infinity to unity.

With the porosity of the undeformed material $\alpha_P = \rho_M/\rho_P$ the current sound speed at an arbitrary compression state can be described through:

$$c = c_P + (c_M - c_P)\left(\frac{\alpha - \alpha_P}{1 - \alpha_P}\right) \tag{4.112}$$

and, thus, the current stiffness of the partly compacted material accordingly.

Influence of Porosity on the Hugoniot State

A characteristic phenomenon observed when porous materials are shock loaded is the convex shape of the $v_S - v_1$ curves and the related deflection of the Hugoniot curve:

Porosity in solids leads to a typical decrease in the observed shock velocity v_S with increasing v_1 for low particle velocities. It is the dispersive nature of the propagated stress waves in the regime just above the compressive strength of porous materials that leads to the observed phenomenon. At a certain v_1 value the shock velocity shows a minimum and starts to rise again with convergence to the slope of the solid materials $v_S - v_1$ characteristic. Riedel and Kawai [325] investigated the phenomenon in mortar by experimental and numerical means. Their proposed stess gauge instrumentation along with the theoretical distinctions of stress waves proposed by Fowles and Williams [130] seems to be a promissing methodolgy for a better understanding of shock waves in inhomogeneous materials.

A porous material with very reproducible mechanical properties is achieved when a matrix is filled with hollow spheres. As an example for that kind of cellular material, epoxy resin filled with glass micro-spheres was investigated by Weirick [416]. Figure 4.19 compares the findings of Weirick with the s$v_S - v_1$ characteristic of mortar reported by Riedel [323]. Hugoniot pressures of the micro-sphere filled epoxy material and of the eneretic material PBX-9407 as listed in Marsh [267] ar given in Figure 4.20. The graphs reflect the aforementioned characteristics of shock loaded porous materials.

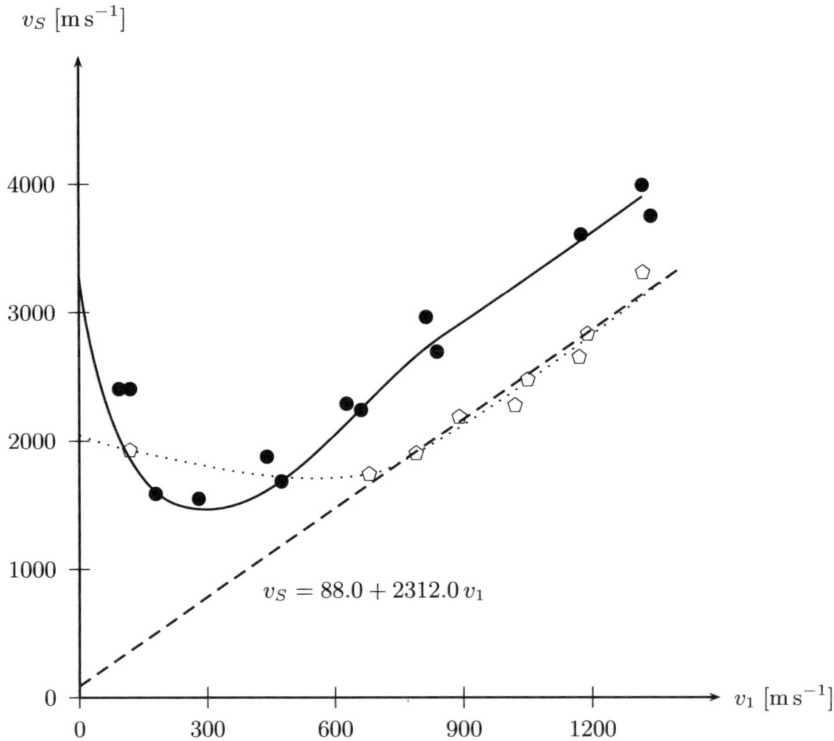

Fig. 4.19. Shock-particle velocity results for Epoxy resin (dashed line) and Epoxy filled to 42 volume percent with glass micro-balloons (⬠) both investigated by Weirick [416] compared to mortar data (•) provided by Riedel [323].

Herrmann's $p - \alpha$ Equation of State

Deformation behaviour of porous materials has long been a topic of scientific investigations. First comprehensive description of shock loaded porous materials including a formulation for a related equation of state were derived more than forty years ago. A first formulation of an EOS was proposed by Herrmann [175]. In order to relate the porous materials' EOS to the one of the solid matrix material, Herrmann invented a porosity and energy dependent formulation, called $p - \alpha$ equation of state:

$$p = p(\frac{V}{\alpha}, e) = p(\phi V, e) \tag{4.113}$$

The distinction of (4.113) to other equations of state is the porosity in the argument of the pressure function. Thus, any existing EOS can be used to

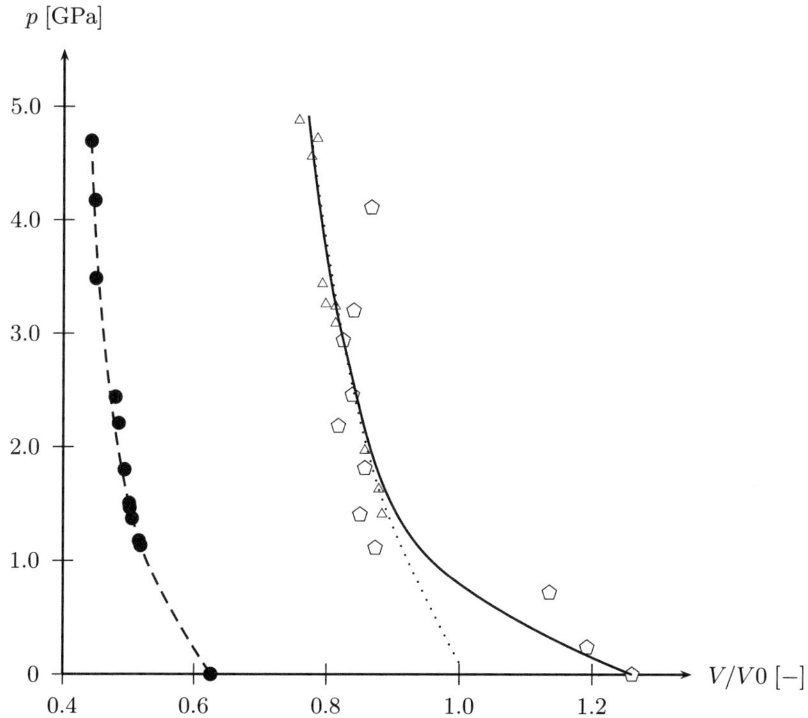

Fig. 4.20. Hugoniot states for pure Epoxy resin (△, dotted), Epoxy filled to 42 volume percent with glass microspheres (⬠, solid) both Weirick [416]), and PBX-9407 (•, dashed) according to data from Marsh [267].

replace the volume dependency by the V/α-dependency.

Equation (4.113) was adjusted by Carrol and Holt [80] to average the pressure in the matrix material over the total porous volume:

$$p = \frac{1}{\alpha} p(\frac{V}{\alpha}, e) = \phi p(\phi V, e) \tag{4.114}$$

The porosity is now a thermodynamic state variable depending on pressure and internal energy:

$$\alpha = \alpha(p, e) \tag{4.115}$$

where the energy dependence is mostly not formulated explicitly due to the lack of data. This gap is usually closed to a sufficient degree of accuracy through an implicit energy dependence carried in the Hugoniot reference curve referred to in the solid materials EOS. This approach is valid as long as no additional energy source terms are in the porous material. Swegle [379] points out that an important advantage of (4.114) over (4.113) is that it avoids unphysically larger bulk sound speeds in the porous material than in the matrix material.

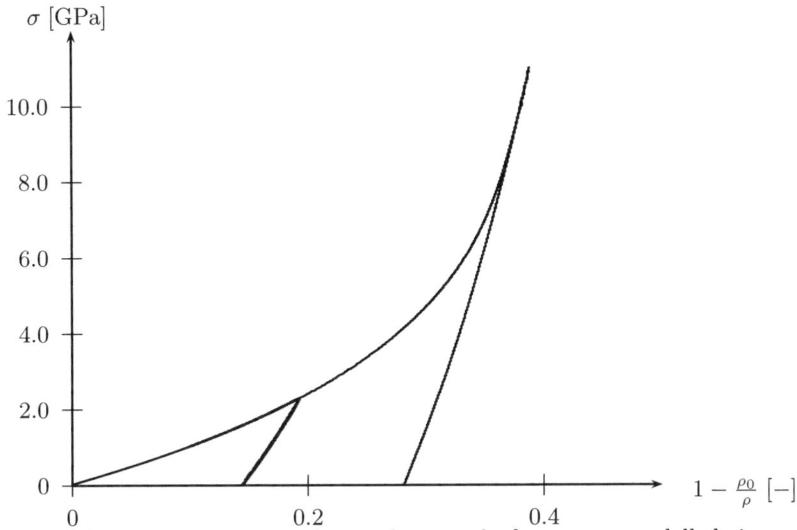

Fig. 4.21. Compression curve and two release paths for cement modelled via a $p-\alpha$ equation of state (Data courtesy of Riedel [323]).

A possible description of the porosity parameter, amongst various other formulations, was proposed by Butcher and Karnes [73] for porous iron:

$$\alpha = 1 + (\alpha_p - 1)\left(\frac{p_C - p}{p_C - p_e}\right)^n \tag{4.116}$$

with a parameter n set to $n = 2$ in [73]. The pressure at initial pore collapse p_e and pressure at the beginning of the compaction phase, denoted by p_C, are further parameters to be defined experimentally. The implicit coupling of pressure, volume, internal energy and porosity demands for an iterative solution of the equations in numerical implementations.

Menikoff-Kober Porous Equation of State

A thermodynamically consistent formulation of a complete equation of state for porous materials was introduced by Menikoff and Kober [274] via a decomposition of the Helmholtz potential into a contribution from the pure solid material ψ_s and potential energy ψ_p associated with the porosity:

$$\psi(V,\,T,\,\phi) = \psi_s(\phi\,V,\,T) + \psi_p(\phi) \tag{4.117}$$

where the porosity is here described via the solid volume fraction $\phi = V_M/V = \rho/\rho_M = \alpha^{-1}$. Alternatively to (4.116), compaction laws for porous materials can also be formulated using ϕ, a parameter p_c denoting the pressure at full compaction, i.e. closure of all pores, and an initial porosity $1-\phi_0$:

$$\phi_{eq}(pV) = 1 - (1-\phi_0)\,e^{-\frac{pV - p_0 V_0}{p_c V_0}} \tag{4.118}$$

To find the porosity ϕ_{eq} of an equilibrium state, Menikoff and Kober minimize the free energy (4.117) to get:

$$pV - \phi_{eq}\,\frac{\mathrm{d}\,\psi_p}{\mathrm{d}\,\phi}(\phi_{eq}) = 0 \tag{4.119}$$

which needs to be solved simultaneously with the equation for equilibrium pressure p_{eq} that is defined according to (3.5) as:

$$p_{eq}(V,\,e) = -\phi_{eq}\,\frac{\partial\,\psi(V_s,\,e_s}{\partial\,V} \tag{4.120}$$

where $V_s = \phi_{eq}V$ and $e_s = e - \psi_p(\phi_{eq})$ are the specific volume and internal energy of the pure solid. Menikoff and Kober point out that equation (4.120) equals the $p-\alpha$ equation, except for the thermodynamically requested distinction between specific internal energy e of the porous material and the one of the solid matrix e_s.

Menikoff [273] uses the matrix material's adiabatic exponent $\gamma_s = \rho_s\,c_s^2\,/P_s$ to describe the ratio of equilibrium and matrix sound speed as:

$$\left(\frac{c_{eq}}{c_s}\right)^2 = 1 - \left(\frac{\gamma_s - 1}{\gamma_s}\right)^2\,\frac{c_s^2}{c_s^2 + \phi_{eq}^2\,\frac{\mathrm{d}^2\psi_p}{\mathrm{d}\phi^2}(\phi_{eq})} \tag{4.121}$$

and, thus, to quantify the decrease in sound speed with increasing porosity. Two further interesting quantities for porous materials formulated in [273] are the Grüneisen parameter:

$$\Gamma = \left(\frac{p_{eq}\,V + \phi_{eq}^2\,\frac{\mathrm{d}^2\psi_p}{\mathrm{d}\phi^2}}{c_s^2 + \phi_{eq}^2\,\frac{\mathrm{d}^2\psi_p}{\mathrm{d}\phi^2}}\right)\,\Gamma_s \tag{4.122}$$

and the related specific heat at constant volume:

$$C_V = \left(\frac{\phi_{eq}^2\,\frac{\mathrm{d}^2\psi_p}{\mathrm{d}\phi^2} + \phi_{eq}\,\frac{\mathrm{d}\psi_p}{\mathrm{d}\phi}}{\phi_{eq}^2\,\frac{\mathrm{d}^2\psi_p}{\mathrm{d}\phi^2} + c_s^2 - \Gamma_s^2\,C_{V\,s}\,T_s}\right)\,\frac{\Gamma_s}{\Gamma}\,C_{V\,s} \tag{4.123}$$

Hayes Equation of State for the Matrix Material

An alternative potential formulation for a complete equation of state was introduced by Hayes [169] and later applied for the description of shock waves in porous high explosives by Sheffield et al. [355] . The matrix material, in this case, is described by the Helmholtz potential:

$$\begin{aligned}\psi(V,T) &= e_0 + p_0\,(V_M - V) - s_0\,T \\ &+ C_V\left[(T-T_0)\left\{1+\frac{\Gamma}{V}\,(V_M - V)\right\} + T\,\ln\frac{T_0}{T}\right] \\ &+ \frac{K_{T0}V_M}{N\,(N-1)}\,\left[\phi^{N-1} - (N-1)\left(1-\phi^{-1}\right) - 1\right]\end{aligned} \quad (4.124)$$

with a constant parameter N, an assumed constant C_V and the isothermal bulk modulus:

$$K_T = K_0\,\phi^N = -\frac{1}{V}\,\frac{\partial p}{\partial V}\bigg|_T \quad (4.125)$$

leading to expressions for the pressure

$$\begin{aligned}p(V,e) &= p_0 + \frac{K_0}{N}\left[\left(\phi^N - 1\right) - \frac{\Gamma_0}{N-1}\,\left(\phi^{N-1} - 1\right) + \Gamma_0\,\left(1-\phi^{-1}\right)\right] \\ &+ \frac{\Gamma_0}{V_M}\,\left[e - e_0 + C_V\,T_0\,\Gamma_0\,\left(1-\phi^{-1}\right) - p_0\,(V_M - V)\right]\end{aligned} \quad (4.126)$$

and the temperature:

$$T(V,e) = T_0 + \frac{V_M}{C_V\,\Gamma_0}\,\left[p(V,e) - p_0 - \frac{K_0}{N}\,\left(\phi^N - 1\right)\right] \quad (4.127)$$

Along with example applications of the Hayes EOS to copper and a discussion of its limitation towards very high pressures $p >> K_0$, Menikoff [273] shows with regard to these formulations that the parameters N and K_0 are related to the coefficients c_0 and S of a linear $v_S - v_1$ relation through:

$$K_0 = \rho_0\,\left(c_0^2 - \Gamma_0^2\,C_V\,T_0\right) \quad (4.128)$$

and

$$N = 4\,S - 1 + (4\,S - \Gamma_0)\,\left(\frac{\Gamma_0^2\,C_V\,T_0}{V_0\,K_0}\right) \quad (4.129)$$

The Hayes equation of state was for example used in combination with Herrmann's $p - \alpha$ model by Setchell and Taylor [352].

4.5.3 Nonlinear Equations of State for Anisotropic Materials

Anderson Model for Anisotropic Solids

According to the conditions for shock wave formation, formulated in (4.31) and (4.32), the nonlinearity in the analytical description of an equation of state is necessary to calculate shock wave evolution and propagation with numerical tools. The characteristic decomposition of the stress tensor $\sigma_{ij} = S_{ij} - p\,\delta_{ij}$ into a deviatoric and a hydrostatic fraction in hydrocodes provides a pragmatic basis to incorporate appropriate descriptions of the hydrostatic pressure p through a thermodynamic equation of state $p = p(\rho,\, e)$.

However, this simple decomposition is only valid for isotropic materials. Anisotropic elasticity is a form of material behavior where deviatoric strain components produce contributions to hydrostatic pressure and volumetric deformation results in deviatoric stresses additive to hydrostatic stresses. Therefore, total hydrostatic pressure must be re-described to account for deviatoric strain contributions. With the decomposition of a total strain tensor according to (2.59) the volumetric contribution for small strains can be written as:

$$\varepsilon^{vol} \approx \frac{1}{3}\,\varepsilon_{ij}\,\delta_{ij} \tag{4.130}$$

and thus, the deviatoric component as

$$\varepsilon_{ij}^{dev} = \varepsilon_{ij} - \frac{1}{3}\,\varepsilon_{ij}\,\delta_{ij} \tag{4.131}$$

The total hydrostatic pressure for linear elastic processes is then:

$$p = p^{el}(\varepsilon^{vol}) + p^{dev}(\varepsilon_{ij}^{dev}) \tag{4.132}$$

and, more general and more adequate for shock wave simulations, the nonlinear equations of state version:

$$p = p^{EOS}(\varepsilon^{vol}) + p^{dev}(\varepsilon_{ij}^{dev}) \tag{4.133}$$

where p^{EOS} can be any, e.g. one of the before discussed, kind of nonlinear equations of state.

Anderson et al. [7] discussed this problem and provided a stepwise solution to be implemented in hydrocodes. Hiermaier et al. [181] implemented the same theory into the commercial hydrocode AUTODYN for application on hypervelocity impact simulations involving orthotropic composite materials. The necessary steps to formulate an approximative description of deviatoric stresses and hydrostatic pressure, including nonlinear equation of state terms, for orthotropic materials are described in the following.

Linear Elastic Orthotropic Stress Increment

Cauchy elastic stress-strain behaviour in an orthotropic material is described by equation (3.38). Accordingly, for a cyclic solution of dynamic processes the following linear elastic stress increments $\delta\,\sigma_{ij}^{el}$ are calculated:

$$\begin{bmatrix} \delta\,\sigma_{xx} \\ \delta\,\sigma_{yy} \\ \delta\,\sigma_{zz} \\ \delta\,\sigma_{yz} \\ \delta\,\sigma_{xz} \\ \delta\,\sigma_{xy} \end{bmatrix}^{el} = \begin{bmatrix} E_{11} & E_{12} & E_{13} & 0 & 0 & 0 \\ E_{21} & E_{22} & E_{23} & 0 & 0 & 0 \\ E_{31} & E_{32} & E_{33} & 0 & 0 & 0 \\ 0 & 0 & 0 & 2G_{23} & 0 & 0 \\ 0 & 0 & 0 & 0 & 2G_{13} & 0 \\ 0 & 0 & 0 & 0 & 0 & 2G_{12} \end{bmatrix} \begin{bmatrix} \delta\,\varepsilon_{xx} + \frac{1}{3}\delta\,\varepsilon^{vol} \\ \delta\,\varepsilon_{yy} + \frac{1}{3}\delta\,\varepsilon^{vol} \\ \delta\,\varepsilon_{zz} + \frac{1}{3}\delta\,\varepsilon^{vol} \\ \delta\,\varepsilon_{yz} \\ \delta\,\varepsilon_{xz} \\ \delta\,\varepsilon_{xy} \end{bmatrix} \tag{4.134}$$

Linear Elastic Pressure Increment

For the hydrostatic pressure increment, expressed as negative mean stress $p = -1/3(\sigma_{xx} + \sigma_{yy} + \sigma_{zz})$, the elastic stress increments result in the following expression:

$$\begin{aligned} \delta\,p^{el} = &-\frac{1}{9}\,(E_{11} + E_{22} + E_{33} + 2\,E_{12} + 2\,E_{23} + 2\,E_{31} +)\;\delta\,\varepsilon_{vol} \\ &-\frac{1}{3}\,(E_{11} + E_{21} + E_{31})\;\delta\,\varepsilon_{11} \\ &-\frac{1}{3}\,(E_{12} + E_{22} + E_{32})\;\delta\,\varepsilon_{22} \\ &-\frac{1}{3}\,(E_{13} + E_{23} + E_{33})\;\delta\,\varepsilon_{33} \end{aligned} \tag{4.135}$$

where, the contribution from volumetric strain is found in the first line and the ones from deviatoric strain components are represented by the second to fourth line. As indicated by the superscript *el*, equation (4.135) can be seen as linear elastic, volumetric stress, i.e. pressure, increment. Thus, to calculate the linear elastic stress deviator increment, the result of (4.135) is subtracted from the total stress increment (4.134):

$$\delta\,S_{ij}^{el} = \delta\,\sigma_{ij} - \delta\,p^{el} \tag{4.136}$$

Nonlinear Pressure Increment

On the other hand, equation (4.135) represents only pressure increments resulting from linear elastic deformations. The nonlinear thermodynamic equation of state element needed for shock wave simulations is not yet part of the theoretical approach. Its implementation can be performed following the suggestion of Anderson et al. [7] by calculating the nonlinear pressure term p^{EOS} in (4.133) with a Mie-Grüneisen formulation. The corresponding pressure increment for a stepwise solution between thermodynamic equilibrium states at step n and $n+1$ is then:

$$\delta\,{}^{n+1}p^{EOS} = p_H(\varepsilon^{vol}) + \Gamma\,\rho\,(e - e_H) - {}^{n}p^{EOS} \tag{4.137}$$

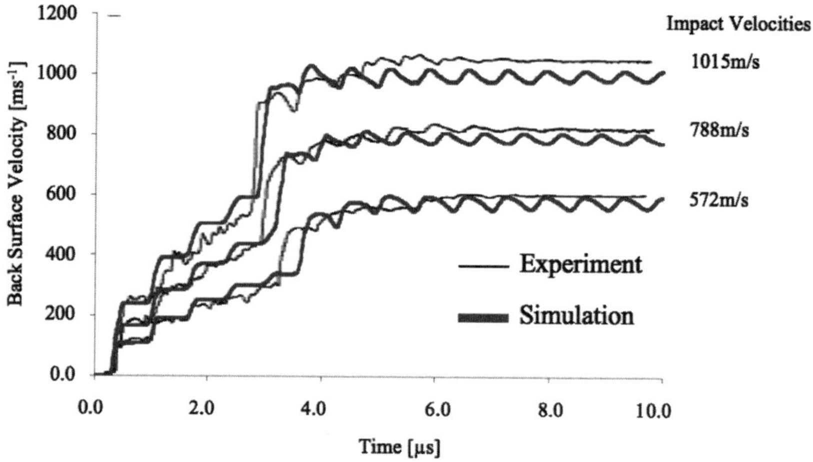

Fig. 4.22. Experimental and numerical results for the back-surface velocities in flyer plate tests at three different impact velocities. The investigated material was Kevlar epoxy resin composite.

Total Pressure Increment and Total Elastic Stress

Equations (4.135) and (4.137), yield an increment in total pressure for general orthotropic materials as:

$$\begin{aligned}\delta\,{}^{n+1}p = \delta\,{}^{n+1}p^{EOS} & \\ & -\frac{1}{3}\,(E_{11}+E_{21}+E_{31})\;\delta\,\varepsilon_{11} \\ & -\frac{1}{3}\,(E_{12}+E_{22}+E_{32})\;\delta\,\varepsilon_{22} \\ & -\frac{1}{3}\,(E_{13}+E_{23}+E_{33})\;\delta\,\varepsilon_{33}\end{aligned} \tag{4.138}$$

Application of (4.134), (4.136) and (4.138), leads to a resulting total elastic stress tensor at step $n+1$ as:

$${}^{n+1}\sigma_{ij} = {}^{n}\sigma_{ij} + \delta\,S_{ij}^{el} - \delta\,{}^{n+1}p\,\delta_{ij} \tag{4.139}$$

This formulation was applied for nonlinear shock compression of composites under hypervelocity impact by Hiermaier et al. [181] in the European Space Agency (ESA) project called Advanced Material Models for Hypervelocity Impact Simulations (AMMHIS). The aim of the project with the cited final report was to provide a numerical model supporting the design of the space debris shield in the European module of the International Space Station

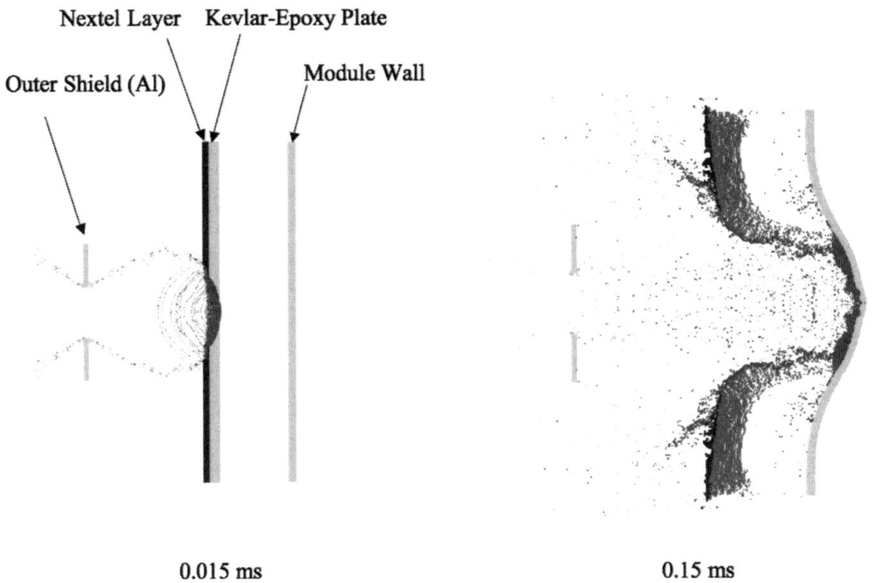

Fig. 4.23. Fragment cloud propagation and deflections of the shield elements at instants of time calculated with the AMMHIS material model implemented in AUTODYN (see Hiermaier et al. [181]).

ISS. The methodology was implemented in the commercial hydrocode AUTODYN is as such available for users' application. Simulation of flyer plate tests, as well as of other tests in the course of material characterization, were performed to validate the model and, thus to find its limitations. Figure 4.22 illustrates the comparison of flyer plate tests on Kevlar epoxy resin composite material at three individual velocities with numerical simulation of the results in terms back-surface velocities.

Validated by comparison of numerical results for the performed material tests, the model was used to design shielding systems against space debris, so called *bumper* or *Whipple shields*. Typically these shields are a combination of several thin aluminum plates serving to break up incoming projectiles. Modern versions are equipped with more advanced materials, such as fiber reinforced polymers or ceramic fabrics. A current standard of these so called *stuffed whipple shields* that is used in similar layouts both in the US and the European modules of the International Space Station is consisting of

- an outer aluminium plate serving to disrupt the projectile via shock wave loading into an expanding fragment cloud,

- a combination of Nextel fabric and Kevlar epoxy resin composite material (the latter only in the European modules) used to further decelerate and disintegrate the impacting fragments
- and finally the module wall of which the integrity after hypervelocity impact on the system is of vital interest for the space station and its crew.

To investigate the protective capabilities of structural designs with varying materials and geometrical dimensions, the formulation described above was used in the equation of state module of the AMMHIS material model. An example result for the fragment cloud propagation and deflections of the shield elements at two time steps is shown in Figure 4.23.

Limitations and Current Deficiencies

Application of the decomposition approach discussed above to hypervelocity impact simulations showed clear improvements in terms of predicting impact damage in shields and fragment clouds. Thus, structural responses in the direct neighbourhood of the impact location are represented with predictive quality.

For a further assessment of the model, three different impact conditions involving aluminum, CFRP and sandwich-structure targets will be investigated next. Figure 4.24 illustrates the geometrical dimensions of the spherical projectiles and the plate targets as well as the impact velocities. Instrumentation with a Laser vibrometer allowed for experimental results including impact induced vibrations of the structures measured in terms of normal velocities. As illustrated in Figure 4.24, the assessment is performed by comparing impact simulations of the orthotropic CFRP target material with an isotropic aluminum target as well as an aluminum-honeycomb stiffened CFRP sandwich target.

Ryan [338] performed experimental and numerical investigation of the CFRP impact as sketched in Figure 4.24-(B). For the numerical simulation of the hypervelocity impact the orthotropic non-linear equation of state material model, described in section 4.5.3, was used. Figure 4.25 illustrates the numerical results for the hypervelocity impact at 4.9 $[km\, s^{-1}]$ on the 0.5 $[mm]$ thick CFRP target. The normal impact initiated shock waves running into the plate, where the propagation speed was directionally dependent on the laminate structure. Comparison with the experimental results in terms of Laser vibrometer based velocity signals from the rear side of the CFRP target indicates that the simulated propagation speeds are correct.

Although the simulations were able to reproduce directional-dependency of the transient wave propagation, significant discrepancies are observed regarding key properties of the transient wave form. Both, amplitudes and wave

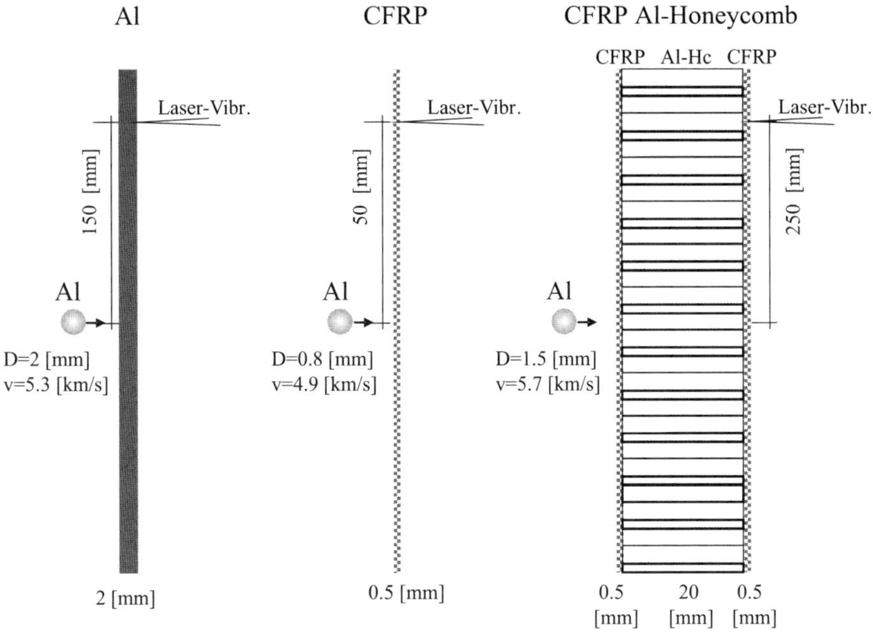

Fig. 4.24. Initial conditions for three normal impact tests involving spherical projectiles and plate targets. A Laser vibrometer was used to measure the vibrational velocities at specified positions remote to the impact location on the back-surfaces of the targets.

lengths of the simulated vibration does not correlate with the experimental results. This discrepancy accounts for the longitudinal, shear and flexural waves.

Hence, Figure 4.26 documents the problems with currently existing material models for shock waves in anisotropic solids. Actually, the problem already starts with the assumptions taken in the framework of experimental parameter derivation in characterization tests. For example, in the course of flyer plate tests a normal stress in the specimen derived according to equation (4.28) is assumed to represent hydrostatic pressure. For the investigation of orthotropic materials this assumption should probably be dropped. At least a quantitative analysis for each material seems to be necessary.

Although the local results in terms of impact damage and fragment clouds can be called predictive[11], the propagating waves cannot be simulated correctly.

[11] For results concerning the local impact damage in the CFRP and CFRP AL-honeycomb targets see also Ryan et al. [339].

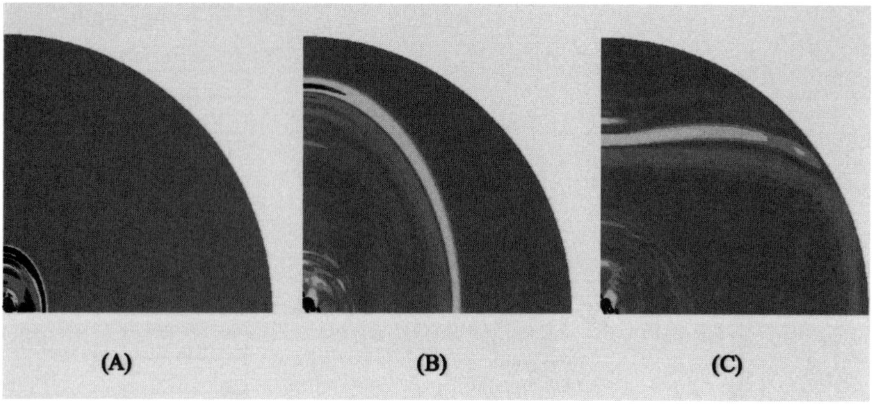

Fig. 4.25. Simulation of the anisotropic wave propagation in a 0.5 [mm] thick CFRP-epoxy composite after a 4.9 [$km\,s^{-1}$] hypervelocity impact of 0.8 [mm] aluminum sphere. The pictures show results from a quarter symmetric simulation with the wave propagation at 3 [μs] (A), 11 [μs] (B) and 18 [μs] (C), respectively. (Reprinted with permission from Ryan [338])

Additional simulation of hypervelocity impact on isotropic or quasi-isotropic targets can help to rule out possible problem sources originating from other code components. Thus, the basic applicability of the existing numerical methods for an investigation of vibrational effects due to hypervelocity impact can be investigated by target variation. Figure 4.27-A illustrates the measured and calculated velocities for the aluminum-aluminum hypervelocity impact. In this case, times of arrival, amplitudes and frequencies over a duration of 30 microseconds is predictively calculated. The example shows that impact induced vibrations in a structure can be simulated with hydrocodes. The set-up which was very similar to the CFRP-impact in terms of impact velocity, target size and measurement location (see Figure 4.24) was simulated with the same code (AUTODYN) and originally performed as a first validation of explicit hydrocodes for that kind of application.

Finally, an impact experiment on a more complex target and the corresponding simulation shall be discussed. As illustrated in Figure 4.24), the sandwich target consists of CFRP face sheets at the front and rear side of an aluminum honeycomb block. The CFRP material and the model used to simulate it are identical to what was used for the pure CFRP target. As can be seen from Figure 4.27-B, the now resulting amplitudes and times of arrival are again very well predicted by the simulation though the frequency in the flexural wave is under-estimated a bit. The problems in modelling the anisotropic CFRP-response are now less prominent since the aluminum honeycomb stiffens the overall structure in an isotropic way.

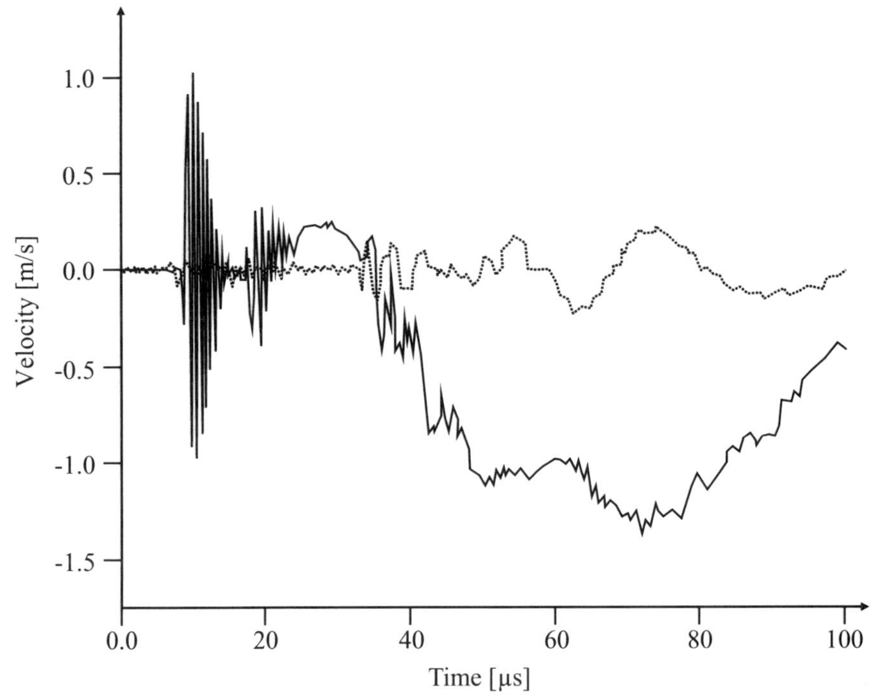

Fig. 4.26. Experimental results (dotted line) and numerical simulation (solid line) using AUTODYN for the velocity histories in a 0.5 [mm] thick CFRP-epoxy composite after a 4.9 [$km\,s^{-1}$] hypervelocity impact of an 0.8 [mm] aluminum sphere. The velocities were measured with a Laser vibrometer at a position 50 [mm] away from the impact locus (Reprinted with permission from Ryan [338]).

In the light of the last two examples explicit hydrocodes can be treated as adequate tool to investigate hypervelocity impact induced shock wave and vibration effects in large structures. The weak link in existing codes are identified to be related to anisotropic material modelling.

An improved simulation of shock wave effects in anisotropic materials including the prediction of remote structural responses requires a more adequate methodology for the shock wave physical assumptions both

- in the experimental parameter derivation
- and in the numerical equation of state formulation and stress tensor decomposition.

A recent contribution towards an improved modelling of shock waves in composite materials was published by Vignjevic et al. [410]. Two alternative decompositions of the stress tensor are proposed and implemented in DYNA3D to simulate flyer plate tests performed on CFRP and reported earlier by Vignjevic et al. [412]. A missing link in the comparison of experimental

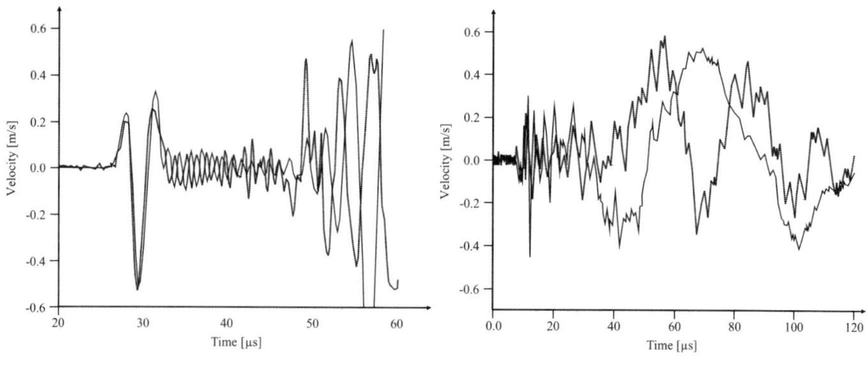

Fig. 4.27. Hypervelocity impact induced vibration of aluminum (A) and CFRP-aluminum honeycomb targets (B). Impact conditions and geometries as introduced in Figure 4.24. Velocities are measured by Laser vibrometer (dotted lines); numerical simulations with AUTODYN provided the calculated velocity histories (solid lines) at the same positions as in the experiments. (Reprinted with permission from Ryan [338].)

and numerical results is the adjusted evaluation of stresses in the experiments. With respect to these measured data in the flyer plate tests, there is still no specific treatment developed referring to the anisotropic material.

Meso-Scale Simulations for EOS Derivations

To improve the existing limitations of equation of state formulations with respect to anisotropic materials, an adequate description and measurement of Hugoniot states in the anisotropic material is needed. A better understanding of measured signals, e.g. resulting from flyer plate tests, can be achieved through parallel numerical modelling of the tests on a meso-scale or even micro-scale level. The meso-scale level includes modelling of components of a composite by the standard continuum mechanical approaches possibly including additional contact, adhesion and interaction descriptions for the interfaces between the components. The micro-scale would concern the formulation of molecular dynamics or atomistic descriptions of the involved materials, which is topic of many research projects in *Multi Scale Material Modelling*[12] .

[12] Interested readers may be referred to introductory literature, e.g. Lu and Kaxiras [254] or Attinger, Koumoutsakos [24]. The topic is also reflected in some conference proceedings on computational material science, for example Gumbsch [153], Steinhauser and Thoma [368], Bulatov et al. [70], Lepinoux et al. [248] or Kirchner, Kubin and Pontikis [226]. An application of molecular dynamics to derive brittle failure models for impact loads is given in Steinhauser et al. [367]

An example for the successful application of meso-scale modelling to derive equation of state data for an inhomogeneous material, i.e. concrete, was given by Riedel [323]. He investigated the components of concrete, cement and aggregate stones, by standard methods of experimental material characterization, specifically in terms of confined and unconfined compression tests at quasi-static and dynamic strain rates as well as flyer-plate tests to derive shock-particle velocity relations. Using the data of the components, Riedel modelled the concrete by explicit discretization of cement and standardized size distributions of aggregate stones. Figure 4.28 shows the discretized 3D sample with a velocity boundary condition and propagating elastic and plastic waves. The resulting calculated $v_S - v_P$ relations are shown in Figure 4.29.

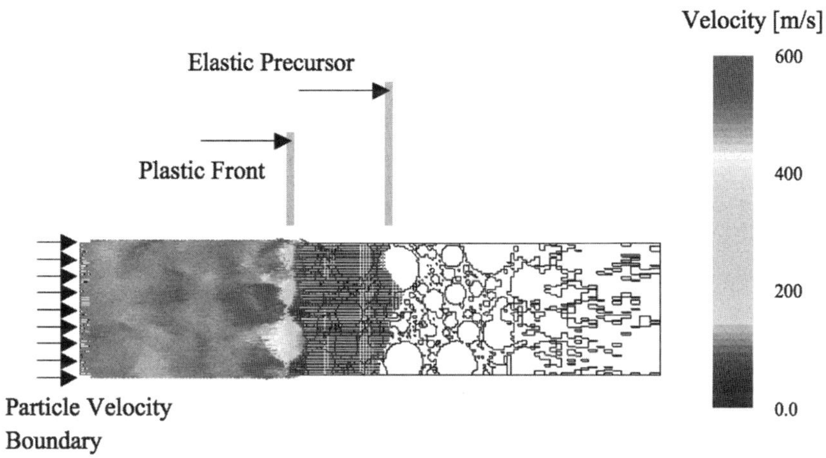

Fig. 4.28. Meso-scale modelled concrete sample loaded by a velocity boundary condition to evoke elastic and plastic wave propagation. The calculated velocities of the shock wave and of the accelerated material behind the shock wave allows to derive Hugoniot states and, thus, a Mie-Grüneisen type equation of state. (Reprinted with permission from Riedel [323])

Comparison to existing experimental data shows the excellent applicability and value of meso-scale simulations of high dynamic material testing methodologies. A similar approach, applied to shock wave propagation in orthotropic materials could deliver an improved understanding of the related Hugoniot states and help provide a new formulation of an equation of state for orthotropic materials.

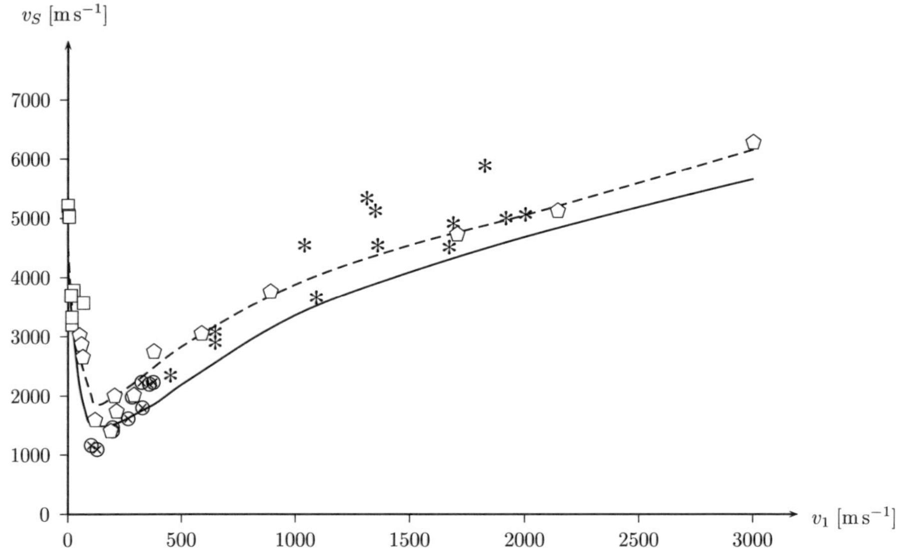

Fig. 4.29. Calculated v_S-v_1 relations for concrete compared to experimental results documented in literature. $\otimes$: Grady [139], $*$: Grady [140], $\square$: Kipp and Chhabildas [225], ⬠ : Gregson [144]. (Reprinted with permission from Riedel [323])

The example calculation of $v_S - v_1$ relations shown above can be observed a first step towards a substantial support from numerical simulation to the determination of equations of state. Other approaches applying methods of molecular dynamics and quantum chemistry have already been taken to provide even deeper theoretical insight in the underlying physics. Successful further examples for the theoretical derivation of equations of state are

- Greeff and Graf [143] for gold,
- Chisolm et al. [91] for aluminium including an investigating concerning the admissibility of neglecting an-harmonic terms, an assumption leading to the Grüneisen parameter Γ (4.67), in the derivation of shock Hugoniots with the Grüneisen theory
- and investigations on the shock induced $\alpha - \omega$ transition in titanium conducted by Greeff, Trinkle and Albers [142]. Using equilibrium free energies for both phases Hugoniot loci were calculated at excellent agreement with experiments. In addition an anomaly assumed by McQueen et al. to be a phase transition could be identified as erroneous interpretation and artifact.

4.6 Discussion of Nonlinear Equations of State for Shock Waves

4.6.1 Summary of Shock Thermodynamics

With the formulations derived in the last sections we have a set of equations that allow for a mathematical treatment of the processes that take place when solid matter is shock loaded. These formulations shall be summarized here along with some suggestions on how to assess their applicability.

Rankine-Hugoniot Equations

Using the conservation for mass, momentum and energy, the conditions ahead and behind a shock wave are related to each other. The resulting Rankine-Hugoniot equations are:

$$\rho_0\ (\ v_S - v_0) = \rho_1\ (\ v_S - v_1) \tag{4.140}$$

$$\rho_1\ (\ v_S - v_1)\ v_1 = p_1 - p_0 \tag{4.141}$$

$$p_1\, v_1 = e_1\ \rho_1\ \ (\ v_S - v_1) - e_0\ \rho_0\ v_S + \frac{1}{2}\ \rho_1\ \ (\ v_S - v_1)\ \ v_1^2 \tag{4.142}$$

For a solution of equations (4.140) - (4.142) one more equation is needed. Either an equation of state $p(\rho,\, e)$ or a relation between shock and particle velocity $v_S(v_1)$ can be used.

Shock - Particle Velocity Relation

Polynomials of the basic form

$$v_S = c_0 + \sum_{i=1}^{n} S_i\ v_1^i \tag{4.143}$$

are used to describe the $v_S - v_1$ relation. Shock behaviour of crystalline materials is well fitted with linear relations, polymers and porous solids and other composite materials require a quadratic description.

Hugoniot State

With the Rankine-Hugoniot equations (4.140) - (4.142) in conjunction with a $v_S - v_1$ relation (4.143) the sum of all thermodynamic states that a specific material reaches under shock loading is described. In the state surface these shock conditions form the Hugoniot curve. Using a linear $v_S - v_1$ relation

$$v_S = c_0 + S\ v_1 \tag{4.144}$$

the Hugoniot states are described by:

$$\rho_1 = \rho_0 \, \frac{c_0 + S\, v_1}{c_0 + v_1(S-1)} \tag{4.145}$$

$$p_1 = p_0 + \rho_0\, c_0^2\, \frac{\eta}{(1 - S\,\eta)^2} \tag{4.146}$$

$$e_1 = e_0 + \frac{\eta}{\rho_0}\, p_1 - \frac{\eta^2}{2}\, \frac{c_0^2}{(1 - S\,\eta)^2} \tag{4.147}$$

Complete Nonlinear Equations of State

According to the combination of first and second law in the Gibbs fundamental equation for a reversible case

$$\mathrm{d}e = -p\,\mathrm{d}V + T\,\mathrm{d}s \tag{4.148}$$

pressure and temperature are defined as

$$p(V,\, s) = -\left.\frac{\partial\, e}{\partial\, V}\right|_s \tag{4.149}$$

and

$$T(V,\, s) = -\left.\frac{\partial\, e}{\partial\, s}\right|_V \tag{4.150}$$

With the specific heat at constant pressure and volume:

$$C_p = T\, \left.\frac{\partial\, s}{\partial\, T}\right|_p \qquad C_V = T\, \left.\frac{\partial\, s}{\partial\, T}\right|_V \quad , \tag{4.151}$$

the coefficient of thermal expansion

$$\beta = \frac{1}{V}\, \left.\frac{\partial\, V}{\partial\, T}\right|_p \quad , \tag{4.152}$$

and the isothermal and isotropic compressibilities

$$K_T = -\frac{1}{V}\, \left.\frac{\partial\, V}{\partial\, p}\right|_T \qquad K_s = -\frac{1}{V}\, \left.\frac{\partial\, V}{\partial\, p}\right|_s \tag{4.153}$$

a more convenient formulation for some second derivatives in equations of state can be found[13]:

- For the Grüneisen parameter we find:

$$\Gamma = V\, \left.\frac{\partial\, p}{\partial\, e}\right|_V = -\frac{V}{T}\, \frac{\partial^2\, e}{\partial\, s\, \partial\, V} = \frac{\beta\, V}{C_V\, K_T} \tag{4.154}$$

[13] See for details Davis [103] as well as Menikoff and Plohr [275]

- the adiabatic exponent becomes:

$$\gamma = \frac{V}{p} \left. \frac{\partial^2 e}{\partial V^2} \right|_s = \frac{1}{p K_s} \tag{4.155}$$

- and the dimensionless specific heat:

$$g = \frac{p V}{T^2} \left. \frac{\partial^2 e}{\partial s^2} \right|_V = \frac{p V}{C_V T} \tag{4.156}$$

The Fundamental Derivative

At the beginning of section 4.4 we already found the fundamental derivative

$$\mathcal{G} = \frac{V^3}{2 c^2} \left. \frac{\partial^2 p}{\partial V^2} \right|_S = 1 + \frac{\rho}{c} \left. \frac{\partial c}{\partial \rho} \right|_S \tag{4.157}$$

to be the criterion required to an equation of state to allow for the existence and stability of shock waves.

The sign of $\mathcal{G}$ determines the existence and stability of shock waves:

- Convex isentropes and, thus, shock waves associated with increasing entropy, are only described by equations of states which strictly obey $\mathcal{G} > 0$. This is the case for all single-phase solids, ideal gases and liquids at normal temperatures[14].
- The limit case of straight isentropes is denoted by $\mathcal{G} = 0$ is for instance the case at the triple point.
- Bethe [58], Zel'dovich [432] and Thompson [389] identified finite areas in van der Waals polytropic gas where the fundamental derivative becomes negative.
- Some exceptional fluids generally show *concave isentropes* and, thus, have the property of a negative $\mathcal{G}$, such as heavy hydrocarbon, heavy fluorocarbons (PP10, PP11), aromatic carbons and syloxanes. These fluids are called *Bethe-Zel'dovich-Thompson* or *BZT fluids*. As a consequence of that anomaly, compressive shock waves become impossible whereas expansion shocks can be initiated.

The Grüneisen Parameter

Further constraints to equations of state have been established that guarantee the monotonicity of properties like the shock wave speed or entropy. Menikoff and Plohr [275] as well as Henderson [172] collect three constraints to the Grüneisen parameter Γ ensuring monotonicity of various properties, given that $\mathcal{G} > 0$:

[14] See for example Henderson [172] for a detailed derivation of the theory, a comprehensive overview on applications as well as material data.

- The *strong constraint* $\Gamma \leq \frac{pV}{e}$ guarantees monotonic decreasing specific volume along a Hugoniot adiabate. Together with $\mathcal{G} > 0$ the criterion for a convex Hugoniot is fulfilled.
- Menikoff and Plohr [275] stated that with the adiabatic exponent γ, a *medium constraint* $\Gamma \leq \gamma + \frac{1}{2}\frac{pV}{e}$ ensures that the specific internal energy is monotonic increasing along a Hugoniot.
- Whereas the *weak constraint* $\Gamma \leq 2\,\gamma$ indicates monotonically increasing pressure and enthalpy with the Hugoniot. A constraint that is obeyed by all known materials.

Schematic illustrations of Hugoniot curves in the $p-V$ and $p-v_1$ planes with geometric shapes that would fulfill or violate the strong, medium and weak constraints are given in Figure 4.30 (see Menikoff and Plohr [275] as well as Henderson for original derivation and more details on these constraints).

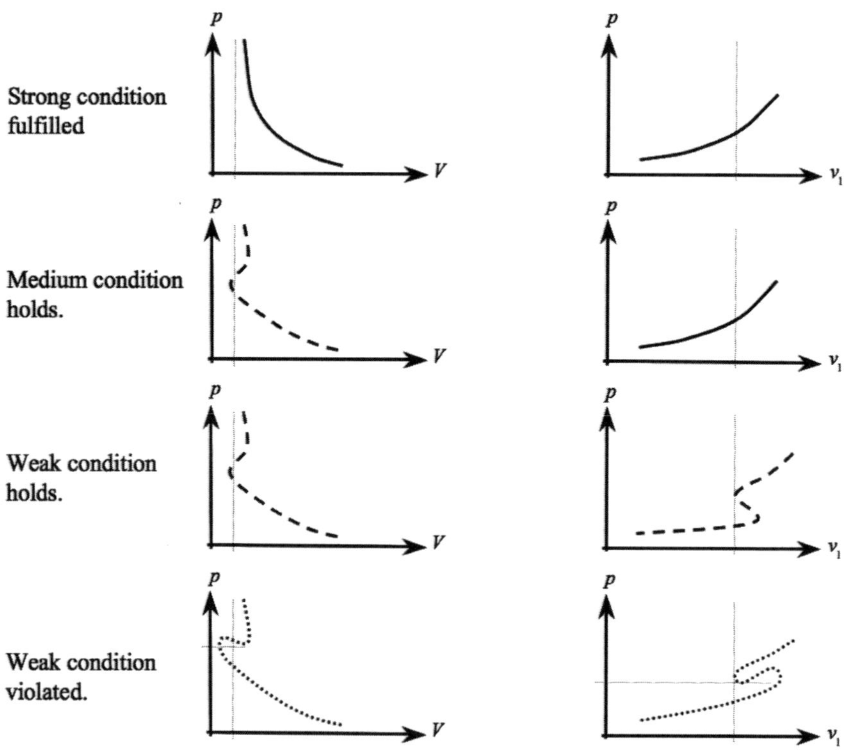

Fig. 4.30. Schematic Hugoniot curves in the $p-V$ and $p-v_1$ planes representing examples for fulfilled or violated constraints of the strong, medium or weak type (Reprinted figure with permission from Menikoff and Plohr [275]. Copyright by the American Physical Society: http://link.aps.org/abstract/RMP/v61/p75).

Finally, Henderson and Menikoff [173] found that the first Bethe criterion (4.33) together with the weak constraint indicates a Hugoniot that contains unique shocks for any elevated pressure, and, that "any pressure increasing shock wave is entropy increasing".

4.6.2 Influence of Nonlinear EOS Formulations on the Calculated Sound Speed

Standard shock equations of state and various polynomial versions are implemented in commercial as well as academic codes for shock wave simulation. To distinguish and to identify the impact of choosing one formulation or parameter set over another one, the resulting pressure and specific internal energy paths in the interesting range of density should be investigated. A reliable measure for the limitations of a EOS formulation is the calculated sound speed with changing density $c(\rho)$. The sound speed is defined through:

$$c^2 = \frac{\mathrm{d}\,p}{\mathrm{d}\,\rho} \tag{4.158}$$

The total differential of p in terms of internal energy and density is

$$\mathrm{d}\,p(\rho,e) = \left.\frac{\partial p}{\partial \rho}\right|_e \mathrm{d}\,\rho + \left.\frac{\partial p}{\partial e}\right|_\rho \mathrm{d}\,e \tag{4.159}$$

For isentropic processes, the change in internal energy is described by

$$\mathrm{d}\,e = -p\,\mathrm{d}\,V = \frac{p}{\rho^2}\,\mathrm{d}\,\rho \tag{4.160}$$

since the change in specific volume $V = \frac{1}{\rho}$ is written in terms of density as:

$$\mathrm{d}\,V = -\frac{1}{\rho^2}\mathrm{d}\,\rho \tag{4.161}$$

For the pressure differential we find:

$$\mathrm{d}\,p(\rho,e) = \left.\frac{\partial p}{\partial \rho}\right|_e \mathrm{d}\,\rho + \frac{p}{\rho^2}\left.\frac{\partial p}{\partial e}\right|_\rho \mathrm{d}\,\rho \tag{4.162}$$

and, thus, for the sound speed the term:

$$c^2 = \frac{\mathrm{d}\,p}{\mathrm{d}\,\rho} = \left.\frac{\partial p}{\partial \rho}\right|_e + \frac{p}{\rho^2}\left.\frac{\partial p}{\partial e}\right|_\rho \tag{4.163}$$

Unfortunately, in case of the Mie-Grüneisen and polynomial equations of state, the values for p, e and c cannot be calculated analytically for a whole path of expansion or compression since the terms for pressure and internal energy are implicitly related to each other. To investigate such a path still outside a hydrocode, where additional factors of influence resulting from discretization et cetera would appear, an iterative solver for equations (4.96) and (4.97) can be written in a few lines of code calculating these data.

In the following, results of an iterative solution will be given for aluminum. The corresponding material data are listed in Table 4.2. The chosen system of units in $cm - g - \mu s$ is common to high pressure and shock compression processes. In particular, the fact that it allows for descriptions of the relevant pressure, energy and sound speed regimes all in the range of -10^1 to 10^1, makes that system of units convenient for our purpose.

Table 4.2. Material Data for Aluminum in Units of $cm - g - \mu s$.

Parameter	Symbol	Value	Unit
Reference Density	ρ_0	2.785	$[g/cm^3]$
Shock Velocity at ρ_0	c_0	0.5328	$[cm/\mu s]$
Slope in Linear $v_S - v_1$ Relation	S	1.338	$[-]$
Initial Grüneisen Gamma	Γ_0	2.03	$[-]$
Coefficients for Polynomial EOS	K_1	0.791	$[Mbar]$
under Compression	K_2	1.325	$[Mbar]$
	K_3	0.8058	$[Mbar]$
Coefficients for Polynomial EOS	$\tilde{K}_1$	0.791	$[Mbar]$
in Expansion	$\tilde{K}_2$	-0.025	$[Mbar]$
Coefficients Describing Energy	B_0	2.03	$[Mbar]$
Dependence in Polynomial EOS	B_1	2.03	$[Mbar]$

Polynomial Equations of State

Cubic Compression and Linear Expansion

As a first formulation, a non-linear polynomial equation of state with a cubic density dependence in compression will be investigated. The energy relation shall be modelled by a term constant in ρ using the equality $B_0 = B_1$ which is equivalent to a constant derivative $\partial p/\partial e$ for the pressure-energy behavior off the reference curve ($e = 0$). For comparison, in a Mie-Grüneisen EOS this would be modeled by a constant $\Gamma(V) = \Gamma_0$. Thus, the compressive path in this model is described by the formulation:

$$p = K_1\mu + K_2\mu^2 + K_3\mu^3 + (\, B_0 + B_1\,\mu\,)\,\rho_0\, e \tag{4.164}$$

The expansion path ($\rho < \rho_0$) shall be formulated by a linear pressure-density relation:

$$p = \tilde{K}_1\mu + \; B_0\,\rho_0\, e = \tilde{K}_1\mu + B_0\,\rho_0\, e \quad . \tag{4.165}$$

An iteratively calculated solution for the variation of pressure, Hugoniot-pressure, specific internal energy, Hugoniot energy and sound speed along an isentropic path in compression and expansion is shown in Figure 4.31.

As theory would expect, the calculated sound speed is not constant in the linear $p - \rho$ expansion regime due to the second term in (4.163). At densities $\rho < 1.76$ the calculated sound speeds are negative and, therefore, in the calculation set to zero. This density marks, thus, a limit of application for this EOS formulation. Also calculated, but not used for the equation of state, are the Hugoniot pressure and energy.

With a $1 - S\,\eta$ term in the denominator, both the Hugoniot pressures and energies have to face singularities with densities approaching

$$\rho = \frac{S\,\rho_0}{S-1} \tag{4.166}$$

which is - in the case of the material data used here - an upper limitation of the formulation at $\rho = 11.025$.

Cubic Compression and Quadratic Expansion

Having found negative sound speeds in the expansion states for densities $\rho < 1.76$, the second example will use a quadratic expansion description along with a changed energy dependence by setting $B_1 = 0$ and

$$B_0 = \frac{1}{\rho_0}\left.\frac{\partial p}{\partial e}\right|_V \tag{4.167}$$

leading to the modified compression model

$$p = K_1\mu + K_2\mu^2 + K_3\mu^3 + B_0\,\rho_0\, e \tag{4.168}$$

To describe the isentropic expansion regime the quadratic formulation of the pressure-density relation given in (4.88) is now used. The results are shown in the graphs of Figure 4.32.

Mie-Grüneisen Equations of State

Constant Grüneisen Γ and Energy Independent Expansion

Applying the shock equation of state (4.77), we first use a constant Grüneisen parameter Γ to follow a most simple approach. The calculated internal energy and pressure values show, however, unphysical drops at density values of

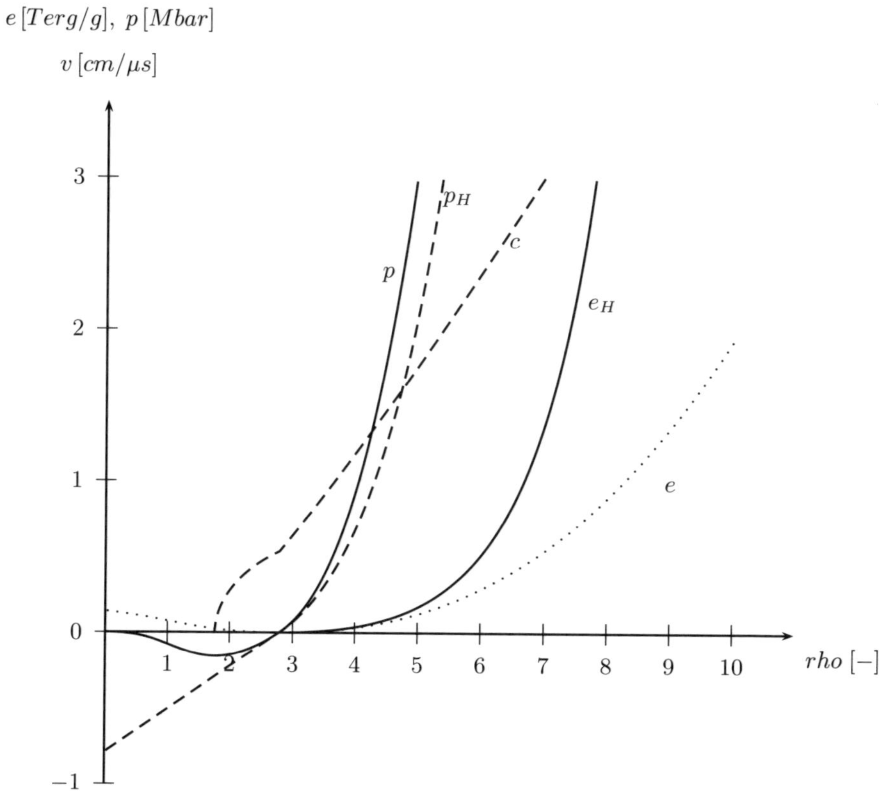

Fig. 4.31. Specific internal energy, pressure, Hugoniot states and sound speed for an isotropic change of density in aluminum through $e_0 = 0$ calculated from a polynomial equation of state with linear and cubic density dependency in expansion and compression, respectively, along with a pressure-energy relation that is constant in ρ. The material data used for the calculations are given in Table 4.2.

$\rho > 5.0$ and $\rho > 6.5$, respectively, which is a consequence of using the constant Grüneisen Γ along with the linear $v_S - v_1$ relation and the results from the third term in the Hugoniot energy (4.79).

The calculated sound speed is constant in expansion since it is only depending on the linear expansion pressure (see Figure 4.33).

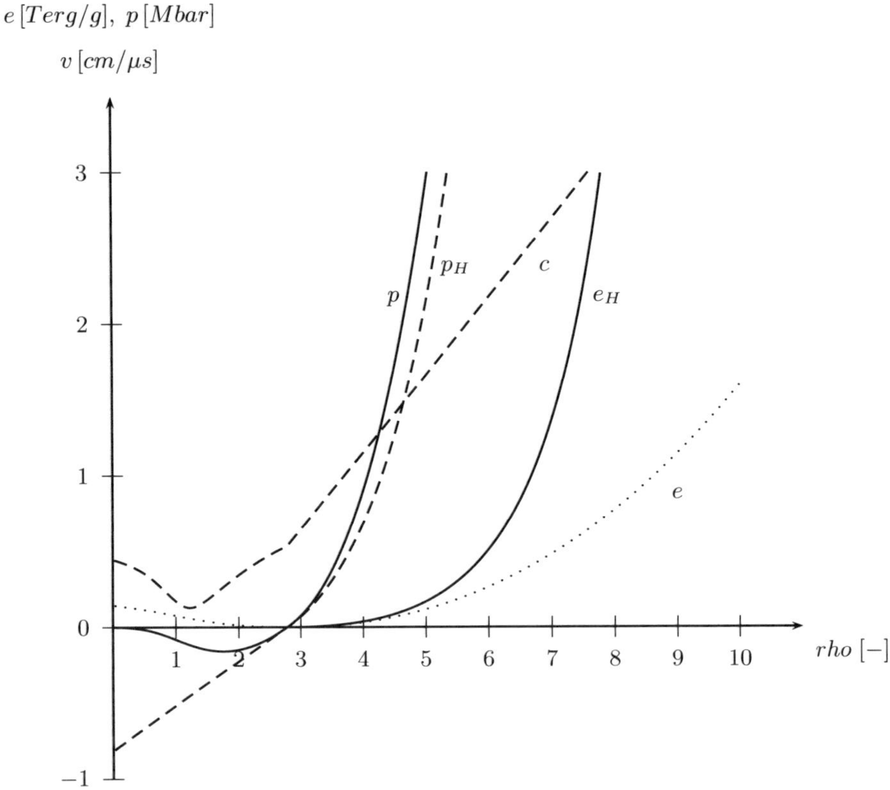

Fig. 4.32. Specific internal energy, pressure, Hugoniot states and sound speed for an isotropic change of density in aluminum through $e_0 = 0$ calculated from a polynomial equation of state with quadratic and cubic density dependency in expansion and compression, respectively, along with a linear pressure energy relation. The material data used for the calculations are given in Table 4.2.

Variable Grüneisen Γ and Energy Dependent Expansion

A variable Grüneisen parameter $\Gamma = \Gamma_0 \, \rho_0 / \rho$ is used in the final formulation. This leads to a physical representation of the pressure and internal energy towards high compressions. The calculated sound speed in compression shows a smaller slope compared to the ones obtained with the constant Γ (see Figure 4.34).

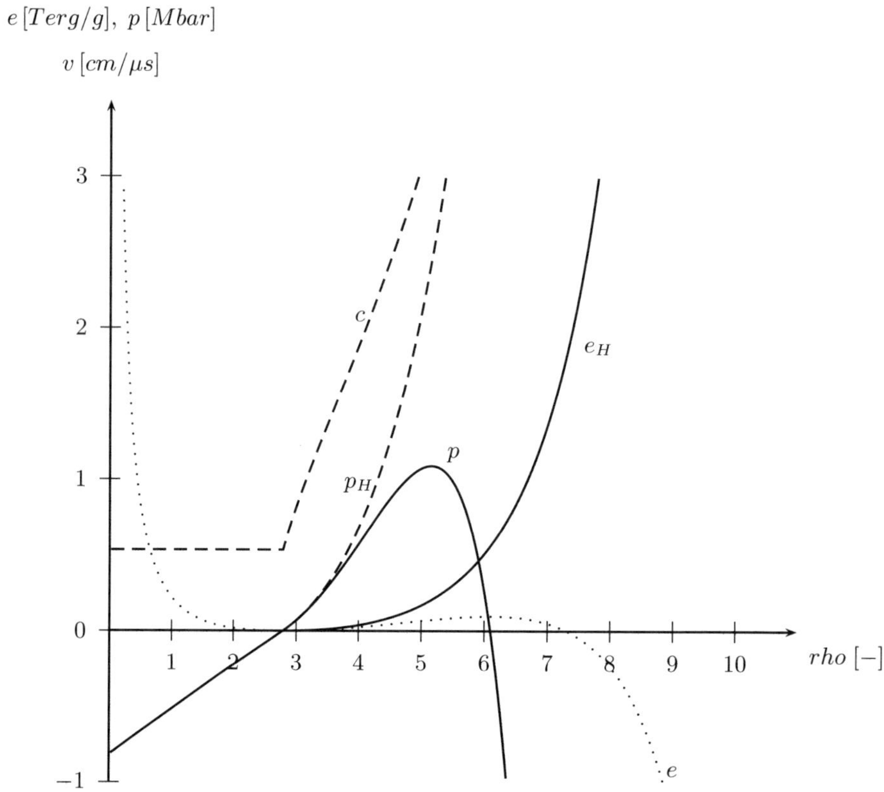

Fig. 4.33. Specific internal energy, pressure, Hugoniot states and sound speed for an isotropic change of density in aluminum through $e_0 = 0$ calculated from a Mie-Grüneisen equation of state with a constant $\Gamma = \Gamma_0$. The material data used for the calculations are given in Table 4.2.

Furthermore, the expansion regime is now formulated with an energy dependence of the sound speed in the last term of equation (4.163) according to:

$$\left.\frac{\partial p}{\partial e}\right|_p = \Gamma \rho \tag{4.169}$$

As a result, the sound speeds calculated in expansion show a similar threshold as seen before in case of the polynomial EOS with linear expansion formulation. This time the critical densities returning negative sound speed is at $\rho = 2.05$. In many codes this energy dependence of the sound speed, in expansion, is not used.

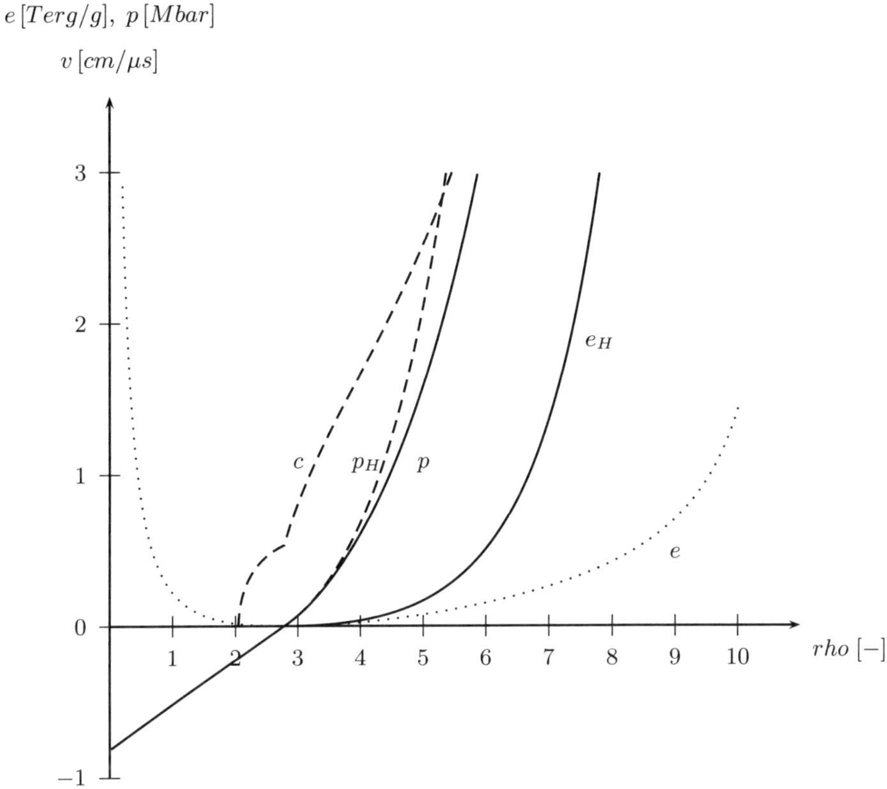

Fig. 4.34. Specific internal energy, pressure, Hugoniot states and sound speed for an isotropic change of density in aluminum through $e_0 = 0$ calculated from a Mie-Grüneisen equation of state with energy dependent sound speed in expansion and a variable Grüneisen $\Gamma = \Gamma_0\,\rho_0/\rho$. The material data used for the calculations are given in Table 4.2.

5

Hydrocodes

Numerical solutions for the time dependent formulations derived in the previous chapters are obtained via discretizations in time and space. A compilation of the manifold ways to discretize the basic equations will be given in the course of this chapter. Specifically for readers with background concerning numerical methods for static solutions, both equilibrium and time dependent formulations with their related solutions will be provided.

As mentioned in the introduction, particular emphasise is put on an adequate choice of discretization techniques for a particular material state. Specifically deformation processes can lead to fundamental changes in the properties of materials. Therefore, adequate choice in this context can mean representation of failure, crack opening, multiple fragmentation and phase changes. Examples for related processes and numerical solutions will be provided at the end of the chapter.

5.1 Modelling of Dynamic Deformation Processes

Crash and impact processes are dominated by transient stress and strain states. Typical duration for the related deformation processes to take place range between milliseconds in impact scenarios and several seconds, e.g. for an automotive structure to deform under crash loads. Depending on the amplitude of the wave effects, a more or less detailed representation of their attenuation is necessary. Low amplitude acoustic waves and strain rates up to several 10^2 $[s^{-1}]$ demand for the consideration of strain rate sensitivity in the materials. The propagation of waves is the observable phenomenon leading to acceleration of matter and, after several reflections, to a specific deformation state. Towards higher strain rates the evolution and propagation of shock waves becomes relevant and a representation of nonlinear thermo-mechanical compressive behavior in terms of equations of state, as discussed in the last chapter, needs to be required. A single wave may now cause failure in structural regions far remote to the position of load initiation.

The strain rate dependent finite deformation of structures is described by the kinematic and constitutive equations discussed before. A solution of the related partial differential equations with the aim of investigating dynamic deformation and energy dissipation needs to include a time resolved description of the process including wave propagation effects. Thus, these equations are to be solved and hence discretized both in space and time. Since analytical or closed form solutions for the complex processes are not at hand, numerical methods have been and are being developed to find approximative solutions. Part of the approximation and core philosophy of numerical methods is the so-called *discretization* of the governing equations, i.e. their selective solution at a finite number of spatial locations and instants of time within the investigated domain. From the particular solutions at discrete locations a subsequent overall continuous solution is reconstructed.

Spatial discretizations of the basic equations may be achieved by various kinds of finite methods to be described in the following chapters. Finite elements (FEM), finite differences (FDM), finite volumes (FVM) or mesh-free methods (MFM) are general categories of numerical methods developed to several different sub-branches for specific applications each. Formulated in Lagrangean or Eulerian kinematics the resulting individual methods are often specifically derived for certain structural components or loading conditions. Examples are finite elements which can be formulated as generally as a numerical method can be or as specific as e.g. plate or shell elements for thin walled structures with two-dimensional stress states.

Concerning *time discretization*, explicit and implicit schemes based on finite difference approximations exist to account for the time dependence of the basic equations. Whereas the explicit formulations are of less computational costs compared to implicit ones, their stability and precision is limited by the time step size. Implicit methods are capable of larger time steps at the same or higher accuracy and its precision can easily be controlled. However, for most dynamic processes under crash or impact conditions an explicit integration scheme is still preferable since a resolution of wave propagation effects demands for extremely short time steps in the order of micro- or nano-seconds which is, thus, eliminating the advantage of implicit methods.

It is the aim of this chapter to give an introduction to the varying schemes for spatial and time discretization, to distinguish their basic approaches and the therefrom resulting advantages as well as limitations. An important aspect derived from experience with the various methods and their application should be emphasised in this book. That is the additional benefit which can be gained from the combined application of different numerical methods. Be it in the case of parallel solutions in two or more physical domains, e.g. for the simulation of fluid-structure-interaction processes where the different regimes should be discretized by Eulerian and Lagrangean grids, respectively. Or maybe when fundamental changes of physical properties, if not even of phases, occur in materials while undergoing loading conditions typical for crash and impact processes. In his area of influence, the author always supported and promoted the combined use and adaptive change of discretizations, where ever useful. Therefore, that idea will also be illustrated in the course of this chapter, and specifically in section 5.9.

5.2 Components of a Hydrocode

Hydrocodes, also called *wave-propagation-codes*, are the typical class of numerical tool for the simulation of crash and impact and at the same time not linked to a specific kind of discretization. Developed in the early 1950's to simulate the physical effects of nuclear weapons, a fluid dynamic approach solving the conservation equations for mass, momentum and energy was chosen. Landmarks in hydrocode development have been set by the Los Alamos National Laboratory (LANL) and the Lawrence Livermore National Laboratory (LLNL). Nowadays available commercial codes for crash and impact simulation all have at least parts of their roots in these codes. The primary application at the time first hydrocodes were developed did not demand for a constitutive equation of shear stresses nor for related failure criteria since the prevailing hydrostatic pressures exceeded the shear strengths by orders of magnitude. Only for later applications for detonation and impact induced shock wave simulations in fluids and structures including structural deformations, deviatoric stress components have been implemented. Over decades models for more and more physical phenomena found implementation into hydrocodes. Therefore, by the nature of their applications and their origins, hydrocodes are also an ideal platform for coupled and adaptive discretizations. Typical representatives of hydrocodes, to name but a few, are ABAQUS, AUTODYN, CTH, DYTRAN, EPIC, HEMP, HULL, LS-DYNA, OURANOS, PAM-SHOCK and RADIOS.

Characteristic elements employed in a hydrocodes are:

- Solution of the conservation equations for mass, momentum and energy.
- Decoupled treatment of the stress tensor in terms of deviatoric and hydrostatic components.
- Formulation of a nonlinear equation of state accounting for shock wave formation and propagation.

- Constitutive equations for elastic and inelastic, rate-dependent material behavior including damage, failure and post-failure behavior.
- Arbitrary spatial and explicit time integration.
- A numerical methodology to capture shock waves, e.g. artificial viscosity or Godunov methods.

5.2.1 Marching Solutions in Time Steps

Marching solutions in time, i.e. stepping forward along the discrete instants in time at which solutions are provided, solve the set of equations at each time step in a specific order. Typically, solutions of the time dependent equation in hydrocodes are organized in the following, or a similar, manner:

A Define the initial conditions for the whole system at a start time $t = t_0$.
B Evaluate the maximum size for a stable time step, i.e. without loosing information or over-predicting propagation speeds.
C Solve the set of discretized equations according to a procedure equal or similar to what is illustrated in Figure 5.1.
D Use the results of **C** to provide new initial conditions for the next time step and continue with **B** until a predefined end-time is reached.

Often cited publications related to hydrocodes and related solution strategies are Wilkins [423], Wilkins [424], Noh [297] and Benson [53]

Before we actually start to collect spatial and time discretization techniques, some general mathematical aspects of partial differential equations concerning their resolvability with numerical tools together with the stability of the solutions need to be addressed. The mathematical character of equations usually applied in computational fluid dynamics, structural dynamics and thermo-mechanics will be discussed. This will lead to basic guidelines to choose finite methods for the solution of the governing partial differential equations in crash and impact simulations.

5.3 Classification of Partial Differential Equations

Linear and quasilinear partial differential equations describe a variety of physical processes. In the course of this evaluation, only linear second order partial differential equations will be analysed. From a physical point of view, the equations we are interested in describe

- either steady state, i.e. time independent situations with static equilibrium,
- or time dependent oscillations and related propagation of waves in fluids and structures,
- or, again time dependent, propagation processes of certain properties including their diffusion.

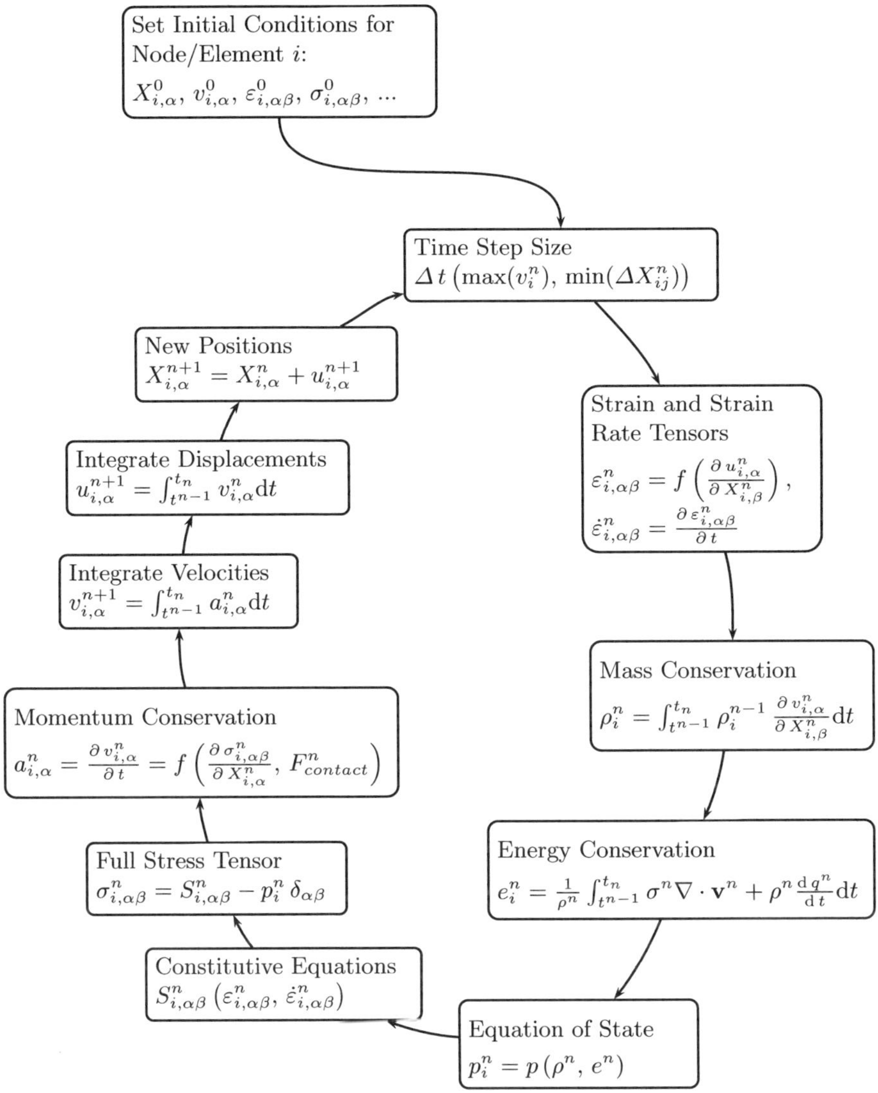

Fig. 5.1. Consecutive solution of equations as to be solved at each discretizing entity i within each time step .

Any partial differential equation has unique solutions in certain areas and non-unique or no solution at all in others. A mathematical approach to identify and separate these areas from each other is the *method of characteristics.* A characteristic represents a border line along which the domain of unique solutions for the independent variables is limited. The meaning and use of characteristics shall be explained with the following linear second order partial differential equation:

$$a\frac{\partial^2 u}{\partial x^2} + 2b\frac{\partial^2 u}{\partial x \partial y} + c\frac{\partial^2 u}{\partial y^2} = 0 \tag{5.1}$$

First, the transformations

$$p = \frac{\partial u}{\partial x} \qquad q = \frac{\partial u}{\partial y} \qquad r = \frac{\partial^2 u}{\partial x^2} \qquad s = \frac{\partial^2 u}{\partial x \partial y} \qquad t = \frac{\partial^2 u}{\partial y^2}$$

shall be used to shape equation (5.1) into a system of first order equations:

$$\begin{bmatrix} a & 2b & c \\ \mathrm{d}x & \mathrm{d}y & 0 \\ 0 & \mathrm{d}x & \mathrm{d}y \end{bmatrix} \begin{bmatrix} r \\ s \\ t \end{bmatrix} = [A]\,[u] = \begin{bmatrix} 0 \\ \mathrm{d}p \\ \mathrm{d}q \end{bmatrix} \tag{5.2}$$

Unique solutions for the system (5.2) exist only if its characteristic equation, represented by the determinant $|A|$, is non-zero:

$$|A| = \begin{vmatrix} a & 2b & c \\ \mathrm{d}x & \mathrm{d}y & 0 \\ 0 & \mathrm{d}x & \mathrm{d}y \end{vmatrix} = a\,\mathrm{d}y^2 - 2b\,\mathrm{d}x\,\mathrm{d}y + c\,\mathrm{d}x^2 \neq 0 \tag{5.3}$$

Characteristics, on the other hand, are defined as locations with non-unique solutions, in other words with determinants $|A|$ equal zero. Thus, characteristics for which

$$|A| = a\,\mathrm{d}y^2 - 2b\,\mathrm{d}x\,\mathrm{d}y + c\,\mathrm{d}x^2 = 0 \tag{5.4}$$

assign the particular directions in the $x - y-$plane along which the partial derivatives are non-unique. Devision by $\mathrm{d}x^2$:

$$|A| = a\frac{\mathrm{d}y^2}{\mathrm{d}x^2} - 2b\frac{\mathrm{d}y}{\mathrm{d}x} + c = 0 \tag{5.5}$$

leads to the wanted directions $\frac{\mathrm{d}y}{\mathrm{d}x}$ as eigenvalues of $|A|$

$$\frac{\mathrm{d}y}{\mathrm{d}x} = \frac{2b \pm \sqrt{4b^2 - 4ac}}{2a} = \frac{b \pm \sqrt{b^2 - ac}}{a} = 0 \tag{5.6}$$

Thus, the discriminant $D = b^2 - ac$ classifies the quality of the solution:

a) For $b^2 - ac > 0$ the characteristics represent two real roots. Two characteristics are running through each point x, y. Equation (5.1) is then called *hyperbolic* .

b) For $b^2 - ac = 0$ the two real roots coincide into a single real one and (5.1) is called *parabolic.*

c) For $b^2 - ac < 0$ there are no real roots. Equation (5.1) is now denoted as *elliptic.*

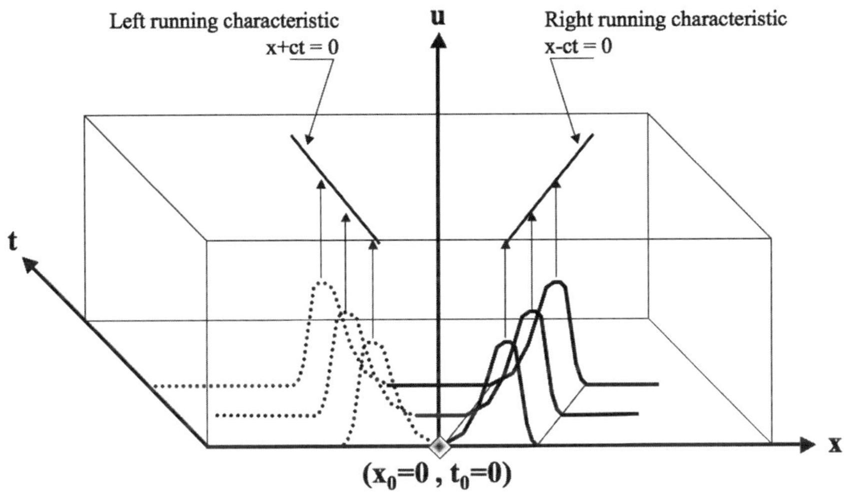

Fig. 5.2. Physical interpretation of the real characteristics of hyperbolic equations: connecting lines between points of constant amplitude of propagating waves.

A physical interpretation of the two real characteristics for hyperbolic equations is shown in Fig. 5.2. Originating from an initiation point x_0, t_0 in a one dimensional description, two waves run into positive an negative x-direction, respectively. In a three-dimensional representation of the wave amplitude over the propagation direction x and time t, the characteristics are lines connecting propagating wave points of constant amplitude.

Each linear second order partial differential equation with constant coefficients in two dimensions can be transformed into a normal form that allows for assigning its type. Accordingly, the normal form or an elliptic equation is:

$$\frac{\partial^2 u}{\partial \eta^2} + \frac{\partial^2 u}{\partial \zeta^2} = f\left(u, \eta, \zeta, \frac{\partial u}{\partial \eta}, \frac{\partial u}{\partial \zeta}\right) \tag{5.7}$$

for a hyperbolic it is:

$$\frac{\partial^2 u}{\partial t^2} - \frac{\partial^2 u}{\partial \eta^2} = g\left(u, \eta, t, \frac{\partial u}{\partial \eta}, \frac{\partial u}{\partial t}\right) \tag{5.8}$$

and for parabolic equations:

$$\frac{\partial^2 u}{\partial \eta^2} = h\left(u, \eta, t, \frac{\partial u}{\partial \eta}, \frac{\partial u}{\partial t}\right) \tag{5.9}$$

Elliptic equations

Steady state problems of mechanical equilibrium, electrostatics or static magnetic problems, are described by *potential equations* of the general three-dimensional form:

$$\Delta u = \frac{\partial^2 u(x,y,z)}{\partial x^2} + \frac{\partial^2 u(x,y,z)}{\partial y^2} + \frac{\partial^2 u(x,y,z)}{\partial z^2} = a \tag{5.10}$$

Performing the classification of partial differential equations illustrated before shows that the potential equation (5.10) is of the elliptic type. The deflection u of a cable or a rope under constant distributed load cold be described by an equation of the type (5.10). In fluid dynamics u denotes a velocity potential. For $a = 0$ in vacuum conditions (5.10) is also called *Laplace equation*; otherwise, such as for $a = 4\pi\rho$ in the case of source terms of density ρ, it is referred to as *Poisson-equation.*
To solve elliptic equations, boundary conditions of the so called *Dirichlet* or of the *Neumann* type are needed along the boundaries ∂G of the investigated domain G:

- *Dirichlet boundary conditions*, also called *geometric conditions*, describe the value of $u(x,y,z)$ along the domain boundary ∂G:

$$u(x,y)|_{\partial G} = \alpha \tag{5.11}$$

- *Neumann* or *natural boundary conditions*, on the other hand, define normal components of the gradient of $u(x,y,z,)$:

$$\left.\frac{\partial u(x,y)}{\partial n}\right|_{\partial G} = \beta \tag{5.12}$$

- and *Robin* boundaries as combination of Dirichlet and Neumann conditions:

$$\left.\frac{\partial u(x,y)}{\partial n}\right|_{\partial G} + \phi\, u(x,y)|_{\partial G} = \gamma \tag{5.13}$$

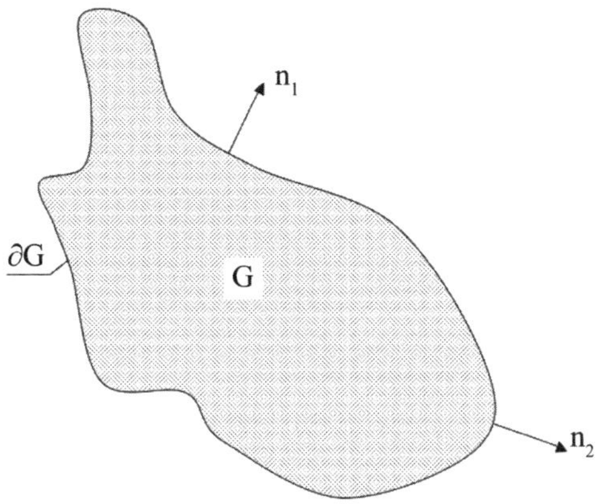

Fig. 5.3. Domain G with boundary ∂G and boundary normals n_i.

The resulting combination of an elliptic equation with Dirichlet, Neumann or Robin boundary conditions is called *boundary value problem* of first, second or third type. If the elliptic equation (5.10) is formulated for a problem in statics with deformations u_i and terms $\frac{\partial u_i}{\partial x_j}$ making up strain components, (5.11) represents displacement boundary conditions whereas Neumann boundaries (5.12) usually prescribe stress components normal to the boundary, e.g. as $\sigma_n = 0$ for free surfaces.

An important aspect for the numerical solution of the discretized equations is the so called *area of influence*. A perturbation, i.e. any event influencing the dependent variable u inside the discretized domain G, affects a certain limited or unlimited spatial region of its neighborhood. In the case of time dependent processes, the affected region may additionally depend on a time interval of influence. Steady state elliptic equations have no time dependence. Initiated at an arbitrary location, the area of influence of a perturbation in an elliptic problem is always identical with the whole domain. Thus, any change in u at any position on G affects the entire system. This is illustrated for a two-dimensional example by Fig. 5.4. The area of influence of a perturbation originating from position C is hatched. It is identical with the entire rectangular domain G defined by $\overline{x_1 x_2}$ and $\overline{y_1 y_2}$.

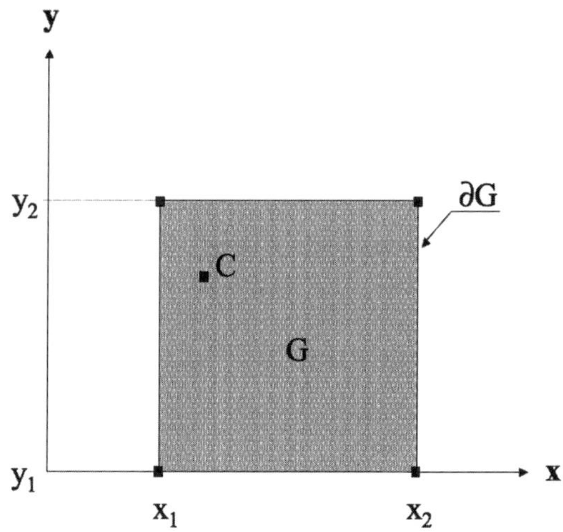

Fig. 5.4. Area of influence of a perturbation at point C for an elliptic problem defined in the domain G.

Hyperbolic equations

If the investigated process is in-stationary, the time dependence is covered by additional terms. Containing second derivatives with respect to time, oscillation and wave propagation processes are of the hyperbolic type. Accordingly, the *wave equation* yields:

$$\triangle u - \frac{1}{c^2}\frac{\partial^2 u}{\partial t^2} = 0 \tag{5.14}$$

Equation (5.14) accounts for acoustic problems with the sound speed c and electro-dynamic or optical problems with c representing the speed of light. A specific one-dimensional version of equation (5.14) describes the motion of a vibrating string:

$$\frac{\partial^2 u}{\partial x^2} - \frac{1}{c^2}\frac{\partial^2 u}{\partial t^2} = 0 \tag{5.15}$$

The propagation of characteristics of hyperbolic equations has already been illustrated in Fig. 5.2. Right and left running characteristics are assigned with $x - ct$ and $x + ct$, respectively. Drumheller [116] derives in detail the *d'Alembert solution* of the wave equation stating that any combination of smooth twice continuously differentiable functions $f(x + ct)$ and $g(x - ct)$ are solutions of the wave equation (5.14):

$$u = f(x + ct) + g(x - ct) \tag{5.16}$$

In hyperbolic problems, a perturbation reaches neighbouring positions in a certain distance Δx after a time interval which is defined by the propagation

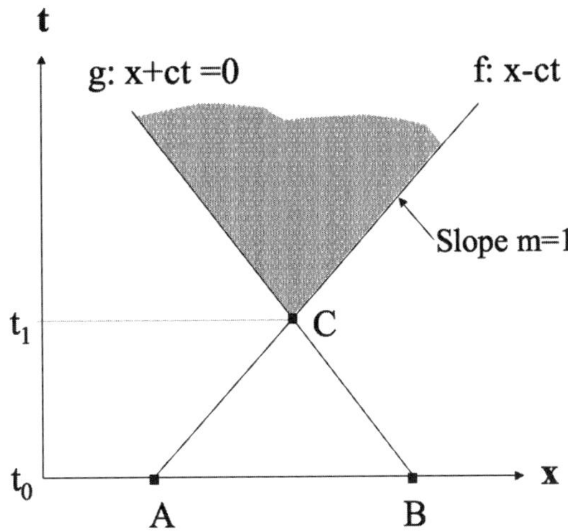

Fig. 5.5. Left and right running characteristics of a hyperbolic equation and area influenced by disturbances originating from a point C.

speed c. Thus, the right and left running characteristics identify the area of influence. The graphical illustration of that trivial definition in Fig. 5.5 shows its non-trivial impact on solution techniques: Within a time duration $\delta t = t_1 - t_0$ an arbitrary point C in space can only be influenced by disturbances originating from neighbouring positions inside the spatial interval $\overline{AB}$. Point C itself, in turn, is able to influence positions hatched in Fig. 5.5. This relation between propagation speed and resulting areas of influence has direct influence on both the stability and accuracy of finite methods introduced in later sections: given initial conditions at t_0, the limited areas of influence allow for a step-by-step marching solution of the equations along the time axis t. Thus, hyperbolic equations need initial conditions

$$u(t = 0) = u_0 \tag{5.17}$$

to be solved and a set of hyperbolic equations plus initial conditions on its domain G is called *initial value problems.*

In fluid dynamics, the conservation equations (2.216) and (2.217), respectively, are formulated for steady or unsteady, compressible or incompressible, viscous or inviscid flow. The resulting governing equations are of different type depending on whether the flow is subsonic or supersonic, as explained in detail by Anderson in Wendt [417]. The *unsteady inviscid compressible* formulations, however, are always of the hyperbolic type. Since they are identical to the formulation used for the simulation of dynamic structural deformation, the stepwise marching solution in time can be applied for our specific problems in computational structural dynamics.

Parabolic equations

An example for time dependent diffusive problems of the parabolic type is the *heat equation* that describes the propagation of temperature through a three dimensional space over time by:

$$\tau \triangle u - \frac{\partial u}{\partial t} = 0 \tag{5.18}$$

where u denotes the temperature and the material dependent thermal diffusivity τ (or temperature conductivity)

$$\tau = \frac{\lambda}{c_P \rho} \tag{5.19}$$

consists of the heat conductivity λ, the mass density ρ and the specific heat capacity at constant pressure c_P. To solve parabolic equations, both initial conditions of the type (5.17) and boundary conditions of either Dirichlet, Neumann or Robin type are needed. Consequently, the problem type is called *initial boundary value problem* of first, second or third type, respectively.

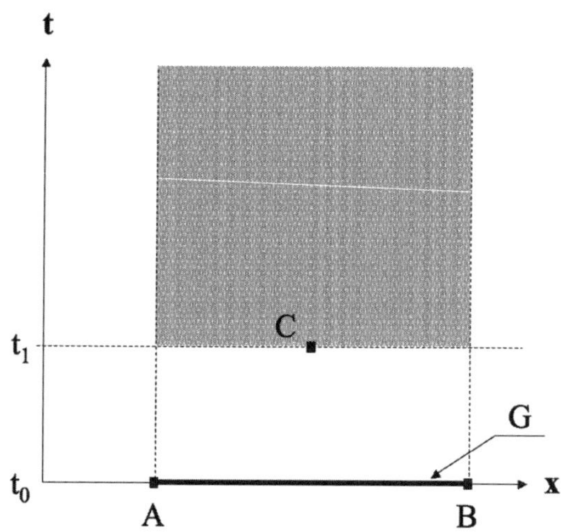

Fig. 5.6. Area of influence over time for a perturbation at point C in a spatially one-dimensional parabolic problem defined in the domain G.

Transport of a contamination of concentration c in a flow environment with the velocity u is another process described by a parabolic equation, the so called *transport equation:*

$$\frac{\partial c}{\partial t} + u \, \nabla c = 0 \tag{5.20}$$

If additional diffusion processes shall be modelled through a diffusion coefficient ϕ, the *convection-diffusion-equation* is used:

$$\frac{\partial c}{\partial t} - \phi \, \triangle c + u \, \nabla c = 0 \tag{5.21}$$

Since in usual applications of crash and impact simulation the propagation of diffusive terms like heat or contamination is not regarded, parabolic equations are of limited importance in that context and will not be investigated further in the course of this book.

5.4 Discretization - The Basic Idea

A repeatedly used term in this chapter will be *discretization.* Its meaning in the context of numerical approximation tools is that equations, formulated to continuously describe a function or functional in space and time, is solved only at certain *discrete locations and instants of time.* Consequently, the solution or an approximation of it, is only known at a *finite* number of discretizing positions which, at the same time, are the nodes forming a grid that represents the structural shape and the timely ladder along which stepwise solutions are calculated. Thats where the names *Finite Differences*, *Finite Elements* or generally *Finite Methods* come from. Some methods, like the Finite Differences, regard exclusively the nodes calculating functions or derivatives there. The Finite Element Method, in contrast, provides solutions for field variables inside the formed cells, here called elements, of the grid as well as solutions for other variables at the nodes' locations.

An introductory example to the fundamental idea of discretizing equations may be given with the numerical integration of a function $f(x)$ between two points A and B. If the analytical solution of building the anti-derivative and inserting the boundary values is not possible or too complex, a numerical approximation can be found through the partitioning of the interesting domain $\overline{AB}$ into a finite number of segments. In the example illustrated in Fig. 5.8, two segments of length h are used with central points x_1 and x_2 at which the functional values y_1 and y_2 are evaluated. The sum over the central values y_i in each segment multiplied by the distances h between the integrating nodes i gives an approximate solution for the integral:

$$\langle \int_A^B f(x) \, \mathrm{d}x \, \rangle = \sum_i f(x_i) h \tag{5.22}$$

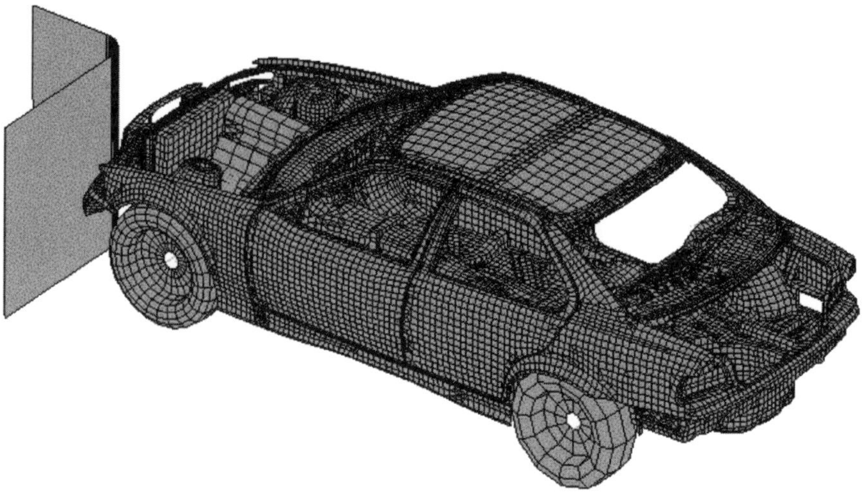

Fig. 5.7. Finite Element discretization of a car to simulate the structural deformations under offset crash.

Each segment could also be called cell or element. The individual products $f(x_i)h$ are local integrals of $f(x)$ in the corresponding cell and the sum (5.22) delivers the solution for the whole domain $\overline{AB}$.

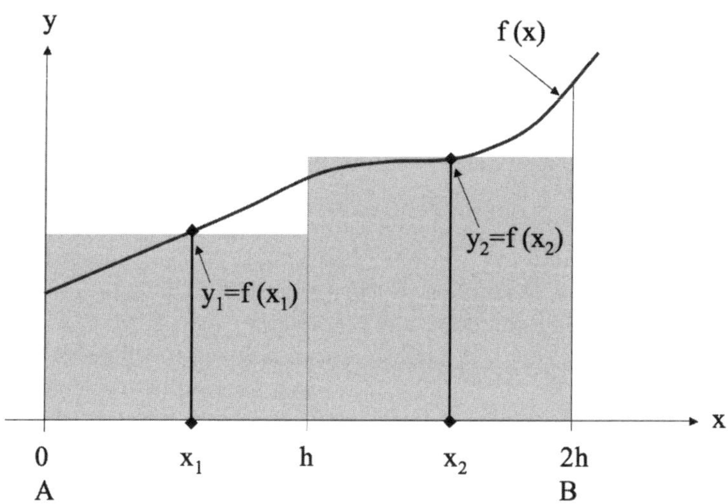

Fig. 5.8. Approximation for the definite integral $\int_A^B f(x)\mathrm{d}x$ using a discretization for the numerical integration.

Depending on the kind of formulation chosen for the kinematics of an investigated continuum (see chapter 2.1.1), two major branches of spatial discretizations can be identified. Application of material coordinates leads to a *body fixed grid*, i.e. the discretizing nodes move according the calculated deformations. Spatial coordinates, on the other hand, keep the nodes fixed in space and observe any material motion or deformation as flux through the cell walls. Mathematically, the existence of flux means additional convective terms in the substantial derivative (2.63). Therefore, the numerical code gets more complex and the computational time to solve the equations increase. Furthermore, at the material side, information on loading history and related accumulation variables are harder to track in Eulerian grids. Also, interfaces, boundaries and complex geometries are easier to model with Lagrangean nodes. Finally, a referential discretization does only need to describe the actual domain, e.g. the structure, whereas a spatial grid must be expanded to all potential positions in space possibly occupied by the structure over the time of the calculation. These are the disadvantages of Eulerian descriptions leading to a preference of Lagrangean coordinates for structural codes. An advantage, on the other hand, is found in the ability of Eulerian grids to describe large deformations without getting into problems of grid entanglement that leads to severe problems in Lagrangean schemes. Combinations of both basic ways of discretization are more and more used to account for interaction problems, like fluid-structure-interaction, where the best of both worlds is used to achieve an optimum total solution.

5.5 Finite Difference Methods

A comparably simple tool for the numerical approximation of ordinary or partial differential equations at discrete locations are the so called *Finite Difference Schemes*. And so is the basic idea which proposes to replace a *differential quotient* by a *difference quotient* which in turn can be evaluated at the discretizing positions. In combination with boundary conditions, an equation or a system of equations can be developed that allows for a derivation of functional values and derivatives at these positions in a system.

Finite Differences for Ordinary Differential Equations

Differential quotients in ordinary differential equations represent the slope in a functional relation between two variables. If we examine a differential quotient $\frac{\mathrm{d}y}{\mathrm{d}x}$ and discretize it in a similar way as illustrated before in Fig. 5.8 it can be understood as the limit value

$$\frac{\mathrm{d}y}{\mathrm{d}x} = \lim_{\Delta x \to 0} \frac{\Delta y}{\Delta x} = \lim_{h \to 0} \frac{y(x+h) - y(x)}{h} \tag{5.23}$$

A discretization like the one shown in Fig. 5.9 allows for three formulations of first order difference quotients approximating the differential quotient (5.23):

a) Forward differences

$$\left.\frac{\mathrm{d}y}{\mathrm{d}x}\right|_i \approx \left.\frac{\Delta y}{\Delta x}\right|_i = \frac{y_{i+1} - y_i}{h} \tag{5.24}$$

b) Backward differences

$$\left.\frac{\mathrm{d}y}{\mathrm{d}x}\right|_i \approx \left.\frac{\Delta y}{\Delta x}\right|_i = \frac{y_i - y_{i-1}}{h} \tag{5.25}$$

c) Central differences

$$\left.\frac{\mathrm{d}y}{\mathrm{d}x}\right|_i \approx \left.\frac{\Delta y}{\Delta x}\right|_i = \frac{y_{i+1} - y_{i-1}}{2\,h} \tag{5.26}$$

Further derivatives can be approximated in an analogous manner, e.g. applying central differences to obtain the second derivative:

$$\left.\frac{\mathrm{d}^2 y}{\mathrm{d}x^2}\right|_i \approx \left.\frac{\Delta y}{\Delta x}\right|_i = \left.\frac{\Delta^2 y}{\Delta x^2}\right|_i = \frac{\frac{y_{i+1}-y_i}{h} - \frac{y_i - y_{i-1}}{h}}{h} \tag{5.27}$$

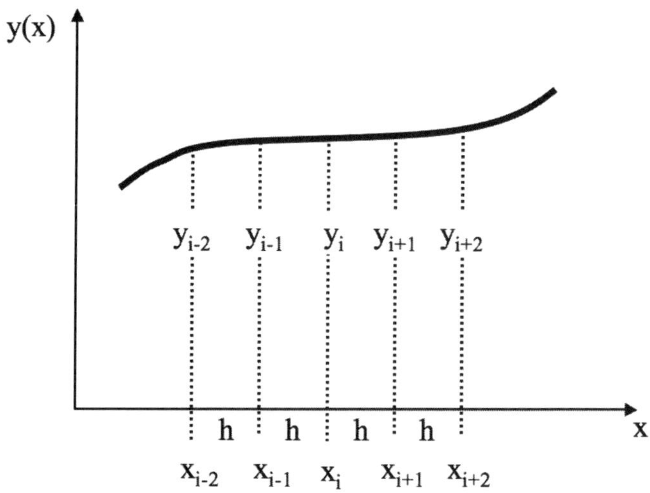

Fig. 5.9. One-dimensional discretization to obtain difference quotients for approximations of differential quotients.

Finite Differences for Partial Differential Equations

The governing equations to describe dynamic deformations of structures, e.g. the conservation equations derived from infinitesimal material elements (2.182), (2.191) and (2.207), are partial differential equations. To obtain solutions for these equations, finite difference schemes can be used to approximate the partial derivatives. A two-dimensional discretizing grid, as for example displayed in Fig. 5.10, shall be used to formulate the corresponding difference quotients. To approximate the value of a function $u_{i+1,j}(x,y)$ at a point $(i+1,j)$, a Taylor series shall be expanded for this variable about the neighbouring point (i,j):

$$u_{i+1,j} = u_{i,j} + \left.\frac{\partial u}{\partial x}\right|_{i,j} h + \left.\frac{\partial^2 u}{\partial x^2}\right|_{i,j} \frac{h^2}{2} + \left.\frac{\partial^3 u}{\partial x^3}\right|_{i,j} \frac{h^3}{6} + \ldots \tag{5.28}$$

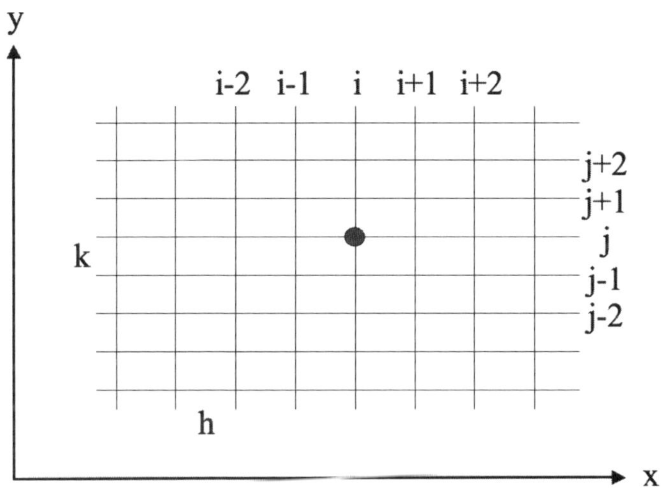

Fig. 5.10. Discretizing grid points (i,j) in the two-dimensional x-y-plane with step sizes h and k, respectively.

Given a convergence of the the step size $h \to 0$, equation (5.28) is exact. That would mean an infinite number of discretizing nodes. Since this is not practically achievable, the Taylor series needs to be truncated after a certain term. A truncation of terms with magnitudes h^{n+1} and higher order results in an approximation of so called $n^t h$ *order accuracy*. In case of equation (5.28) a first order approximation would be

$$u_{i+1,j} = u_{i,j} + \left.\frac{\partial u}{\partial x}\right|_{i,j} h + O(h^2) \quad , \tag{5.29}$$

where $O(h^2)$ indicates that contributions with h-terms of second and higher order are truncated. Equation 5.29 can be rearranged to obtain an expression for $\left.\frac{\partial u}{\partial x}\right|_{i,j}$:

$$\left.\frac{\partial u}{\partial x}\right|_{i,j} = \frac{u_{i+1,j} - u_{i,j}}{h} - O(h) \tag{5.30}$$

which, in turn, is a first order forward difference expression for the partial derivative $\left.\frac{\partial u}{\partial x}\right|_{i,j}$ and, thus, equivalent to (5.24).
Expansion of a Taylor series at position $(i-1,j)$, instead, and again about (i,j)

$$u_{i-1,j} = u_{i,j} + \left.\frac{\partial u}{\partial x}\right|_{i,j} h + \left.\frac{\partial^2 u}{\partial x^2}\right|_{i,j} \frac{h^2}{2} + \left.\frac{\partial^3 u}{\partial x^3}\right|_{i,j} \frac{h^3}{6} + ... \tag{5.31}$$

leads to the first order backward difference formulation

$$\left.\frac{\partial u}{\partial x}\right|_{i,j} = \frac{u_{i,j} - u_{i-1,j}}{h} - O(h) \tag{5.32}$$

A central difference formulation is achieved if the backward scheme 5.32 is subtracted from the forward differences 5.30 to get:

$$\left.\frac{\partial u}{\partial x}\right|_{i,j} = \frac{u_{i+1,j} - u_{i-1,j}}{2h} - O(h^2) \quad . \tag{5.33}$$

Compared to the first order forward and backward differences, the central scheme 5.33 is of second order accuracy as indicated by the symbol $O(h^2)$. Of the same order of accuracy in two dimensions is the following central second difference approximation for a mixed derivative:

$$\left.\frac{\partial^2 u}{\partial x\,\partial y}\right|_{i,j} = \frac{u_{i+1,j+1} + u_{i-1,j-1} - u_{i+1,j-1} - u_{i-1,j+1}}{4hk} - O(h^2, k^2) \quad . \tag{5.34}$$

More detailed introductions to the basics of FDM can be found with Anderson [9], Hirsch [187], Mitchell [281] or Strikwerda [376].

5.5.1 Time Integration with Finite Difference Schemes

As we have seen so far, finite differences are a numerical method to solve partial differential equations by the transformation of differential terms into difference quotients. The basic equations that we intend to solve contain derivatives with respect to spatial coordinates and with respect to time, e.g. the conservation equations for mass, momentum and energy (2.217) neglecting source terms:

$$\frac{\partial \mathbf{u}}{\partial t} + \nabla \psi = 0 \tag{5.35}$$

Similar to the spatial discretization, though independent to the actual choice of the numerical method in space, time can be discretized via finite differences to solve non steady equations like (5.35). With initial conditions defined on the whole spatial domain at time t_0, a marching solution shall be formulated with an arbitrary but finite number n_t of solution steps in time. Application of finite differences to the first term in (5.35) delivers particular solutions at discrete times $t_0 + i\,\Delta t$, where $i = 1, 2, ..., n_t$ is the number of the time step. To distinguish discrete points in space from discrete times the lower right index will be kept for spatial positions in the discretizing grid and the upper right index will indicate the position in time. Thus, $u_{i,j,k}^{n+1}$ identifies the value of variable $u(x_{i,j,k}, t^{n+1})$ at a spatial grid position i, j, k and at time interval $n + 1$.

First, we consider only the first term of (5.35): $\frac{\partial \mathbf{u}}{\partial t}$. At cycle n, the forward differences (5.30) with a time step size Δt yield

$$\left.\frac{\partial u}{\partial t}\right|_i^n = \frac{u_i^{n+1} - u_i^n}{\Delta t} - O(\Delta t) \tag{5.36}$$

and with the isolated term u_i^{n+1} a solution for the next time step:

$$u_i^{n+1} = u_i^n + \Delta t \left.\frac{\partial u}{\partial t}\right|_i^n + O(\Delta t) \tag{5.37}$$

that demonstrates the *explicit* character of forward differences in the context of time integration: all terms at the right hand side of (5.37) are known and, thus, allow for a direct calculation of the new values in the next time step. The situation is different if we use the backward differences (5.32) :

$$\left.\frac{\partial u}{\partial t}\right|_i^{n+1} = \frac{u_i^{n+1} - u_i^n}{\Delta t} - O(\Delta t) \tag{5.38}$$

leading to

$$u_i^{n+1} = u_i^n + \Delta t \left.\frac{\partial u}{\partial t}\right|_i^{n+1} + O(\Delta t) \quad . \tag{5.39}$$

The fact that now the time derivative at the right hand side is taken at time step $n + 1$ makes equation (5.39) *implicit*, i.e. iterative methods are needed to solve it. Consequences and limitations that result from choosing either the explicit or the implicit solution will be discussed in chapter 5.5.2.

Since we are interested in a solution of the whole equation (5.35), a combined solution for the time dependent and the spatial term is needed. Of course, the already derives difference schemes can be used also for the spatial term. Various options for the actual formulation arise when a time step must be assigned with the spatial terms. That choice has impact on the accuracy of the approximation as well as on whether the total discretized is of explicit or implicit character, as the following examples will show.

Single Step Schemes

To start with, finite difference formulations will be chosen that build differences between direct neighbours in both the spatial and the time discretization. A first example is formulated with forward differences for the time and backward differences for the space, i.e. one step schemes in both domains. For a one-dimensional version of (5.35), that results in the so called *Upwind* or *Forward Euler method*:

$$\frac{u_i^{n+1} - u_i^n}{\Delta t} + O(\Delta t) + \frac{\psi_i^n - \psi_{i-1}^n}{\Delta x} + O(\Delta x) = 0 \tag{5.40}$$

For the formulation of the spatial derivative in (5.40) all ψ-terms were taken at time step n, i.e. the current ones which are all known. As shown before the character of the scheme (5.40) is therefore explicit. For any explicit scheme, Courant, Friedrichs and Lewy [98] have shown that it can only be stable if the width of the time step Δt is limited by the ratio of maximum propagation speeds of disturbances, i.e. in our context wave and material speeds, and the minimum size of discretizing cells in space. For the maximum time step size that is stable the *Courant-Friedrichs-Lewy condition* (CFL) means that

$$\Delta t_{max} \leq \Delta t_{CFL} \sim \frac{v_{max}}{\Delta x_{min}} \tag{5.41}$$

The meaning of the CFL condition will be illustrated in chapter 5.5.2. For now we note that both high velocities (speeds of sound or motion) and small discretizing cells in space result in small time steps and, thus, increase the computational cost since the number of time steps needed to calculate a particular conditions at a specific time increases.

To establish a marching solution in time, values of the dependent variables at the new time step $n + t$ must be calculated which is achieved from (5.40) by isolating u_i^{n+1}:

$$u_i^{n+1} = u_i^n - \Delta t_{max} \frac{\psi_i^n - \psi_{i-1}^n}{\Delta x} \tag{5.42}$$

If the ψ-terms in the Upwind scheme (5.40) are evaluated at the new time step $n + 1$ an implicit version of the Euler method is formulated, the so called *Backward Euler* scheme:

$$\frac{u_i^{n+1} - u_i^n}{\Delta t} + O(\Delta t) + \frac{\psi_i^{n+1} - \psi_{i-1}^{n+1}}{\Delta x} + O(\Delta x) = 0 \tag{5.43}$$

In the *backward Euler* scheme, three terms are contained that are to be simultaneously solved for the next time step $n+1$. To solve the resulting system of equations an algorithm for tridiagonal matrices is needed.
In analogy to the explicit Upwind scheme (5.40) there exists an again explicit *Downwind scheme* taking the opposite neighbor in space to formulate the one-step spatial derivative, i.e. combining a forward difference scheme in time with a forward scheme in space:

$$\frac{u_i^{n+1} - u_i^n}{\Delta t} + O(\Delta t) + \frac{\psi_{i+1}^n - \psi_i^n}{\Delta x} + O(\Delta x) = 0 \quad . \tag{5.44}$$

Since (5.44) is also explicit, the CFL condition (5.41) applies again when u_i^{n+1} is to be calculated

$$u_i^{n+1} = u_i^n - \Delta t_{max} \frac{\psi_{i+1}^n - \psi_i^n}{\Delta x} \tag{5.45}$$

However, the downwind scheme (5.44) turns out to be unconditionally instable (see e.g. Hirsch [187]). In other words no matter how short the time step is chosen, it will always tend to become instable.

Multi Step Schemes

The same unconditional instability as in (5.44) would result if besides the forward scheme in time a centered difference scheme was chosen for the spatial component:

$$\frac{u_i^{n+1} - u_i^n}{\Delta t} + O(\Delta t) + \frac{\psi_{i+1}^n - \psi_{i-1}^n}{2\Delta x} + O(\Delta x^2) = 0 \quad . \tag{5.46}$$

Equation (5.46) is considered here as multi-step scheme since it contains both the right and the left neighbor in the centered spatial formulation. Its instability problem can be resolved if the spatial derivatives are calculated at the new time step $n+1$:

$$\frac{u_i^{n+1} - u_i^n}{\Delta t} + O(\Delta t) + \frac{\psi_{i+1}^{n+1} - \psi_{i-1}^{n+1}}{2\Delta x} + O(\Delta x^2) - 0 \quad . \tag{5.47}$$

leading to an implicit solution:

$$u_i^{n+1} = u_i^n - \Delta t_{max} \frac{\psi_{i+1}^{n+1} - \psi_{i-1}^{n+1}}{2\Delta x} \tag{5.48}$$

An alternative way to stabilize (5.46) is found if the centered differences for the spatial derivative are also centered in time leading to the so called *Crank-Nicolson scheme*:

$$\frac{u_i^{n+1} - u_i^n}{\Delta t} - \frac{\psi_{i+1}^n - 2\psi_i^n + \psi_{i-1}^n + \psi_{i+1}^{n+1} - 2\psi_i^{n+1} + \psi_{i-1}^{n+1}}{2\Delta x^2} = 0 \tag{5.49}$$

with its implicit solution:

$$u_i^{n+1} = u_i^n + \Delta t \frac{\psi_{i+1}^n - 2\psi_i^n + \psi_{i-1}^n + \psi_{i+1}^{n+1} - 2\psi_i^{n+1} + \psi_{i-1}^{n+1}}{2\Delta x^2} \quad . \tag{5.50}$$

Both the formulation of equation (5.47) and the Crank-Nicolson scheme are perfect examples on how influential the precise choice of locations in space and time is concerning the accuracy and stability of approximations.

An often used multi-step method that consists of centered differences for the time as well as centered differences for the space discretization is the *Leap frog scheme*:

$$\frac{u_i^{n+1} - u_i^{n-1}}{2\Delta t} + O(\Delta t^2) + \frac{\psi_{i+1}^n - \psi_{i-1}^n}{2\Delta x} + O(\Delta x^2) = 0 \quad . \tag{5.51}$$

with the corresponding explicit solution:

$$u_i^{n+1} = u_i^{n-1} - \Delta t_{max} \frac{\psi_{i+1}^n - \psi_{i-1}^n}{\Delta x} \tag{5.52}$$

A more complex explicit solution can be found in the so called *Lax-Wendroff scheme* also known as *predictor-corrector scheme*. The basic idea is to approximate the Taylor series developed around $u(x,t)$:

$$u(x, t + \Delta t) = u(x,t) + \Delta t \frac{\partial u}{\partial t}(x,t) + \frac{\Delta t^2}{2}\frac{\partial^2 u}{\partial t^2}(x,t) + O(\Delta t^3) \tag{5.53}$$

using the investigated equation (5.35)

$$\frac{\partial \mathbf{u}}{\partial t} = -\nabla \psi \tag{5.54}$$

for the time derivatives at the right hand side of (5.53):

$$u(x, t + \Delta t) = u(x,t) - \Delta t \frac{\partial \psi}{\partial x}(x,t) + \frac{\Delta t^2}{2}\frac{\partial^2 \psi}{\partial x^2}(x,t) + O(\Delta t^3) \tag{5.55}$$

which are finally approximated by centered differences to:

$$u_i^{n+1} = u_i^n - \Delta t \frac{\psi_{i+1}^n - \psi_{i-1}^n}{2\Delta x} + \Delta t^2 \frac{\psi_{i+1}^n - 2\,\psi_i^n + \psi_{i-1}^n}{2\,\Delta x^2} + O(\Delta t^2) + O(\Delta x^2) \tag{5.56}$$

The name *predictor-corrector* scheme is used for that method since an alternative derivation takes the time and spatial derivatives at the half time steps to identify a *predictor*:

$$\hat{u}_{i+1/2}^{n+1/2} = \frac{1}{2}\left(\psi_i^n + \psi_{i+1}^n\right) - \Delta t \left(\frac{\psi_{i+1}^n - \psi_i^n}{2\,\Delta x}\right) \tag{5.57}$$

and a subsequent *corrector* step:

$$u_i^{n+1} = u_i^n - \Delta t \left(\frac{\hat{u}_{i+1/2}^{n+1/2} - \hat{u}_{i-1/2}^{n+1/2}}{\Delta x}\right) \tag{5.58}$$

5.5.2 Explicit or Implicit Time Integration Schemes?

Basic aspects of both implicit an explicit schemes have become evident in the course of the derivations in the last chapter. Accuracy as well as stability of the resulting numerical integration depends on the choice of discretizing positions in time and space. The unconditional stability of implicit methods is an advantage over the conditional stability of explicit ones. The CFL condition (5.41) for stable schemes limits the time step length. That limit is directly linked to the areas of influence discussed earlier to classify partial equations in chapter 5.5.1.

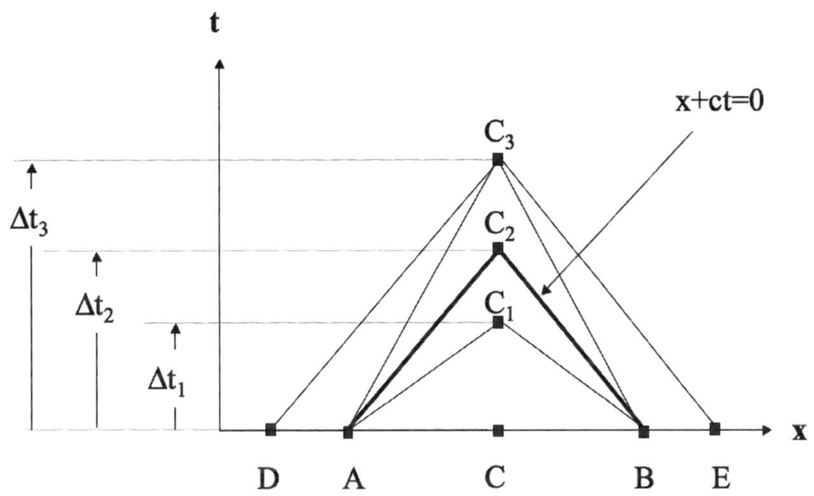

Fig. 5.11. Illustration of the CFL condition and the relation to the regions of influence.

Figure 5.11 illustrates that link: The scheme under consideration is one dimensional in space. Solutions are of interest a location (C) at a future time $t + \Delta t$ by a spatial discretization that takes into regard the neighbors (A) and (B). Right and left running characteristics originating at (A) and (B), respectively, coincide at (C_2), i.e. position of (C) at $t+\Delta t_2$. With a given local sound speed c, the CFL criterion defines Δt_2 as maximum time step. For any longer time steps, e.g. Δt_3, spatial positions outside the interval $(A)\bar{(B)}$ would influence the solution at C_3. These influences, however, would not be covered by the spatial discretization and the numerical scheme would turn instable since the area of influence was underestimated in that case.

On the other hand, any time step shorter than Δt_2 would overestimate the area of influence. As evident also from Figure 5.11, a higher propagation speed would be needed to influence position (C) from neighbor (A) and (B) with in a time step as short as e.g. Δt_1. That example illustrates a common problem in the application of explicit methods for Lagrangean discretizations as usually applied in computational solid mechanics. Other than in Eulerian grids for computational fluid dynamics, Lagrangean meshes typically contain cells of different size. For the time step calculation according to the CFL criterion the smallest discretizing cell in the grid defines the time step. That means that there are always larger cells for which the zone of interest is overestimated. Though the scheme remains stable under these conditions, the overestimation of the zone of influence results in a systematic error and, thus, reduced accuracy.

The discussed example shows the major problem of explicit methods with small cells sizes and specifically with large variations in size of cells with a spatial grid for which an overall time step is calculated. A way of that dilemma may be found in local time stepping methods, e.g. with so called local subcycling. Proposed methods that use different time steps at different nodes are for example reported in Belytschko et al. [50], Neal and Belytschko [292] or Tamma and D'Costa [384]. Some of them still struggle with problems of momentum conservation at time step interfaces. Others with so called statistical stability, i.e. narrow bands of stable time steps (see for example Daniel [101] or Klisimski and Mostrom [227]).

However, many applications involving complex structures of multiple components can only be discretized with reasonable total numbers of cells if the cells vary in size. Specifically if local effects during a deformation process shall be investigated by simulation, e.g. if local failure mechanisms are of interest, selected zones need enhanced grid resolution. In order to achieve satisfying accuracy with reasonable computation times more intelligent time integration schemes are needed.

Alternatively, implicit methods are at hand which are unconditionally stable and, thus, allow for larger time steps. However, the necessary time resolution to investigate wave propagation effects is usually the reason why that advantage gets lost. Consequently, its enhanced computational efforts make the implicit schemes less attractive for applications to dynamic problems in computational solid mechanics.

5.6 Finite Volume Method

5.6.1 Basic Concept of Finite Volume Methods

A common numerical methodology in computational fluid dynamics is the *finite volume method*. Its particular applicability to shock waves in fluids and, thus, for also fluid-structure interactions makes it attractive for dynamic problems. As in the case of the aforementioned finite difference schemes, finite volume methods solve the basic partial differential equations as algebraic equations. However, other than a standard finite difference scheme the conservative, i.e. integral, formulations of the conservation equations (2.216) are solved.

Conservative Discretization

In its integral form, conservation of a quantity U in a domain Ω is described by source terms Q and flux terms F as:

$$\int_\Omega \frac{\partial U}{\partial t}\,\mathrm{d}\Omega = \int_\Omega Q\,\mathrm{d}\Omega - \int_\Omega \nabla\cdot F\,\mathrm{d}\Omega \tag{5.59}$$

This means that in the absence of source terms the conserved quantity U is only depending on the gradient of the fluxes and, moreover, recalling the divergence theorem:

$$\int_\Omega \nabla F\,\mathrm{d}\Omega = \int_S F\cdot\mathbf{n}\,\mathrm{d}S \tag{5.60}$$

only depending on the fluxes through the surface S of the domain.

Concerning the conservation equations in arbitrary sub-domains ω_i of Ω, the balance is formulated by the statement that fluxes in adjacent domains are equal and of opposite sign. Thus, inside a control volume dicretize by cells, he fluxes between the cells have to compensate each other in a way that the sum of all sub-domain flux integrals equals the flux integral of the total domain:

$$\sum_i \int_{\omega_i} \nabla\cdot F\,\mathrm{d}\omega_i = \int_\Omega \nabla\cdot F\,\mathrm{d}\Omega \tag{5.61}$$

A detailed discussion of discretizing either the non-conservative equations or the conservative ones is given in Hirsch [187]. The exemplary discretizations with central differences shows that non-conservative formulations produce internal source terms of unphysical nature. The quantity of these errors are investigated by Taylor series expansions and found to be in the range of second order influence. Thus, with respect to steady-state flow problems the error can be neglected. In the case of transient problems, however, the perturbations become a significant problem. Examples for numerical simulations of shock wave effects with both non-conservative and conservative formulations will be illustrated in section 5.10.

Types of Control Volumes

There are various types and formulations of finite volume methods. Basically, the investigated domain is subdevided into discretizing cells and nodes. On that grid of cells, control volumes are defined. Using the known values of the state variables at the node locations, the conservative balance equations are formulated, where the nodes may be located at the cell corners or in their centers (see Figure 5.12). The shape of the cells, in two as in three dimensions, can vary using an arbitrary number of edges.

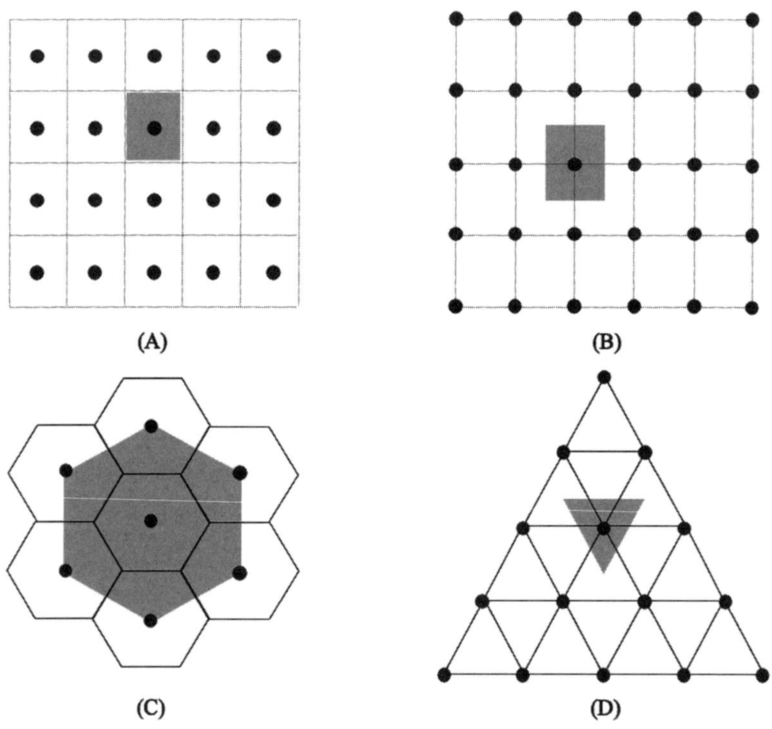

Fig. 5.12. Various ways of finite volume discretizations (control volumes marked in grey).

Some versions of how control volumes can be defined are illustrated in Figure 5.12. A cell-centered control volume is performed in Figure 5.12-(A). With the same cell shape and resolution a corner-centered type of control volumes can be implemented, as shown in Figure 5.12-(B). Also corner-centered, but with triangular instead of rectangular cells is the option of Figure 5.12-(D).

An overlapping cell-centered definition on base of hexagonal cells is implemented in Figure 5.12-(C).

Approximation of Balance Equations

A finite volume approximation to conservation of a balanced specific quantity u in a discretizing volume V

$$\int_V \frac{\partial u}{\partial t}\, \mathrm{d}V + \int_V \nabla \cdot f \, \mathrm{d}V = \int_V q \, \mathrm{d}V \tag{5.62}$$

is found after applying the divergence theorem to the flux term which yields a flux formulation corresponding to surfaces between adjacent volumes :

$$\int_V \nabla \cdot f \, \mathrm{d}V = \int_S f \cdot n \, \mathrm{d}S \tag{5.63}$$

Using the surface normals on the interfacial surfaces, the integral over all flux terms in a control volume can be described by the sum of all fluxes through the volume surfaces:

$$\int_V \nabla \cdot f \, \mathrm{d}V = \sum_k f_i \cdot n_{k,i} s_{k,i} \tag{5.64}$$

where f_i is the i-direction component of the flux in a volume V_k with evaluation point x_k (preferably the center of gravity). The resulting approximation yields:

$$\int_V \frac{\partial u}{\partial t}\, \mathrm{d}V + \sum_i f_i \cdot n_{k,i} s_{k,i} = \int_V q \, \mathrm{d}V \tag{5.65}$$

The configuration illustrated in Figure 5.13 represents a cell-centered formulation with evaluation points x_k in volumes V_k. A distinctive feature between finite volume methods is the particular formulation of the fluxes f_i. In order to guarantee a conservative scheme, the fluxes through a surface $s_{k,i}$ from adjacent volumes need to be equal. Therefore, an averaging algorithm is often applied which may either be found from:

$$f = \frac{1}{2}\left(f_1 + f_2\right) \tag{5.66}$$

or from averaging the arguments:

$$f = f\left(\frac{1}{2}\left(u_1 + u_2\right)\right) \tag{5.67}$$

With an averaging of that type and the sum $\sum_k f_i \cdot n_{k,i}\, s_{k,i}$ in equation (5.64) representing total flux through surface $s_{k,i}$, a conservative discretization of

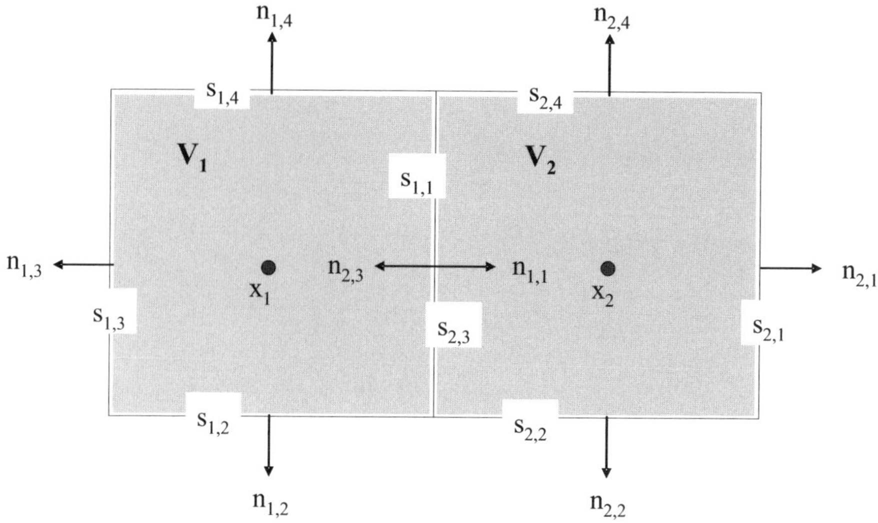

Fig. 5.13. Finite volume discretization with volumes V_i, normal vectors $n_{k,i}$, edge lengths $s_{k,i}$ and nodes x_i.

the fluxes is achieved.

A particular version of the finite volume methods that is capable of treating the propagation of discontinuous perturbations in terms of shock waves is the *Godunov scheme* [136] which performs the following solution steps:

- Assumption of piecewise constant or linear distribution of a solution in the control volumes at a certain time instant. Depending on these assumptions, a scheme of first, second or higher order results depending on whether the distributions are piecewise constant, linear or quadratic, respectively.
- Evaluation of the state variables for the next time step through direct formulation and solution of the Riemann problem across the interfacial surfaces between adjacent volumes.
- Averaging of the state variables calculated from all neighbouring volumes in the last step.

Hirsch [187] describes the originally presented version of the Godunov scheme with piecewise constant distributions of the flux terms and exact solution of the Riemann problem. In addition, approximative Riemann solvers are discussed, e.g. solvers introduced by Osher and Roe, and discussed with respect to their applicability.

Results of an example application of a second order finite volume Godunov scheme are illustrated in section 5.10. Detailed introductions to the theory of various finite volume schemes can be found in Hirsch [187] and [188], Toro [392], van Kan and Segal [224] or Wendt [417].

5.7 Finite Element Method

The *Finite Element Method* (FEM) is probably the most commonly known numerical discretization technique and the term FEM itself is often used as a synonym for computer based simulation. Its wide spectrum of available element types and the sophisticated mathematical formulation and assessment of these elements is unique amongst all existing descretization methodologies. This is partly due to the matrix formulations of the equations and the related access of schematized computational methods to solve them. The early development of the method started in the fifties with the presumably first application of the method by Turner et al. [402] in 1956. Having experienced the limitations of the *direct stiffness method* for the solution of boundary value problems in beam structures, Turner et al. applied the Ritz method [329] to a number of structural components, the later elements, to calculate flutter induced deflections in delta-wing structures of Boing airplanes. Only later Clough [93], one of the co-authors of Turner [402], invented the term *finite element method* in 1960.

Its precise mathematical relation to the Ritz method, which was developed already at the turn of the nineteenth century, became clear first after its initial development and several applications in civil and aeronautical engineering. That understanding of the finite element method as a generalized Ritz method, applied to the finite number of elements instead of the total structure as in the Ritz method, enhanced the capabilities of the method and provided a more profound mathematical basis. It also allowed for a more elegant solution of any kind of equilibrium problems, be it of mechanical, electro-static or steady-state fluid-dynamical kind.

With the application of the Galerkin method [132] the range of applicability of the finite element method became again wider. The introduction of further weak forms, e.g. the weighted residuals, opened the door to solutions of non-equilibrium problems.

Today, numerous commercial and and non-commercial FE codes exist. To name but a few commercial ones, ABAQUS, ADINA, ANSYS, LS-DYNA and NASTRAN shall be mentioned here. At the same time manifold introductory and in depth describing literature on finite elements has been written over the last five decades. Examples are Bathe [37], Hughes [201], Zienkiewicz et al. [435]. It would be far beyond the scope of this book to provide another introduction to the details of the finite element method. However, the basic concept shall be described briefly since that is the minimum of what is needed to assess the potential and the limitations of finite elements compared to the other discretization techniques described in this chapter.

5.7.1 Solutions of the Euler-Lagrange Equation

In the course of section 2.3 we found that the Euler-Lagrange equation

$$\frac{\partial F}{\partial y} - \frac{\mathrm{d}}{\mathrm{d}\,x}\left(\frac{\partial F}{\partial y'}\right) = 0 \tag{5.68}$$

prescribes a criterion for solutions y that minimize the functional form

$$\Pi = \int_{x_0}^{x_1} F\,(x,\, y,\, y')\,\mathrm{d}\,x \tag{5.69}$$

Relating this minimization task to a structural problem of static equilibrium, the searched function y represents the deformation state u for which equilibrium is satisfied. With the example of a hanging beam we have seen a simple version of formulating the Euler-Lagrange equation allowing for exact solutions, which in not too complex systems can even be performed by hand.

In general, however, real applications involving more complex structural elements than straight beams cannot be treated in a closed form solution of that kind. Therefore approximations are needed some of which are briefly mentioned in the following. These basic methodologies form the fundamental theories for the different existing finite element formulations.

Ritz Method

A well known concept to solve boundary value problems of the type (2.218) in its strong form was proposed at the beginning of the twentieth century by Ritz [329]. According to the Ritz method an approximation for the deformation state $\hat{u}(x)$ at equilibrium is formulated as

$$\hat{u}(x) = \sum_{i}^{n} a_i\,\varphi_i(x) \tag{5.70}$$

with n known linear independent admissible trial functions φ_i and n unknown trial parameters a_i. The attribute admissible to the trial functions requires that the geometric boundary conditions (2.219) are satisfied.

With the approximation (5.70) the variation of the functional form (2.150) is now only depending on the unknown parameters a_i. Accordingly, the minimization problem describing equilibrium reduces to

$$\frac{\partial \bar{\Pi}}{\partial a_i} = \int_{x_0}^{x_1} \left(\frac{\partial F}{\partial \hat{u}(x)} \frac{\partial \hat{u}(x)}{\partial a_i} + \frac{\partial F}{\partial \hat{u}'(x)} \frac{\partial \hat{u}'(x)}{\partial a_i} \right) \mathrm{d}\, x = 0 \tag{5.71}$$

and with

$$\hat{u}'(x) = \sum_i^n a_i \, \varphi_i'(x) \tag{5.72}$$

we find the following system of n linear independent equations for n unknown trial parameters a_i:

$$\frac{\partial \bar{\Pi}}{\partial a_i} = \int_{x_0}^{x_1} \left(\frac{\partial F}{\partial \hat{u}(x)} \varphi(x) + \frac{\partial F}{\partial \hat{u}'(x)} \varphi_i'(x) \right) \mathrm{d}\, x = 0 \tag{5.73}$$

Methods of Weighted Residuals

With the Ritz method the solution (5.73) of a minimization problem was found via the formulation of the approximations (5.70) for the deformation state $\hat{u}$ and through demanding for a stationary value of the approximated functional $\bar{\Pi}$. An approach that is restricted to minimization problems.

An alternative, more general, approach is the *method of weighted residuals*. Instead of directly addressing the Euler-Lagrange equations, the variational form (2.160) of the functional is approximated:

$$\delta\, \Pi(u) = \int_\Omega \delta\, u \left[\frac{\partial F}{\partial u} - \frac{\mathrm{d}}{\mathrm{d}\, x} \left(\frac{\partial F}{\partial u'} \right) \right] \mathrm{d}\, x = 0 \tag{5.74}$$

which can be written as:

$$\delta\, \Pi(u) = \int_\Omega \delta\, u \, \{ L\, [u(x)] - f(x) \} \; \mathrm{d}\, x = 0 \tag{5.75}$$

For the displacements u the basic approximation (5.70) known from the Ritz approach is kept, yielding:

$$\delta\, \Pi(\hat{u}) = \int_\Omega \delta\, u \, \{ L\, [\hat{u}(x)] - f(x) \} \; \mathrm{d}\, x = 0 \tag{5.76}$$

The new aspect compared to the Ritz method is that now the discrepancy between the approximate solution $L\,[\hat{u}(x)]$ and the right hand side $f(x)$ is investigated. That deviation is called *residual*, denoted by R. It compares the difference in the integrand:

$$R\,(\hat{u}) = L\,[\hat{u}(x)] - f(x) = L \left[\sum_i^n a_i \, \varphi_i(x) \right] - f(x) \tag{5.77}$$

In order to find a best approximation, i.e. a minimization of the residual, a weak form (2.221) of the residual is formulated:

$$\int_{\Omega} R(\hat{u})\; w_i(x)\, \mathrm{d}x = 0 \tag{5.78}$$

demanding for the integrated residual to vanish on the domain Ω.

By the particular choice of the weighting functions $w_i(x)$ various types of weighted residual methods are defined:

- Formulation (5.78) yields the *Bubnow-Galerkin* method:

$$\int_{\Omega} \left\{ L \left[\sum_{i}^{n} a_i\, \varphi_i(x) \right] - f(x) \right\} w_i(x)\, \mathrm{d}\Omega = 0 \tag{5.79}$$

- The *Petrow-Galerkin* method is characterized by choosing the identical functions for the trial functions $\varphi(x)$ and for the weighting functions $w(x)$:

$$\int_{\Omega} \left\{ L \left[\sum_{i}^{n} a_i\, \varphi_i(x) \right] - f(x) \right\} \varphi(x)\, \mathrm{d}\Omega = 0 \tag{5.80}$$

- If Dirac functions $w_i(x) = \delta(x - x_i)$ are used the *collocation method* is formulated which reduces the residual to zero at the locations x_i
- In case of the *least squares method*, the weighting functions are identified as derivatives of the residual itself $w_x = \partial R / \partial a_i$ with respect to the unknown trial parameters a_i which yields:

$$\begin{aligned} \frac{\partial}{\partial a_i} \int_{\Omega} R^2(\hat{u})\, \mathrm{d}\Omega &= \frac{\partial}{\partial a_i} \int_{\Omega} \left\{ L \left[\sum_{i}^{n} a_i\, \varphi_i(x) \right] - f(x) \right\}^2 \mathrm{d}\Omega \\ &= \int_{\Omega} R(\hat{u})\, L \left[\sum_{i}^{n} \varphi_i(x) \right] \mathrm{d}\Omega = 0 \end{aligned} \tag{5.81}$$

The discussed solution techniques for the Euler-Lagrange equations can be discretized with finite elements. Discretization means that the particular solutions are formulated for a finite number of elements which, together, describe the total structure of interest. Application of these solution techniques to elements with n_e nodes will be topic of the following subsections.

5.7.2 Ritz Version of Finite Elements

A basic version of the finite element method, formulated as generalized Ritz method, describes static equilibrium as variational problem. It provides a solution in terms of a minimum in the governing functional, e.g. the Euler-Lagrange equation of the particular structural element. In contrast to the classical Ritz method however, not the total structure is the basis for the Euler-Lagrange equation to be approximated, but a finite number of elements forming the structure. A detail that makes a significant difference, since now it will be much easier to find many (a finite number of) solutions for the discretizing elements instead of finding one single solution valid for the whole structure. The essential difference is

- the disintegration of an arbitrarily complex structure into many structural elements, each of which has a well known mechanical behaviour,
- the subsequent determination of solutions for each individual element
- and, finally, the construction of an overall solution from the combination of the elementary solutions.

In analogy to the Ritz method a variational problem $\delta \Pi = 0$ is formulated to describe equilibrium and the independent variables are approximated by trial functions which. When applied to deformation in an element u with nodal degrees of freedom u_i, the trial functions are called *shape functions.* Their application provides the benefit of utilizing the well known convergence characteristics of the Ritz method[1]

In contrast to the Ritz method however, finite element solutions are constructed by formulating the governing equations and its approximations on the element level first. For equilibrium problems this means that internal and external energy terms are formulated for each element.

To describe the internal energy Π_{int} of an element with volume V the strain energy is formulated by:

$$\Pi_{int} = \frac{1}{2} \int \sigma_{ij} \, \varepsilon_{ij} \, \mathrm{d}V = \frac{1}{2} \int \varepsilon_{ij} \, E_{ijkl} \, \varepsilon_{kl} \, \mathrm{d}V \tag{5.82}$$

Depending on the chosen strain tensor, a particular strain-displacement relation

$$\varepsilon_{ij} = b_{ij} \, u_k$$
$$\text{or}$$
$$\varepsilon = \mathbf{b} \, \mathbf{u} \tag{5.83}$$

is used to get

$$\Pi_{int} = \frac{1}{2} \int \mathbf{u}^T \, \mathbf{b}^T \, \mathbf{E} \, \mathbf{b} \, \mathbf{u} \, \mathrm{d}V \tag{5.84}$$

On the other hand, the external energy for an element with n_N nodes and n_{ndf} nodal degrees of freedom is expressed by the negative product of all force components $\{_j$ at all nodes of the element times the corresponding nodal displacement components u_j:

$$\Pi_{ext} = - \sum_{j=1}^{n_N * n_{ndf}} \{_j \, u_j \tag{5.85}$$

[1] Convergence characteristics of the Ritz method are for example discussed in Birkhoff et al. [60], Chatelin and Lemordant [85], Ciarlet et al. [92] or Mertins [272].

Using a nodal force vector $\mathcal{F}_i$ describing force vectors at node i and a corresponding nodal displacement vector $\mathbf{u}_i$, each of dimension $n_N * n_{ndf}$, the external energy yields:

$$\Pi_{ext} = -\mathcal{F}_i\,\mathbf{u}_i \tag{5.86}$$

leading to a total potential energy of:

$$\Pi = \frac{1}{2}\int \mathbf{u}^T\,\mathbf{b}^T\,\mathbf{E}\,\mathbf{b}\,\mathbf{u}\,\mathrm{d}V - \mathcal{F}_i\,\mathbf{u}_i \tag{5.87}$$

Since the total potential energy (5.87) formulates the strains in terms of displacements, the minimization problem is reduced to require

$$\frac{\partial\,\Pi}{\partial\,\mathbf{u}} = 0 \tag{5.88}$$

Moreover, the displacements in an element are, according to the Ritz method, approximated by the shape functions from the nodal displacements $\mathbf{u}_i$:

$$\hat{\mathbf{u}}(\mathbf{x}) = \varphi_i(\mathbf{x})\,\mathbf{u}_i \tag{5.89}$$

meaning that the total potential is now a direct function of the nodal displacements:

$$\Pi(\mathbf{u}_i) = \frac{1}{2}\int \mathbf{u}_i\,\varphi_i^T\,\mathbf{b}^T\,\mathbf{E}\,\mathbf{b}\,\varphi_i\,\mathbf{u}_i\,\mathrm{d}V - \mathcal{F}_i\,\mathbf{u}_i \tag{5.90}$$

With the matrix of spatial derivatives $\mathbf{B}$ applied to the shape functions:

$$\mathbf{B} = \mathbf{b}\,\varphi \tag{5.91}$$

we find the following formulation:

$$\Pi(\mathbf{u}_i) = \frac{1}{2}\int \mathbf{u}_i\,\mathbf{B}^T\,\mathbf{E}\,\mathbf{B}\,\mathbf{u}_i\,\mathrm{d}V - \mathcal{F}_i\,\mathbf{u}_i \tag{5.92}$$

Therefore, equilibrium is now described by the partial derivatives of the total potential energy with respect to the unknown nodal displacemens $\mathbf{u}_i$:

$$\frac{\partial\,\Pi}{\partial\,\mathbf{u}_i} = \frac{\partial\,\Pi}{\partial\,\hat{\mathbf{u}}}\,\frac{\partial\,\hat{\mathbf{u}}}{\partial\,\mathbf{u}_i} = \frac{\partial\,\Pi}{\partial\,\hat{\mathbf{u}}}\,\varphi_i \tag{5.93}$$

and thus for

$$\frac{\partial\,\Pi}{\partial\,\mathbf{u}_i} = \int \mathbf{B}^T\,\mathbf{E}\,\mathbf{B}\,\mathbf{u}_i\,\mathrm{d}V - \mathcal{F}_i = 0 \tag{5.94}$$

With the invention of the term

$$\mathbf{K} = \int \varphi_i^T\,\mathbf{b}^T\,\mathbf{E}\,\mathbf{b}\,\varphi_i\,\mathrm{d}V = \int \mathbf{B}^T\,\mathbf{E}\,\mathbf{B}\,\mathrm{d}V \tag{5.95}$$

called *stiffness matrix*, equilibrium in the element is described by:

$$\mathbf{K}\,\mathbf{u}_i = \mathcal{F}_i \tag{5.96}$$

relating the nodal displacements to the respective forces via the stiffness matrix.

For a solution of the finite element formulation for equilibrium (5.96) according to the Ritz version, the following components are needed:

- Approximative descriptions of the displacement using shape functions
- A numerical integration scheme to derive the stiffness matrix (5.95)
- A solver for the resulting system of algebraic equations

These components shall be derived in the subsections 5.7.4 and 5.7.5, respectively.

5.7.3 Finite Elements for Dynamic Problems

With the generalization of the finite element method that was introduced by the various weak forms its applicability to non-equilibrium problems was invented. In addition, Hamilton's principle (2.244) applied to the equations of motion (2.228) provides a formulation that allows for variational approaches to solve eigenfrequency problems with finite element approximations. With updated Lagrangean formulations solutions of the conservation equations with finite elements are an alternative to finite difference or finite volume methodologies[2]. In the following, a discrete version of the updated Lagrangean version of the momentum balance will be provided. This in turn is the basis of implementations in hydrocodes for investigations of dynamic deformation problems including shock wave propagation with finite elements.

Discretized Weak Form of the Equation of Motion

To approximate the equation of motion, total and updated Lagrangean formulations were introduced with equations (2.232) and (2.236), respectively. The most convenient form used in explicit codes is the updated Lagrangean version (2.236):

$$\int_\Omega \rho\,\delta\,\mathbf{v}\,\frac{\partial\,\mathbf{v}}{\partial\,t}\,\mathrm{d}\,\Omega + \int_\Omega \frac{\partial\,\delta\,\mathbf{v}}{\partial\,\mathbf{x}}\,\sigma\,\mathrm{d}\,\Omega - \int_\Omega \rho\,\delta\,\mathbf{v}\,\mathcal{B}\,\mathrm{d}\,\Omega - \int_\Omega \delta\,\mathbf{v}\,(\mathbf{n}\,\sigma)\,\mathrm{d}\,\Gamma = 0 \tag{5.97}$$

Since the virtual velocities $\delta\,\mathbf{v}$ are stationary, they can by approximate by the shape functions φ_i in the same way as the displacements:

$$\delta\,\mathbf{v} = \delta\,\mathbf{v}_i\,\varphi_i \tag{5.98}$$

[2] Donea and Huerta [111] as well as Dick [107] are examples for applications of Galerkin methods to computational fluid dynamics.

Thus, a discrete version of the weak form (5.97) is obtained with the formulation:

$$\int \rho\, \varphi_j^T\, \varphi_i \,\mathrm{d}V \,\frac{\partial\, \mathbf{v}_i}{\partial\, t} + \int \sigma\, \frac{\partial\, \varphi_i}{\partial\, \mathbf{x}} \,\mathrm{d}V - \int \rho\, \varphi_i\, \mathcal{B}_i \,\mathrm{d}V - \int \varphi_i \mathbf{n}\, \sigma \,\mathrm{d}V = 0 \quad (5.99)$$

In brief we can write

$$\mathbf{M}\, \frac{\partial\, \mathbf{v}_i}{\partial\, t} + \mathbf{\Pi}_{int} - \mathbf{\Pi}_{ext} = 0 \tag{5.100}$$

where

$$\mathbf{M} = \int \rho_0\, \varphi_j^T\, \varphi_i \,\mathrm{d}V \tag{5.101}$$

is the elements *consistent mass matrix*[3]. Consistent means in this context that $\mathbf{M}$ is achieved with the same shape functions as the stiffness matrix $\mathbf{K}$. For some applications, specifically in the context of explicit time integration[4] it is advantageous to apply a *diagonal mass matrix.* $\mathbf{M}^D$. A straight forward way to establish the diagonal form, besides direct distribution of the total element mass to nodes, is found in summing rows of the consistent matrix:

$$M_{ii}^D = \sum_j M_{ij} \tag{5.102}$$

The diagonal mass matrix (5.102) is also called *lumped mass matrix.* There are several other ways to derive diagonal forms of the mass matrix. A good overview can be found in Cook [96].

Equation (5.99) is often called the *semi-discretized momentum equation*[5] since it is not yet discretized in time.

Discretized Hamilton Principle

For the solution of eigenfrequency problems the principle of least action was derived with equation (2.244):

$$\int_{\Omega_0} \rho_0\, \delta\, \mathbf{u}\, \frac{\partial^2\, \mathbf{u}}{\partial\, t^2} \,\mathrm{d}\Omega_0 + \int_{\Omega_0} \delta\varepsilon^T\, \mathbf{P}\, \varepsilon \,\mathrm{d}\Omega_0 - \int_{\Omega_0} \rho_0 \delta\, \mathbf{u}^T\, \mathcal{B} \,\mathrm{d}\Omega_0 - \int_{\Gamma_0} \mathbf{n}\, \mathbf{P} \,\mathrm{d}\,\Gamma_0 = 0 \tag{5.103}$$

for which the related discrete version in an element of initial volume V_0 can be expressed as:

---

3 Consistent mass matrices were originally introduced by Archer [12] and [13].

4 Computing the accelerations $\mathbf{a} = \mathbf{F}/\mathbf{M}$ from the dynamic force vector $\mathbf{F}$ and a diagonal mass matrix is by far more efficient than using a consistent mass matrix.

5 Belytschko et al. [47] give detailed derivations of the weak form momentum conservation in total and updated Lagrange formulations.

$$\int_{V_0} \rho_0\, \varphi_j^T\, \varphi_i \,\mathrm{d}\, V_0\, \frac{\partial^2 \mathbf{u}_i}{\partial\, t^2} + \int_{V_0} \mathbf{B}^T\, \mathbf{E}\, \mathbf{B} \,\mathrm{d}\, V_0\, \mathbf{u}_i - \int_{V_0} \varphi_i\, \mathcal{B}_i \,\mathrm{d}\, V_0 - \int_{S_0} \varphi_i\, \mathbf{f}_i \,\mathrm{d}\, S_0 = 0 \tag{5.104}$$

With the already introduced terms for a mass matrix **M**, stiffness matrix **K** and a total force vector $\mathcal{F}_i$ the discrete equation (5.104) can be written as:

$$\mathbf{M}\, \frac{\partial^2 \mathbf{u}_i}{\partial\, t^2} + \mathbf{K}\, \mathbf{u}_i = \mathcal{F}_i \tag{5.105}$$

If additional damping forces are to be considered, a velocity dependent term is needed which is facilitated by a damping matrix **D** to yield:

$$\mathbf{M}\, \frac{\partial^2 \mathbf{u}_i}{\partial\, t^2} + \mathbf{D}\, \frac{\partial\, \mathbf{u}_i}{\partial\, t} + \mathbf{K}\, \mathbf{u}_i = \mathcal{F}_i \tag{5.106}$$

where a combination of weighted mass and stiffness components is often used to describe Rayleigh type damping:

$$\mathbf{D} = \alpha\, \mathbf{M} + \beta\, \mathbf{k} \tag{5.107}$$

where α and β are the scalar weighting terms.

5.7.4 Shape Functions

Requirements for Shape Functions

Shape functions for finite elements may be of exact nature, e.g. if they are a solution of the Euler-Lagrange equation of the element. However, this is only possible in the case of very simple formulations like in beam elements. For more complex elements, for example shell elements, the shape functions are an approximation of the deformation distribution over the element. Some general properties need to be required for the shape functions in an element which, for the sake of simplicity, shall for now be a one-dimensional element of n_e nodes with the local coordinates $\xi \in [-1,\, 1]$ (see Figure 5.14):

- Shape functions $\varphi_i(\xi)$ must be continuous functions on the domain ξ of the element.
- For each node i of an element the shape function has to describe the unique relation:

$$\varphi_i(\xi = \xi_j) = \begin{cases} 1 \text{ for } i = j \\ 0 \text{ for } i \neq j \end{cases} \tag{5.108}$$

  With (5.108) *continuity* of the shape functions across element domain borders is satisfied.
- At each position in the element, the sum of all shape functions needs to equal 1:

$$\sum_{i=1}^{n_e} \varphi_i \equiv 1 \tag{5.109}$$

  This requirement is often entitled as *completeness requirement.*

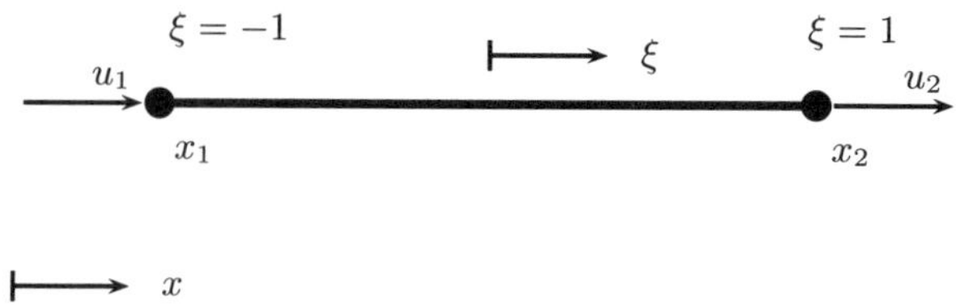

Fig. 5.14. One-dimensional two-node element with global (x) and local (ξ) coordinate system.

Simple Elements

As a consequence from the requirements (5.108) and (5.109), polynomial shape functions are of order $n-1$ along an element edge descretized by n nodes. In other words, the displacements in a one-dimensional element, as illustrated in Figure 5.14, with $n = 2$ nodes are described by two linear shape functions φ_i. An arbitrary linear distribution of longitudinal displacement in the element

$$u(x) = a_0 + a_1 x \tag{5.110}$$

is accordingly approximated using the nodal degrees of freedom u_1 and u_2 with the linear combination of shape functions φ_1 and φ_2 by:

$$\tilde{u}(x) = \varphi_1\, u_1 + \varphi_2\, u_2 = \frac{1}{2}\,(1-\xi)\; u_1 + \frac{1}{2}\,(1+\xi)\; u_2 \tag{5.111}$$

since the relation between local and global coordinates is expressed by:

$$x = \frac{1}{2}\,(1+\xi) \tag{5.112}$$

With the local coordinates $\xi_1 = -1$ and $\xi_2 = 1$ the found shape functions φ_i can be generalized to the form:

$$\varphi_i(\xi) = \frac{1}{2}\,(1 + \xi_i\, \xi) \tag{5.113}$$

Extension of the linear approximation to a quadratic one is achieved by a third node, placed in the middle between the two edge nodes. The resulting quadratic shape functions for the three nodes are:

$$\varphi = \begin{bmatrix} \varphi_1 \\ \varphi_2 \\ \varphi_3 \end{bmatrix} = \begin{bmatrix} \frac{1}{2}\,\xi\,(\xi-1) \\ 1-\xi^2 \\ \frac{1}{2}\,\xi\,(\xi+1) \end{bmatrix} \tag{5.114}$$

Deformations in two- and three-dimensional elements are approximated by analogous shape functions. To give an impression of the related formulations, shape functions for some elements are formulated next.

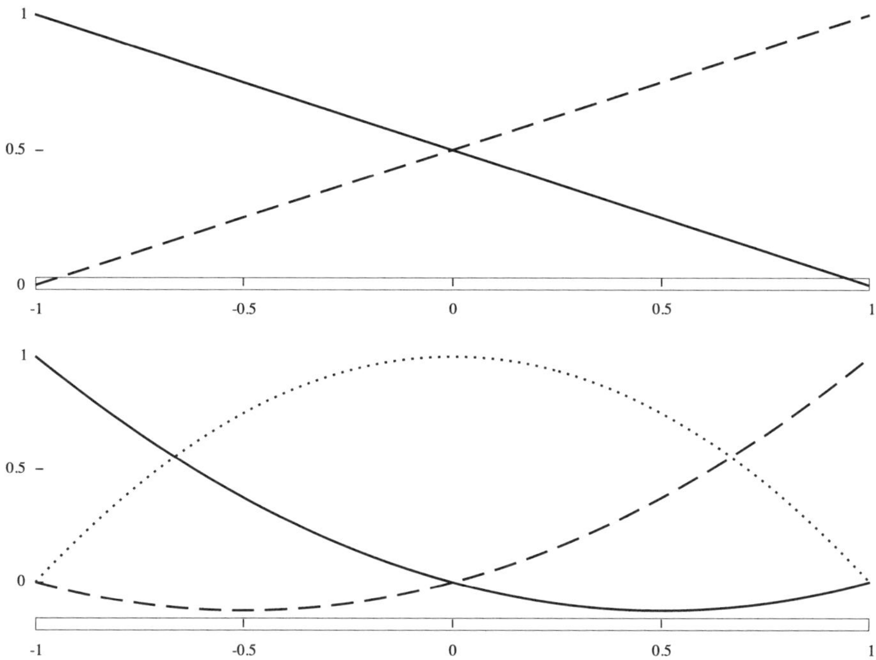

Fig. 5.15. Linear (top) and quadratic (down) shape functions for a one-dimensional element.

Isoparametric Elements

Straight edges of elements allow for a not too complex formulation of shape functions, as we have seen in the one-dimensional case. Curvilinear edges in two or three dimensions, however, make it much more complex to formulate proper functions. Moreover, a determination of shape functions for every particular geometrical shape would make the finite element method way too inefficient. To avoid this effort and to be able to use standard shape functions and solution procedures for whole classes of element types, a transformation of arbitrary element shapes to related standardized elements of unit dimensions and shape was invented[6]. Thus, what is needed is a transformation of the original nodal coordinates into a unit, or as illustrated in Figure 5.18, into a bi-unit configuration.

We will now use the coordinates ξ, η for a unit element with straight edges, as for example the triangular element in Figure 5.16 (B). The characteristic nature of *isoparametric elements* is that the transformation between original and unit configuration is described by the shape function. Thus, the shape of

[6] Specifically the isoparametric transformation was introduced by Taig [383] and by Irons [205]

the unit element is described in terms of its original coordinates by the same functions which are used to describe the distribution of the displacements.

Accordingly, in two dimensions the configuration of a n_e-nodes element is transformed between the global $\mathbf{x} = (x, y)$ coordinates and the so called *parent domain* of $\xi = (\xi, \eta)$ coordinates by:

$$\mathbf{x} = \sum_{i=1}^{n_e} x_i \, \varphi_i(\xi) \tag{5.115}$$

The shape functions in terms of ξ, η-coordinates are defined by the relations:

$$\begin{aligned} x(\xi, \eta) &= \sum_{i=1}^{n_e} a_i \, \varphi_i \\ y(\xi, \eta) &= \sum_{i=1}^{n_e} b_i \, \varphi_i \end{aligned} \tag{5.116}$$

where the necessary parameters a_i and b_i are to be identified through a consistency condition for the transformation which demands that the nodal coordinates $\mathbf{x}_i$ and ξ_i in both configurations coincide at the respective nodes:

$$\begin{aligned} x(\xi_i, \eta_i) &= x_i \\ y(\xi_i, \eta_i) &= y_i \end{aligned} \tag{5.117}$$

In a unit or bi-unit isoparametric element this leads to a further requirement to the shape functions in the parent domain:

$$\varphi_i(\xi_b) = \delta_{ab} \tag{5.118}$$

This yields a definition equation for the parameters a_i and b_i and, after substitution into (5.116), the shape functions $\varphi_i(\xi, \eta)$. The procedure will become apparent with the example of a linear triangular isoparametric element to be discussed next.

Linear and Quadratic Triangles

To illustrate the coordinate transformation, an isoparametric version of a triangular element with three nodes, illustrated in Figure 5.16 (A), shall be developed. In that element, displacements are described by linear shape functions:

$$u(x, y) = \varphi^T \, \mathbf{u}_i \quad \text{and} \quad v(x, y) = \varphi^T \, \mathbf{v}_i \tag{5.119}$$

with the nodal displacements

$$\mathbf{u}_i = \begin{bmatrix} u_1 \\ u_2 \\ u_3 \end{bmatrix} \quad \text{and} \quad \mathbf{v}_i = \begin{bmatrix} v_1 \\ v_2 \\ v_3 \end{bmatrix} \tag{5.120}$$

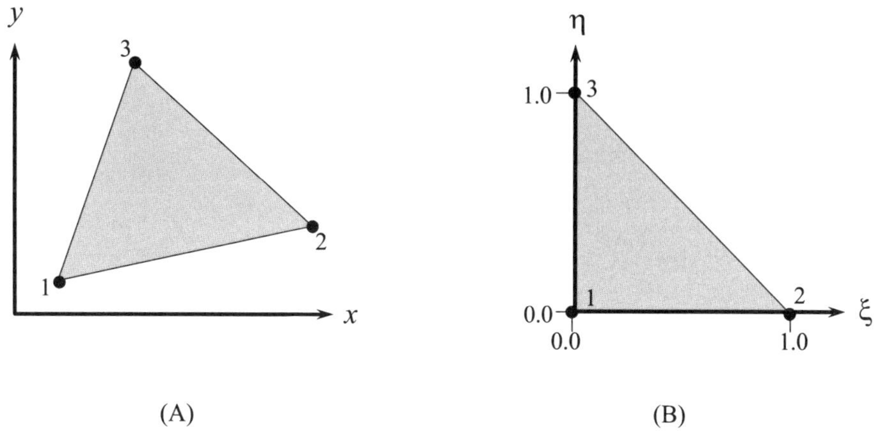

Fig. 5.16. Linear triangular element with straight edges in its original (A) and isoparametrically transformed (B) configurations.

and the shape functions $\bar{\varphi}_i$ in the x, y domain

$$\bar{\varphi}^T(x, y) = \left[\bar{\varphi}_1 \; \bar{\varphi}_2 \; \bar{\varphi}_3\right] = \left[1 \; x \; y\right] \tag{5.121}$$

To transform the element into the unit configuration, transformation (5.115) is applied with the shape functions $\bar{\varphi}$ in terms of ξ and η yielding the linear coordinate transformation:

$$x = a_1 + a_2\xi + a_3\eta \tag{5.122}$$

$$y = b_1 + b_2\xi + b_3\eta \tag{5.123}$$

The parameters a_i and b_i are identified by the unity requirement (5.118) and the coincidence requirement (5.117) of the nodal coordinates in both systems. Together these conditions are formulated by:

$$\begin{bmatrix} x_1 \\ x_2 \\ x_3 \end{bmatrix} = \begin{bmatrix} 1 & 0 & 0 \\ 1 & 1 & 0 \\ 1 & 0 & 1 \end{bmatrix} \begin{bmatrix} a_1 \\ a_2 \\ a_3 \end{bmatrix}$$

$$\begin{bmatrix} y_1 \\ y_2 \\ y_3 \end{bmatrix} = \begin{bmatrix} 1 & 0 & 0 \\ 1 & 1 & 0 \\ 1 & 0 & 1 \end{bmatrix} \begin{bmatrix} b_1 \\ b_2 \\ b_3 \end{bmatrix} \tag{5.124}$$

$$\mathbf{x}_i = \mathbf{A}\,\mathbf{a}_i$$

$$\mathbf{y}_i = \mathbf{A}\,\mathbf{b}_i$$

With the matrix $\mathbf{A}$ the relation between the shape functions in the two domains is determined as:

$$\varphi(\xi,\eta) = \mathbf{A}^{-T}\,\bar{\varphi}(\xi,\eta) \tag{5.125}$$

leading to the corresponding linear shape functions in the unit element configuration:

$$\varphi = \begin{bmatrix} 1-\xi-\eta \\ \xi \\ \eta \end{bmatrix} \tag{5.126}$$

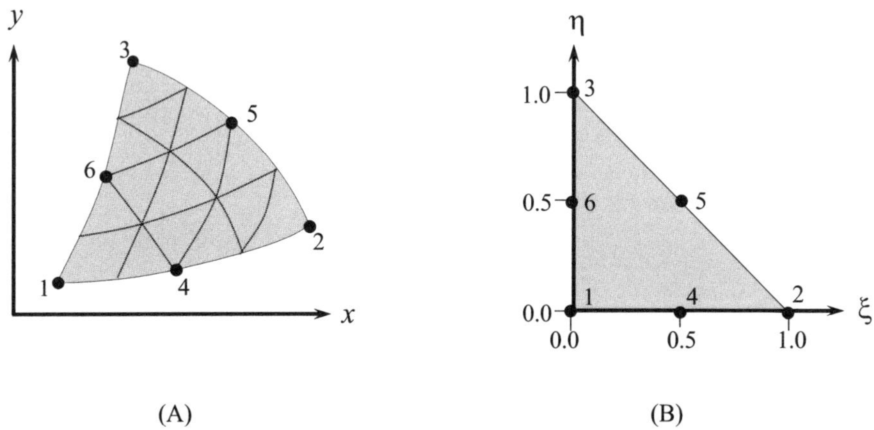

Fig. 5.17. Quadratic triangular element with curvilinear edges in its original (A) and isoparametrically transformed (B) configurations.

With intermediate nodes, located between the corner nodes of the triangular element, a quadratic description of displacements in the element is achieved. Since the strain field inside an element is calculated from relations containing spatial derivatives of the displacement, a linear displacement distribution would yield a constant strain value in the element and, by invention of a constitutive law, also constant stresses. Accordingly, a quadratic shape allows to describe a linear distribution of strain and stress in an element. Basically it can be stated, that the higher the gradients in strain and stress in a certain region of the discretized structure is, the more adequate is a higher order shape function. However, since the computational effort is significantly lower to solving the same equations for a linear compared to quadratic or higher order formulations, it can be more efficient to apply a higher resolution in the discretization than to use higher order elements.

Nevertheless, we will of course discuss higher order elements. the first one shall be the mentioned quadratic triangular element. For the configuration illustrated in Figure 5.17 the following shape functions are found after application of the same procedure that lead to the linear shape functions (5.126):

$$\varphi = \begin{bmatrix} \varphi_1 \\ \varphi_2 \\ \varphi_3 \\ \varphi_4 \\ \varphi_5 \\ \varphi_6 \end{bmatrix} = \begin{bmatrix} (1-\xi-\eta)\,(1-2\xi-2\eta) \\ 2\xi^2-\xi \\ 2\eta^2-\eta \\ 4\,(1-\xi-\eta)\,\xi \\ 4\,\xi\,\eta \\ 4\,\eta\,(1-\xi-\eta) \end{bmatrix} \tag{5.127}$$

Quadrilateral Elements

Figure 5.18 illustrates a quadratic quadrilateral element in the two domains x, y and ξ, η. If the nodes 5, 6, 7, 8 and 9 are omitted the same nomenclature can be used for its bilinear version. Moreover, omission of node 9 only describes another quadratic element. Quadratic quadrilaterals with less than nine nodes are called *serendipity* since it was perceived as sort of an unexpected gift to be able to formulate quadratic two-dimensional elements with reduced effort.

Note that in contrast to the unit triangular that we discussed before, we now have a so called bi-unit isoparametric coordinate system that is centered in the middle of the element with coordinates $\xi = \pm 1$ and $\eta = \pm 1$ at the corners of the element.

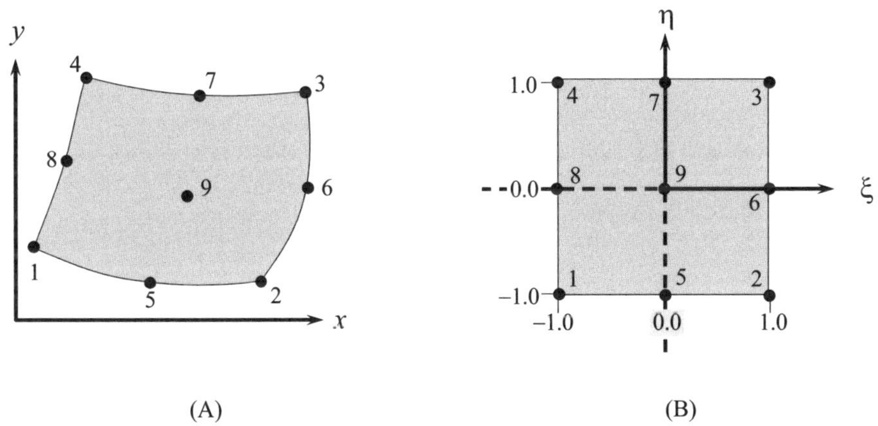

Fig. 5.18. Quadratic quadrilateral element with curvi-linear edges in its original (A) and isoparametrically transformed (B) configurations.

Before we derive shape functions for some quadrilaterals, a certain basic kind of shape functions should be introduced, called *Lagrange polynomials*[7]. The usual denomination for a Lagrange type polynomial shape function $l_i^r(\xi)$

[7] See for example Hughes [201]

uses a lower index indicating the node number and an upper index denoting the order of the polynomial. The general definition for a $r-$th order Lagrange polynomial as shape function in a n_e-node element at node i is:

$$l_i^r(\xi) = \frac{\prod_{\alpha=1,\,\alpha\neq i}^{n_e}(\xi - \xi_\alpha)}{\prod_{\alpha=1,\,\alpha\neq i}^{n_e}(\xi_i - \xi_\alpha)} \tag{5.128}$$

With definition (5.128) linear Lagrange polynomials are:

$$l_1^1(\xi) = \frac{1}{2}\,(1-\xi) \tag{5.129}$$

$$l_2^1(\xi) = \frac{1}{2}\,(1+\xi) \tag{5.130}$$

whereas the quadratic polynomials at the three nodes along an element edge are:

$$l_1^2(\xi) = \frac{1}{2}\,\xi\,(\xi-1) \tag{5.131}$$

$$l_2^2(\xi) = \left(1-\xi^2\right) \tag{5.132}$$

$$l_1^2(\xi) = \frac{1}{2}\,\xi\,(\xi+1) \tag{5.133}$$

Utilizing the linear Lagrange polynomials, a bilinear quadrilateral element can be described as a two-dimensional extension of a linear beam element. The combination of linear polynomials for the two directions yields:

$$\varphi_i(\xi,\,\eta) = l_j^1(\xi)\,l_k^1(\eta) \tag{5.134}$$

where the combinations of indices j and k at each of the four nodes is defined by:

$$\begin{aligned} &\text{Node 1}: \quad j=1 \quad k=1\\ &\text{Node 2}: \quad j=2 \quad k=1\\ &\text{Node 3}: \quad j=2 \quad k=2\\ &\text{Node 4}: \quad j=1 \quad k=2 \end{aligned}$$

The fully formulated shape functions are accordingly:

$$\begin{aligned} \varphi_1(\xi,\,\eta) &= \frac{1}{4}\,(1-\xi)\,(1-\eta)\\ \varphi_2(\xi,\,\eta) &= \frac{1}{4}\,(1+\xi)\,(1-\eta)\\ \varphi_3(\xi,\,\eta) &= \frac{1}{4}\,(1+\xi)\,(1+\eta)\\ \varphi_4(\xi,\,\eta) &= \frac{1}{4}\,(1-\xi)\,(1+\eta) \end{aligned}$$

On the other hand, a quadrilateral element with quadratic shape functions and related nine nodes as shown in Figure 5.18 is able to describe curvilinear edges. The Lagrange polynomials for the shape functions are encoded in the same manner as before by (5.134) for linear quadrilaterals. The quadratic scheme is:

$$\varphi_i(\xi,\, \eta) = l_j^2(\xi)\, l_k^2(\eta) \tag{5.135}$$

with adjusted index relations:

Node 1 :	$j = 1$	$k = 1$	Node 5 :	$j = 2$	$k = 1$
Node 2 :	$j = 3$	$k = 1$	Node 6 :	$j = 3$	$k = 2$
Node 3 :	$j = 3$	$k = 3$	Node 7 :	$j = 2$	$k = 3$
Node 4 :	$j = 1$	$k = 3$	Node 8 :	$j = 1$	$k = 2$
Node 9 :	$j = 2$	$k = 2$			

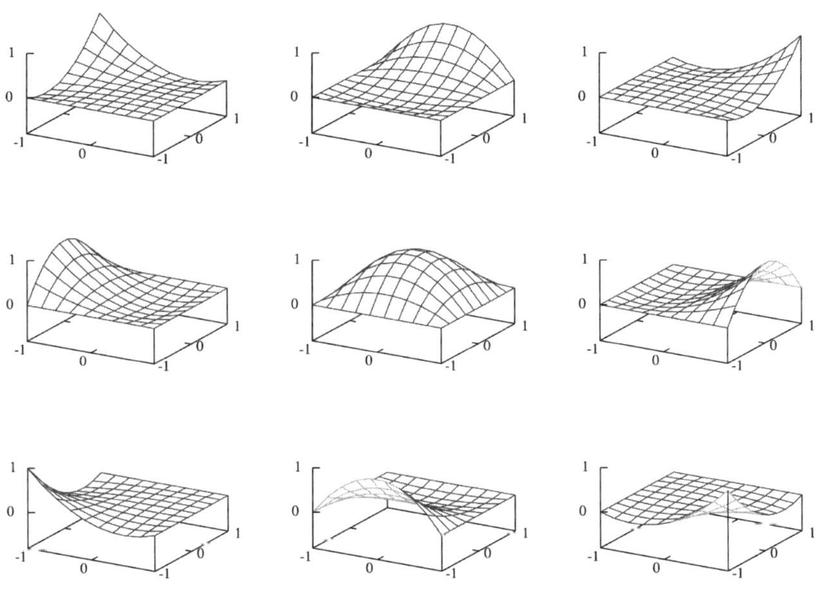

Fig. 5.19. Quadratic shape functions for a nine-node quadrilateral element.

The shape functions for the nine node quadrilateral are illustrated in Figure 5.19. More information on serendipity elements with one or more nodes of the quadratic nine-node Lagrange element omitted, as well as on the degeneration of quadrilaterals to become triangular elements is for example provided in Hughes [201].

Three-Dimensional Elements

As a simple example for three dimensional elements, the eight node hexahedral tri-linear element, shall be mentioned. The isoparametric ξ, η, ζ-coordinate system of the brick element is centered in the cube's center. The shape functions are:

$$\varphi_i(\xi, \eta, \zeta) = \frac{1}{8} (1 + \xi_i \xi) (1 + \eta_i \eta) (1 + \zeta_i \zeta) \tag{5.136}$$

An extension to quadratic formulations in the brick element takes 27 nodes and is generated in the analogous way as seen for the quadratic quadrilateral.

5.7.5 Stiffness Matrices, Mass Matrices and Numerical Solution

Finite element solutions are developed via the formulation of discrete equations for elements, e.g. the equilibrium form in (5.94) or the semi-discrete momentum conservation (5.99), and the subsequent combination of these to equations describing the behaviour of the total structure. Before that recombination can be performed, the involved element based stiffness and mass matrices need to be evaluated.

The partial derivatives in the strain-displacement formulations (5.83) are to be taken with respect to the global coordinates (x, y, z), whereas the element coordinates and, thus, the displacements are available in the local system (ξ, η, ζ). The same coordinate transformation in derivatives is needed to evaluate the size of the element volume dV.

Moreover, for the calculation of the integral terms in the consistent mass matrix (5.101) as well as in the stiffness matrix (5.95) numerical integration methods are needed. Therefore,

- an inverse Jacobian matrix (2.16) and, thus, its determinant is needed to transform the local coordinates into global ones
- and a numerical integration scheme is needed to evaluate the matrix equations.

Jacobian Matrix and Resulting Stiffness Matrices

The general procedure shall be illustrated by the linear triangular element (Figure 5.16-(B)) for which the shape functions were identified in (5.126) to be:

$$\varphi = \begin{bmatrix} 1 - \xi - \eta \\ \xi \\ \eta \end{bmatrix} \tag{5.137}$$

In the given two-dimensional case, the Jacobian matrix $\mathbf{J}$ relating derivatives with respect to local and global coordinates to each other is used:

$$\begin{bmatrix} \frac{\partial}{\partial \xi} \\ \frac{\partial}{\partial \eta} \end{bmatrix} = \begin{bmatrix} \frac{\partial}{\partial x} \\ \frac{\partial}{\partial y} \end{bmatrix} \begin{bmatrix} \frac{\partial x}{\partial \xi} & \frac{\partial y}{\partial \xi} \\ \frac{\partial x}{\partial \eta} & \frac{\partial y}{\partial \eta} \end{bmatrix} = \begin{bmatrix} \frac{\partial}{\partial x} \\ \frac{\partial}{\partial y} \end{bmatrix} \mathbf{J} \tag{5.138}$$

Its inverse describes the opposite transformation:

$$\begin{bmatrix} \frac{\partial}{\partial x} \\ \frac{\partial}{\partial y} \end{bmatrix} = \mathbf{J}^{-1} \begin{bmatrix} \frac{\partial}{\partial \xi} \\ \frac{\partial}{\partial \eta} \end{bmatrix} = \frac{1}{J} \begin{bmatrix} \dot{y} & -y' \\ -\dot{x} & x' \end{bmatrix} \begin{bmatrix} \frac{\partial}{\partial \xi} \\ \frac{\partial}{\partial \eta} \end{bmatrix} \tag{5.139}$$

wherein

$$J = det\mathbf{J} = x'\,\dot{y} - y'\,\dot{x} \tag{5.140}$$

is the *Jacobian determinant*, dots $(\dot{\ })$ symbolize derivatives $\frac{\partial}{\partial \xi}$ and apostrophes $(\ )'$ denote $\frac{\partial}{\partial \eta}$, respectively.

For a linear strain tensor, the strain tensor components are expressed as

$$\begin{aligned} \varepsilon_{xx} &= \frac{\partial u_x}{\partial x} = \frac{1}{J}\left(\dot{y}\,u_x' - y'\,\dot{u}_x\right) \\ \varepsilon_{yy} &= \frac{\partial u_y}{\partial y} = \frac{1}{J}\left(-\dot{x}\,u_y' + x'\,\dot{u}_y\right) \\ \varepsilon_{xy} &= \frac{1}{2}\left(\frac{\partial u_x}{\partial y} + \frac{\partial u_y}{\partial x}\right) = \frac{1}{2J}\left(-\dot{x}\,u_x' + x'\,\dot{u}_x\right) + \frac{1}{2J}\left(\dot{y}\,u_y' - y'\,\dot{u}_y\right) \end{aligned} \tag{5.141}$$

The resulting strain-displacement relation is expressed by

$$\mathbf{b} = [\mathbf{b}_x\ \mathbf{b}_y] = \begin{bmatrix} \frac{\partial \varepsilon_{xx}}{\partial u_x} & \\ & \frac{\partial \varepsilon_{yy}}{\partial u_y} \\ \frac{1}{2}\frac{\partial \varepsilon_{xy}}{\partial u_x} & \frac{1}{2}\frac{\partial \varepsilon_{xy}}{\partial u_y} \end{bmatrix} \tag{5.142}$$

In equation (5.141) the inverse Jacobian $\mathbf{J}$ is used to derive the global derivatives. That necessary inversion is possible solely for a non-singular Jacobian determinant J (5.140). That restriction has significant influence on the admissible deformation states of an element. For example, internal angles of 180° or more are impossible.

Implementation of the strains (5.141) into (5.142) and application of the shape functions leads to a description of the strain-displacement relation of:

$$\mathbf{B} = \frac{1}{J} \begin{bmatrix} \dot{y}\,\varphi'^T - y'\,\dot{\varphi}^T & \\ & -\dot{x}\,\varphi'^T + x'\,\dot{\varphi}^T \\ \frac{1}{2}\left(-\dot{x}\,\varphi'^T + x'\,\dot{\varphi}^T\right) & \frac{1}{2}\left(\dot{y}\,\varphi'^T - y'\,\dot{\varphi}^T\right) \end{bmatrix} = \frac{1}{J}\,[\mathbf{B}_x\ \mathbf{B}_y] \tag{5.143}$$

With the abbreviations:

$$\beta_1 = \dot{y}\,\varphi' - y'\,\dot{\varphi} \qquad \beta_2 = -\dot{x}\,\varphi' + x'\,\dot{\varphi}$$
$$\bar{\beta}_1 = \dot{y}\,\varphi'^T - y'\,\dot{\varphi}^T \qquad \bar{\beta}_2 = -\dot{x}\,\varphi'^T + x'\,\dot{\varphi}^T$$

we can write:

$$\mathbf{B} = \frac{1}{J}\begin{bmatrix} \bar{\beta}_1 & \\ & \bar{\beta}_2 \\ \bar{\beta}_2 & \bar{\beta}_1 \end{bmatrix} \tag{5.144}$$

Having applied the coordinate transformation to the derivatives, the stiffness matrix for the linear triangle

$$\mathbf{K} = \int \mathbf{B}^T\,\mathbf{E}\,\mathbf{B}\,\mathrm{d}V \tag{5.145}$$

can be formulated with the thickness of the triangle t and the differential surface element

$$\mathrm{d}x\,\mathrm{d}y = J\;\mathrm{d}\xi\,\mathrm{d}\eta \tag{5.146}$$

through the relation:

$$\begin{aligned} K &= t\int\limits_{\xi=0}^{1}\int\limits_{\eta=0}^{1}\frac{1}{J^2}\begin{bmatrix}\mathbf{B}_x^T\,E\,\mathbf{B}_x & \mathbf{B}_x^T\,E\,\mathbf{B}_y \\ sym. & \mathbf{B}_y^T\,E\,\mathbf{B}_y\end{bmatrix} J\;\mathrm{d}\eta\;\mathrm{d}\xi \\ &= t\int\limits_{\xi=0}^{1}\int\limits_{\eta=0}^{1}\begin{bmatrix}\beta_1 E_{11}\bar{\beta}_1 + \beta_2 E_{33}\bar{\beta}_2 & \beta_1 E_{12}\bar{\beta}_2 + \beta_2 E_{33}\bar{\beta}_1 \\ sym. & \beta_2 E_{22}\bar{\beta}_2 + \beta_1 E_{33}\bar{\beta}_1\end{bmatrix}\frac{1}{J}\;\mathrm{d}\eta\;\mathrm{d}\xi \end{aligned} \tag{5.147}$$

Numerical Integration

To solve equations (5.96) or (5.100) with the derived mass and stiffness matrices, integrals of the general shape

$$I\left(\Upsilon(\xi,\,\eta)\right) = \int\limits_{\xi}\int\limits_{\eta}\Upsilon(\xi,\,\eta)\;\frac{1}{J}\;\mathrm{d}\eta\,\mathrm{d}\xi \tag{5.148}$$

need to be calculated. Numerical integration methods are used to approximate the integral terms. The basic concept of numerical integration is to replace the integrand function $\Upsilon(\xi,\,\eta)$ by a polynomial evaluated at h locations. The quality of the numerical integration scheme depends on the number of locations at which the polynomial is evaluated and on the order of the integrand function. If a Gauss quadrature is applied, a polynomial of order $2\,h-1$ is calculated exactly with at least h integrating locations. In the two-dimensional case discussed before, a Gauss quadrature yields approximations as weighted sums of function $\Upsilon(\xi_g, \eta_g)$ values:

$$\int_{\xi}\int_{\eta} \Upsilon(\xi, \eta) \frac{1}{J} \,\mathrm{d}\eta \,\mathrm{d}\xi \cong \sum_{g=1}^{h} \Upsilon(\xi_g, \eta_g) \frac{1}{J} w_g \tag{5.149}$$

using h locations (ξ_g, η_g) at which the functions values are calculated and multiplied by weighs w_g. The locations and weighs for various types of topologies are well documented in literature[8]. Data for a four-point Gauss integration in triangular and in quadrilateral elements are illustrated in Figure 5.20.

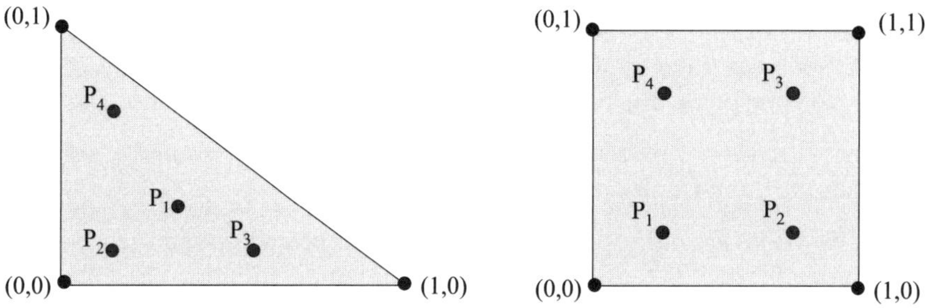

Positions:

$P_1(\xi_1,\eta_1),\ P_2(\xi_2,\eta_2),\ P_3(\xi_3,\eta_3),\ P_4(\xi_4,\eta_4)$

$\xi_1 = \eta_1 = \frac{1}{3},\ \xi_2 = \eta_2 = \frac{1}{5},$

$\xi_3 = \frac{3}{5},\ \eta_3 = \frac{1}{5},\ \xi_4 = \frac{1}{5},\ \eta_4 = \frac{3}{5}$

Weights:

$w_1 = -\frac{27}{96},\ w_2 = w_3 = w_4 = \frac{25}{96}$

Positions:

$P_1(\xi_1,\eta_1),\ P_2(\xi_2,\eta_2),\ P_3(\xi_3,\eta_3),\ P_4(\xi_4,\eta_4)$

$\xi_1 = \xi_4 = \eta_1 = \eta_2 = 0.211325,$

$\xi_2 = \xi_3 = \eta_3 = \eta_4 = 0.788675$

Weights:

$w_1 = w_2 = w_3 = w_4 = \frac{1}{4}$

Fig. 5.20. Locations and weights for four-point Gaussian quadrature in triangular and quadrilateral elements.

5.7.6 Shell Elements

With the continuum theory of thin shells an indispensable tool is available for efficient calculation of the mechanics in thin walled structured. Technical applications of thin two-dimensional structural elements are manifold and range

[8] See for example Bathe [37] or Stroud and Secrest [378].

from automobile car bodies over fuselages and wings of airplanes to building structures in civil engineering.

The dramatic reduction in computational effort through the application of shell elements is achieved by the following factors:

- Shell elements model a two-dimensional plane stress configuration which saves the out-of-plane nodes, the related degrees of freedom and, thus, *reduces the number of equations* to be solved.
- With the through-thickness direction omitted in the calculation, the smallest spatial dimension has no influence on the calculation of the *time step*. This may lead to time step increases by an order of magnitude and thus reduces the total number of cycles to be calculated until the final instant of time which is of interest is achieved.
- A further gain in computational efficiency is provided through a *reduced integration* using a single Gaussian point, albeit in three or more layers.

Along with these advantages, related drawbacks arise from the assumptions taken in the formulation of shell elements:

- Shell elements are only applicable to shell structures although their computational efficiency is tempting to expand the understanding of shell structures to quasi-shells which sometimes are not covered by the related kinematic assumptions.
- Stress wave transmission in through-thickness directions is not possible.
- Reduced integration leads to *zero energy modes*, i.e. deformation patterns which are not associated with a strain energy. Additional numerical damping mechanisms (*hourglass damping*)
- Kinematic constraints lead to overestimated strain energies

The finite element formulation of shells can basically be derived through formulating the weak form of the equilibrium or the momentum equation for a shell. Since that approach is complex, specifically for nonlinear shells, a simpler approach is usually followed for finite elements which imposes the kinematic constraints on the shape functions of a continuum element.

Two basic branches of the thin shell theory are implemented in shell finite elements. One is the *Kirchhoff-Love* type of shells for which the kinematics assume that a normal to the mid-surface remains *straight and normal* under deformation of the shell. This assumptions does not admit transverse shear in the element, not even for non-constant moments where it should be non-zero to guarantee equilibrium. The alternative theory, called *Mindlin-Reissner* shell theory, assumes normals to the mid-surface to remain straight and, thus, admit transverse shear. However, as a result of the Mindlin-Reissner formulation, the transverse shear components are constant through the thickness of the shell and thus also inconsistent with equilibrium since it should vanish at the top and bottom surfaces. The resulting error in terms of overestimated

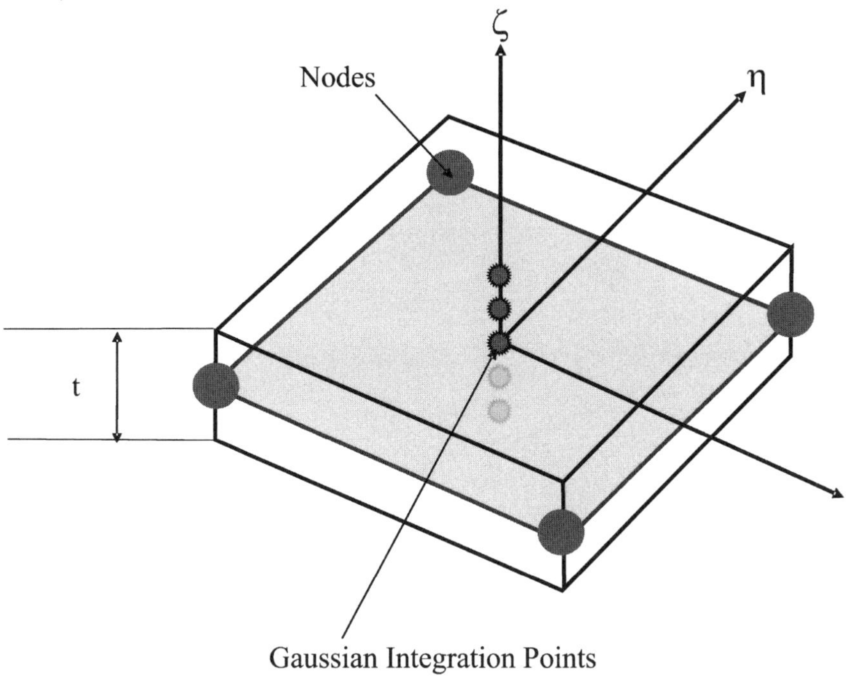

Fig. 5.21. Shell element of thickness t with reduced integration at one Gaussian integration point evaluated at five layers.

shear energy is often reduced by shear correction factors.

Whereas the Kirchhoff-Love shells are preferably applied to thin shells, Mindlin-Reissner formulations are better suited for sandwich and composite structures[9].

An implementation that is available in most commercial and academic codes is the shell element introduced by Hughes and Liu [202]. Even more attractive, since more efficient, and most widely used is the Belytschko-Lin-Tsay element [46]. Its enhanced efficiency originates from an optimized quadrature that reduced the necessary mathematical operations for a reduced integrated Hughes-Liu shell element from 4050 to 725 (Hallquist [158]). This was achieved by the calculation of strain rates in a local coordinate system in two components: the first is contribution from the mid-surface motion, the second is element rotation. In a Mindlin-Reissner shell element with non-plane cross-surfaces under shear deformation, the velocity at an arbitrary position in the

[9] Belytschko et al. [47] give a detailed description of the continuum mechanics of shell elements, of the various formulations and of the inconsistencies and idiosyncrasies of the theoretical basics. The latter is outlined in [47], Sec. 9.6.5.

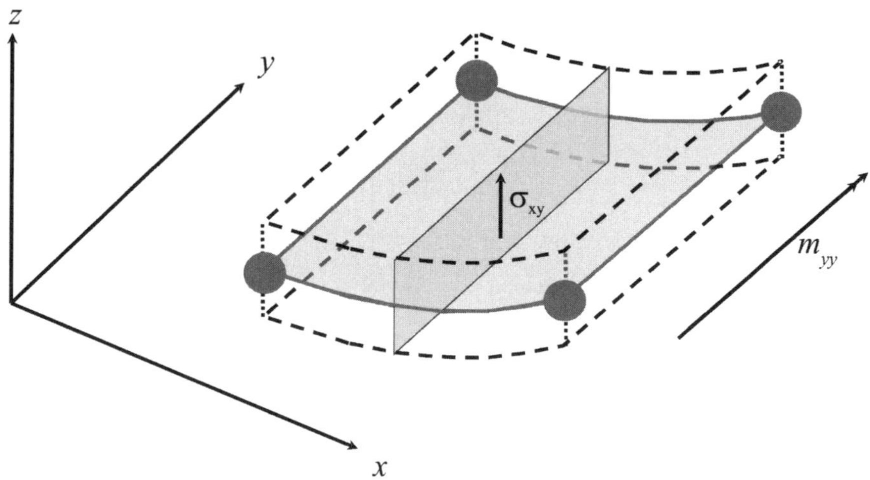

Fig. 5.22. Shell element under constant bending moment m_{xx} and transverse shear component σ_{xz} which should vanish for constant moments and be non-zero for a moment that is not constant. In addition, it should vanish at the top and bottom surfaces. However, σ_{xz} is constant in Mindlin-Reissner shells (leading to shear locking) and vanishes totally in Kirchhoff-Love shells.

shell with distance z to the mid-surface plane is described as sum of mid-plane velocity $\mathbf{v}^m$ and rotational velocity Θ:

$$\mathbf{v} = \mathbf{v}^m - z \times \Theta \tag{5.150}$$

For the time derivative of the strain displacement relation, (5.150) yields the strain rates:

$$\dot{\varepsilon}_{ij} = \frac{1}{2}\left(\frac{\partial \dot{u}_i}{\partial x_j} + \frac{\partial \dot{u}_j}{\partial x_i}\right) + \frac{z}{2}\left(\frac{\partial \dot{\Theta}_i}{\partial x_j} + \frac{\partial \dot{\Theta}_j}{\partial x_i}\right) \tag{5.151}$$

A correct calculation of moments and membrane forces under elastic-plastic loading conditions becomes possible if the integration at the single Gaussian point is performed at multiple layers in the shell thickness directions. Hourglass damping methodologies for the Belytschko-Lin-Tsay element were proposed by Flanagan and Belytschko [126] as well as by Engelmann and Whirley [121] an implemented in many codes.

Another critical point with shell elements is their tendency to exhibit the so called *shear and membrane locking*. Effectively these effects result in an underestimation of displacement due to incorrect representations of specific deformation states. In the case of shear locking, the normality constraint on the mid-surface appears as a penalty term in the energy and becomes larger with decreasing shell thicknesses. Shear locking is lessened through reduced integration. The membrane locking on the other hand, evolves from the inability of shell elements to represent extensional modes of deformation. It is of particular importance for the simulation of buckling effects. A method to avoid shear and membrane locking is the assumption of strain states in an element. A detailed discussion of shear and membrane locking as well as of methods to circumvent it can be found in Belytschko et al. [47], Sec. 9.7. Basic mathematical understanding of locking effects is provided in Arnold [16] as well as by Babuška and Suri [29], [30].

5.7.7 Finite Element Methodologies for Discontinuities

Inherent to dynamic loading conditions are discontinuities of various kinds. Shock waves are an example for discontinuities in the solution that needs special treatment if the simulation shall predict its propagation, shape and amplitude as precisely as possible. This topic will be discussed in section 5.10.

Crack Representation with Finite Elements

Another type of discontinuity arises when loading conditions lead to micro- or macroscopic failure of the material associated with the evolution of shear bands, pores or cracks. Shear bands are a deformation state which is characterized by a very local *strain discontinuity*. A discontinuity in displacement is present at a stress free crack tip or at a mode II or III crack opening. It goes along with an infinity value of the associated strain (see Figure 5.23). Whereas the first type is called *weak discontinuity*, the crack formation represents a so called *strong discontinuity*.

Crack opening and propagation is a general problem in finite element solutions since the evolving new surfaces with normal stress free boundary and the additional crack opening displacement need special attention. The crack opening displacement is sometimes modelled by a continuous smeared element displacement (see Figure 5.24-(A)). An approach which, however, does not account for the new free surface. If processes including the generation of free surfaces, e.g. multiple fragmentation, are intended to be modelled with smeared displacement solutions, a specific threshold strain is usually chosen at which the affected element is taken out of the calculation. The free surfaces and the crack opening are then approximated by the neighbour elements. However, realistic crack patterns are hardly represented by that approximation (see Figure 5.24-(B)). The crack tip is typically blunt to the degree of the element size. On the same scale is the initial crack width modelled.

A better crack representation is achieved if an element separation is enforced through node split as illustrated in Figure 5.24-(C). The resulting crack

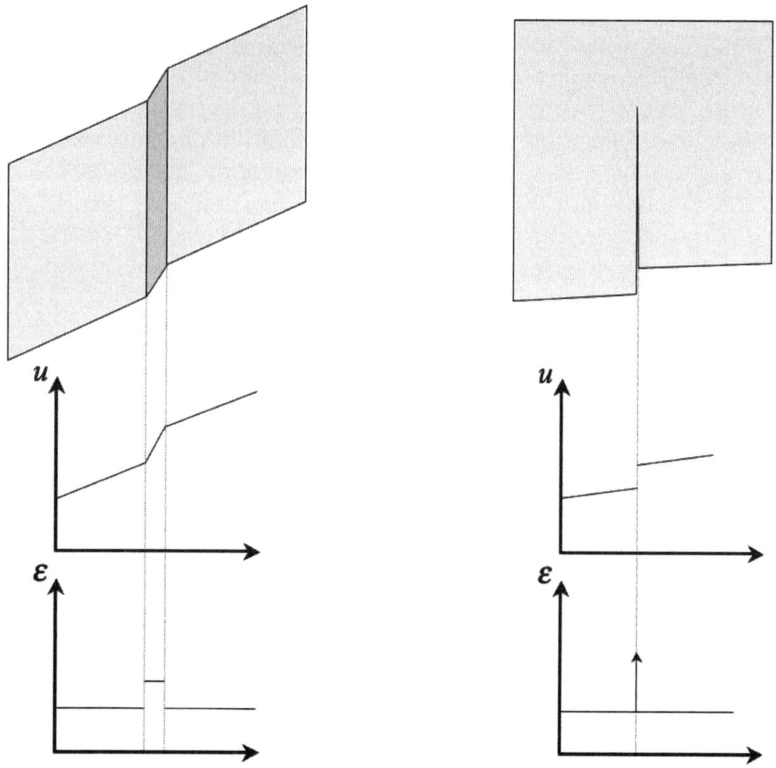

Fig. 5.23. Weak (left) and strong right discontinuities in strain and displacement, respectively.

tip is sharp but the exact positioning of the new nodes on the crack surface can be complex. The complexity arises partly from the fact that contact algorithms need to be initiated at these surfaces to allow for pressure contact under reverse loading conditions. Moreover the crack pattern is depending on the predefined element boundaries which allows a good representation of the cracks only if they are aligned with the mesh.

To allow for arbitrary crack paths, a successive adaption of the grid, i.e. an automatic mesh refinement, can be performed as the crack propagates. Depending on related fracture mechanical models, the direction and velocity of the crack propagation can be modelled by adjusted element boundaries (Figure 5.24-(D)). The continuous re-meshing is of course an additional time consuming computational effort. But the alternatives are either course or not really less expensive. Remarkable achievements in two and three dimensional crack modelling with automatic mesh refinement have been shown by the

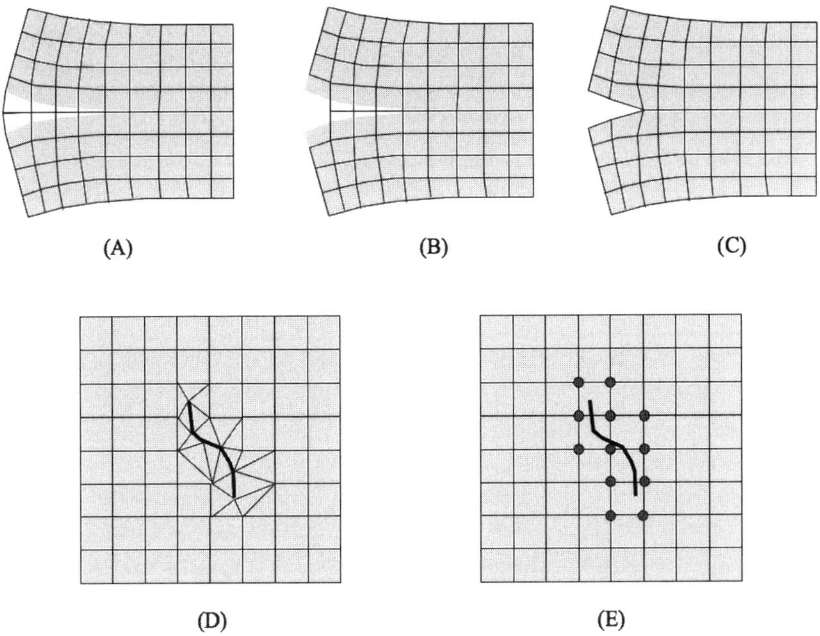

Fig. 5.24. Crack treatment with finite elements: (A) Smeared representation of crack opening displacement, (B) Element elimination at certain strain threshold, (C) Node split and generation of free surface for cracks aligned with the mesh, (D) Grid adaption for explicit crack modelling and (E) Enriched elements for non-aligned cracks using enrichment functions and nodes with additional degrees of freedom.

Ingraffea group, particularly published in Martha et al. [268] or in Carter et al. [79]. The limitations of mesh refinement methods are reached under dynamic multiple fragmentation, specifically with multiple crack intersections.

EED and X-FEM – Enrichment of Elements by Discontinuous Fields

More recent developments for finite element applications to crack propagation processes follow the idea of modelling cracks independent of the element boundaries. In the case of *Elements with Embedded Discontinuities* (EED), additional local degrees of freedom facilitate the crack representation. To describe the resulting equilibrium an additional condition is needed which is found in the crack model. So called *enrichment functions* Ψ_{EED} are used to modify the standard shape functions:

$$\Psi_{EED} = \Psi(\mathbf{x}) - \varphi(\mathbf{x}) \tag{5.152}$$

The resulting conditions to element geometries, specifically in three dimensions and for multiple discontinuities, lead to complex handling of the method. Otherwise the robustness and convergence of the methodology is lost (see Jirásek, Belytschko [209]).

An alternative approach, called *Extended Finite Element Method* (X-FEM), bases on the partition of unity method which was introduced by Babuška and Melenk [28] and is capable of discontinuities of both the strong and weak type. Belytschko and Black [42] introduced a X-FEM method for curved cracks in two-dimensional problems by mapping the straight crack enriched field. An improved version, applicable to three-dimensional crack patterns and long cracks, was introduced by Moës et al. [288] separately modelling the crack faces away from the crack tip and the near tip zones. Basically, a single crack is approximated with the finite element displacements and shape functions by:

$$\mathbf{u}^h(\mathbf{x}) = \sum_{i=1}^{n_N} \mathbf{u}_i\, \varphi_i(\mathbf{x}) + \underbrace{\sum_{i=1}^{n_{CD}} \mathbf{a}_i\, \varphi_i(\mathbf{x})\, H(\mathbf{x})}_{\text{Crack Discontinuity}}$$
$$+ \underbrace{\sum_{i=1}^{n_{T1}} \varphi_i(\mathbf{x}) \left(\sum_{j=1}^{4} \mathbf{b}_{i,1}^j\, F_1^j(\mathbf{x}) \right)}_{\text{First Crack Tip}} + \underbrace{\sum_{i=1}^{n_{T2}} \varphi_i(\mathbf{x}) \left(\sum_{j=1}^{4} \mathbf{b}_{i,2}^j\, F_2^j(\mathbf{x}) \right)}_{\text{Second Crack Tip}} \tag{5.153}$$

where n_N is the total number of nodes in the mesh, the nodal degrees of freedom for a specific node i are denoted by $\mathbf{u}_i$, the shape functions by φ_i. The number of nodes along the crack discontinuity with shape functions intersecting the crack but not containing either one of its tips, is here referred to as n_D. The additional degrees of freedom for associated with those nodes are denoted by $\mathbf{a}_i$. The number of nodes with shape functions support containing a crack tip is n_{T1} and n_{T2} for the first and second crack tip, respectively. The related new degrees of freedom are $\mathbf{b}_{i,1}^j$ and $\mathbf{b}_{i,2}^j$.

The key aspect of the method is represented by

- the discontinuous function $H(\mathbf{x})$ which is constant on each side of the crack, e.g. -1 on the one and 1 on the other crack face
- and the near-tip functions F_1^j and F_2^j which are given in Moës et al. [288] for a two-dimensional case with four node elements ($j = 1, \ldots, 4$) by

$$\left\{F_1^j(\mathbf{x})\right\} = \left\{\sqrt{r}\,\sin\left(\frac{\theta}{2}\right),\ \sqrt{r}\,\cos\left(\frac{\theta}{2}\right),\right.$$
$$\left.\sqrt{r}\,\sin\left(\frac{\theta}{2}\right)\sin(\theta),\ \sqrt{r}\,\cos\left(\frac{\theta}{2}\right)\sin(\theta)\right\} \tag{5.154}$$

An important further capability of the X-FEM approach was introduced by Daux et al. [102] providing an extension to the three dimensional case and to multiply branched as well as intersecting cracks. To many standard applications of hydrocodes multiple fragmentation of structures is an inherent processes. Therefore, this latter extension of the extended finite element method is extremely valuable in the branch of impact and crash simulations. In addition, the power of X-FEM was stretched by Chessa and Belytschko [87] showing the applicability of the method to arbitrary discontinuities including the Rankine-Hugoniot jump conditions. With that application X-FEM became an alternative approach to the Godunov-type finite volume methods as well as to the invention of artificial viscosities, both representing existing standard solution techniques for shock wave propagation in fluids and solids.

The need for discretization methodologies with the capability of modelling multiple fragmentation was one major motivation for the development of meshfree methods which will be discussed in the next section.

5.8 Meshfree Methods

5.8.1 Motivation to Develop Meshfree Methods

Smooth particle hydrodynamics (SPH), a discretization methodology that works without any grid or mesh in the sense that neighbouring nodes of an initial configuration stay neighbours throughout the calculation, was originally developed for the simulation of astrophysical problems. The evolution, motion and possible collision of interstellar gas clouds were typical applications of its early implementations. Why should one try to adopt a numerical method designed for gas or particle dynamics to structural mechanics when other methodologies like finite differences or finite elements are already established over several decades? Why go for a new relatively unknown method, after generations of engineers and mathematicians dedicated their professional life to provide a sophisticated, stable and in its mathematical character well known tool?

The reason was the limited capability of existing methods and the strongly increasing complexity of them if these limitations shall be overcome. Finite elements, for example, are in its genuine version not capable of fragmentation effects. Nevertheless it is not impossible to model cracks, e.g. with the extended finite element method discussed before. However, the complexity of the algorithm grows with the additional functionalities and therewith also the tendency of the method to become less reliable. Increasing complexity goes along with increasing risk and decreasing robustness[10].

Meshfree methods, on the other hand, were adapted to structural dynamics by the pioneering work of Libersky and Petschek [251] when they implemented constitutive equations into SPH. Its simplicity and at the same time perfectly adequate qualities for large deformations possibly ranging to total structural disintegration convinced many scientists and software developers of the methods potential. Though it turned out soon that SPH carried along inherent shortcomings of mathematical nature that required urgent examination, it was continuously further developed and improved.

The significantly increased effort in research work on particle methods also facilitated developments towards totally new directions. With the Element-Free-Galerkin (EFG) method Belytschko et al. [48] developed a particle scheme that is not totally working without a grid and not capable of applications the original particle methods were focused on. But it provides a serious alternative, even to extended finite element methods, for fracture mechanical investigations under static loading conditions. Rabczuk and Belytschko [316] presented the application of a total Lagrangean EFG formulation coupled with finite elements that is able to predict multiple and intersecting cracks in reinforced concrete slabs. The coupling concerned a representation of concrete by EFG particles and a modelling of the steel rebars using finite elements.

It seems like, in the past, a certain kind of inertia lead to a relatedness in the application of numerical tools towards one or another discretization technique. However, that kind of solidarity is neither reasonable nor useful when a wide spectrum of deformations, material conditions and interactions of different kind shall be investigated by numerical methods. The tool that solves the mathematical problem with the highest precision at comparably lowest effort should always be preferred. And that accounts not only for the various classes of problems in their initial configurations. Fundamental changes in the physical consistency of materials during a deformation process require not only for an equation of state that reflects the phase change. A discretization that hardly describes evolving new free surfaces cannot be the tool of choice for massive dynamic fragmentation processes. Similarly, large deformation processes with strong dependence on a precise tracking of material interfaces can hardly be simulated with Eulerian methods. The complexity of Eulerian methods rises enormously if a related multi-material tracking in individual cells is activated. And the resulting accuracy, at least in the case of structural applications, does most often not justify the effort.

---

[10] Inter-relations between complexity, risk and robustness of systems is for example investigated in Marczyk, Deshpande [265]

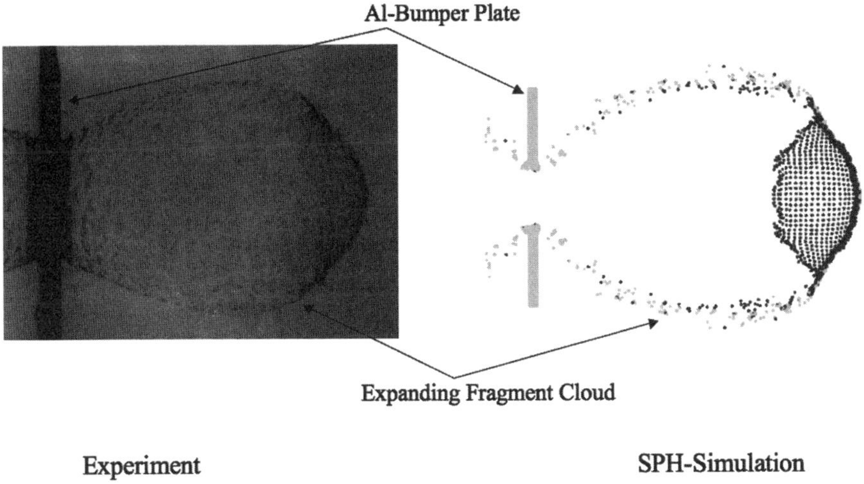

Fig. 5.25. Fragmentation of a 1.0 [cm] diameter Aluminum projectile after a 6.18 [km/s] hypervelocity impact on a 0.4 [cm] thick Aluminum plate. Experimental picture (left) taken by Flash X-Ray instrumentation. Simulation using the 2D axis-symmetric SPH version of AUTODYN.

Therefore, advanced application of numerical methods means right choice for an adequate discretization technique. And advanced provision of methods in a software package must mean that discretization methods can be combined and adaptively changed during the simulation, when ever needed.

Reflecting the most challenging tasks in the simulation of dynamic processes, two major sources of motivation can be identified in support of using meshfree methods instead or in combination with grid-based approximation methodologies:

- Simulation of **fundamental disintegration processes** in materials, i.e. deformation processes involving *multiple fragmentation* and possibly *phase changes* of solid materials to liquid or gaseous phases.
- **Fluid-Structure-Interaction** with complex and over time significantly changing interfaces between the fluid and the structural domain.

A standard example for the first motivation, fundamental structural disintegration, is the whole spectrum of hypervelocity impact processes. After perforation of a shielding plate, an impacting solid projectile is shattered by shock wave effects in a way that it propagates on as a dusty cloud of fragments. A typical expansion state of such a fragment cloud is illustrated in Figure 5.25. Depending on the impact velocity, the projectile and target materials and the geometrical dimensions of the involved structures, the cloud can consist of solid, liquid, gaseous or mixed phase constituents. Deformation states as for example the expanding fragment cloud illustrated in Figure 5.33 are far beyond the range of what can be covered with grid based Lagrangean methods. Also, Eulerian methods fail for application due to their deficiencies in keeping the loading history and material interfaces.

Topological changes of interfaces in fluid-structure-interaction simulations are the second challenge and motivation named above to use meshfree methods. For example, when the structure undergoes failure processes in the course of the interaction opening new free surfaces or ventings, the interaction schemes often face problems. Again, the fragment cloud propagating into a pressurized gas vessel represents an example for that problem type. A more common situation is the simulation of airbag inflation where meshfree methods like the Finite Pointset Method (FPM, Kuhnert [237]) are meanwhile implemented in commercial codes. Besides existing approaches to model airbag deployment using Arbitrary-Lagrangean-Eulerian (ALE) methods[11] and specifically used for inflation simulations[12].

5.8.2 Evolution and Maturing of Meshfree Methods

Meshfree particle methods exist already for several decades, its continuum mechanical branch since the early nineties. Other particle methods, which were not really meshfree but served as a meso-scopic representation of small particles in fluid flow, have been known before. The roots of these schemes go back to Francis Harlow's (Harlow [162], Harlow [163], or Evans and Harlow [122]) Particle in Cell (PIC) method.

Since its first introduction to solve continuum mechanical problems, particle methods were developed towards different directions. One basic determination can be made with respect to the type of approximation that is implemented in the various meshfree methods:

- Early particle methods, like SPH, are based on *kernel approximations*, with a so-called kernel function that is a weighting function as will be explained later. In the case of these kernel approximation methods, the implementation of the related weak form operation takes place directly with the numerical integration of the function.
- Later developed *field approximation* methods, for example Element-Free-Galerkin (EFG, Belytschko et al. [48]) or Meshless Local Petrov-Galerkin Methods (MLPG, Atluri and Zhu [21], Atluri, Shen [22]), perform the weak form operation in the course of creating the discrete equations.

---

[11] An ALE application to airbag deployment simulations is for example found in Marklund and Nilsson [266].

[12] Michalke et al. [278] investigate FPM for the simulation of airbag inflation.

Motivated by the need for a discretization technique for large deformation and fragmentation problems, SPH was totally meshfree from the beginning. After the identification of numerical problems inherent with the original SPH methodology, e.g. its lack in stability and consistency which will be discussed in the next subsection, the mentioned alternative particle methods with weak form descriptions in the discrete equations were developed. As a consequence of that change in philosophy, a background mesh was needed to perform the integration. In the case of EFG it is a global mesh whereas MLPG needs local background meshes for integration. Obviously, these new branches in particle methods, invented by scientist with a profound background in the numerical mathematics of finite elements, offer the stability and consistency that SPH had not. On the other hand, the core intention of particle methods to describe finite deformations including fragmentation was at least partly lost as well. Moreover, the local conservation of linear and angular momentum could not be satisfied any more, e.g. in the case of Moving Least Squares version of SPH introduced by Dilts [108], [109].

Therefore, recent re-evaluations of SPH based on a solid mathematical investigation as performed by Randles and Libersky [312], Monaghan [285], Belytschko et al. [43] and Rabczuk et al. [316] were essential steps forward in the development of SPH to becoming a mature, stable and consistent numerical method.

Since the large deformation processes including fragmentation and phase changes are the focal interest of this book, the particle method of choice to be discussed further is SPH. This is not in disrespect to other particle methods but due to the limited space that can be dedicated to finite methods in this book. A description of the basic SPH solution strategy in its original version, faced numerical problems and found remedies as well as currently existing robust SPH versions shall be introduced next.

5.8.3 Smoothed Particle Hydrodynamics

Integral Representation of a Function and its Approximation

Having evolved from a probabilistic particle method in molecular dynamics[13], Smoothed Particle Hydrodynamics was introduced by Lucy [255] as well as by Gingold and Monaghan [135] to solve the conservation equations for compressible and viscous fluids in astrophysical applications. An early example for the simulation of shock waves in fluids using SPH is Monaghan and Gingold [286]. A logical extension of the method to solids required an implementation of strength of materials. With the work of Libersky and Petschek [251] this

[13] Molecular dynamics: Stillinger and Rahmann [374]; Statistical Physics: Binder [59]

important step was performed and application of SPH to dynamic finite deformations in solids became possible for the first time. An early example of applying the SPH method to impact problems is found with Stellingwerf and Wingate [369].

The development of the basic numerical method starts with the integral representation of an arbitrary function $u(\mathbf{x})$:

$$u(\mathbf{x}) = \int_{\Omega} u(\mathbf{x}')\, \delta(\mathbf{x} - \mathbf{x}')\, \mathrm{d}\mathbf{x}' \tag{5.155}$$

which is exact if $u(\mathbf{x})$ is defined on the domain Ω and if the Dirac function defined by:

$$\delta(\mathbf{x} - \mathbf{x}') = \begin{cases} 1 & for\ \mathbf{x} = \mathbf{x}' \\ 0 & for\ \mathbf{x} \neq \mathbf{x}' \end{cases} \tag{5.156}$$

is used. Though it is exact, the integral expression (5.155) is not particularly useful for continuum mechanical solutions. Therefore, an approximation of the Dirac delta by an adequate continuous "smoothing" function is used to approximate equation (5.155). A possible estimation $\langle u(\mathbf{x})\rangle$ is found if a weighting function[14] $W(\mathbf{x} - \mathbf{x}')$ is used instead of the Dirac function:

$$u(\mathbf{x}) = \int_{\Omega} u(\mathbf{x}')\, W(\mathbf{x} - \mathbf{x}')\, \mathrm{d}\mathbf{x}' \tag{5.157}$$

This approximation leads us to the "smooth" character of Smoothed Particle Hydrodynamics as we will see in the following.

Original SPH Kernel Approximation

SPH Kernel Estimate of Functions

In its original version, the discretized SPH estimation of a scalar function $\langle u(\mathbf{x}_i)\rangle$ at the location of a discrete particle i is performed by an Eulerian approximation function $W(\mathbf{r}/h)$, which in the SPH context is usually called *kernel function.* It weights the functional values at neighbouring particles j which are located in a distance $\mathbf{r} = |\mathbf{x}_i - \mathbf{x}_j|$. The support domain is defined by a maximum distance $2\,h$ to which the kernel function is scaled. The scaling $\mathbf{r}/h$, for example, facilitates an approximate particle spacing of 1 particle per h (see Figure 5.26 for illustration).

The SPH kernel estimate $< u >$ is constructed by the integral:

[14] Lucy [255] denoted the weighting function W as *broadening function* which describes the influence of u to a certain surrounding region.

$$\langle u(\mathbf{x}_i,\, t)\rangle = \int_{\Omega} u(\mathbf{x}_j,\, t)\, W(|\mathbf{x}_i - \mathbf{x}_j|/h)\, \mathrm{d}\mathbf{x}_j \tag{5.158}$$

If we identify m_j/ρ_j as the differential volume $\mathrm{d}\mathbf{x}_j$, we can formulate the discrete version of (5.158) with the related summation:

$$\langle u_i\rangle = \sum_{j}^{n_N} u_j\, W(|\mathbf{x}_i - \mathbf{x}_j|/h)\, \frac{m_j}{\rho_j} \tag{5.159}$$

The number n_N of neighbour particles j building the support for the estimations (5.159) and (5.165) is defined by the domain of influence of the kernel function, i.e. by the smoothing length h. It can vary from particle to particle and at each individual particle from time step to time step. We will drop this upper limit of the summation in the following for convenience.

With the two-dimensional illustration of the kernel approximation in Figure 5.26 it becomes apparent that the nature of a particle method can be described as overlapping patches. A characteristic that reminds of the *Partition of Unity* (PU) method[15]. In fact, the close relation of many particle and finite element approaches to the PU method as root has been shown for example in Babuška et al. [27] or Atluri and Shen [20].

SPH Kernel Estimate of Derivatives

To derive an estimation for the spatial derivative of a function, the same integral representation as for the function estimate (5.158) can be formulated for the derivative $\nabla \cdot u(\mathbf{x}_i,\, t)$:

$$\langle \nabla \cdot u(\mathbf{x}_i,\, t)\rangle = \int_{\Omega} \nabla \cdot u(\mathbf{x}_j,\, t)\, W(|\mathbf{x}_i - \mathbf{x}_j|/h)\, \mathrm{d}\mathbf{x}_j \tag{5.160}$$

Following the traditional SPH path of approximating the derivative[16], the integrand of (5.160) can be written as:

$$\begin{aligned}\nabla \cdot u(\mathbf{x}_j,\, t)\;\; W\;(|\mathbf{x}_i - \mathbf{x}_j|/h) = \\ \nabla \cdot \left[u(\mathbf{x}_j,\, t)\, W(|\mathbf{x}_i - \mathbf{x}_j|/h)\right] - u(\mathbf{x}_j,\, t) \cdot \nabla W(|\mathbf{x}_i - \mathbf{x}_j|/h)\end{aligned} \tag{5.161}$$

which means for the approximation (5.160) after integration by parts that:

---

15 The partition of unity method was introduced by Babuška and Melenk [28] and serves for constructing mathematical proof on nonlinear manifolds.

16 For example introduced as basis for general particle methods in hydrodynamics by Monaghan [283].

$$\langle \nabla \cdot u(\mathbf{x}_i,\, t) \rangle = \int\limits_{\Omega} \nabla \cdot [u(\mathbf{x}_j,\, t) \;\; W\;(|\mathbf{x}_i - \mathbf{x}_j|/h)]\; \mathrm{d}\mathbf{x}_j \\ - \int\limits_{\Omega} u(\mathbf{x}_j,\, t) \cdot \nabla W(|\mathbf{x}_i - \mathbf{x}_j|/h)\; \mathrm{d}\mathbf{x}_j \tag{5.162}$$

Applying the divergence theorem to the first integral, (5.162) can be reformulated as a surface integral over the boundary S of the problem domain Ω with a unit normal vector $\mathbf{n}$ to it:

$$\langle \nabla \cdot u(\mathbf{x}_i,\, t) \rangle = \int\limits_{S} [u(\mathbf{x}_j,\, t) \;\; W\;(|\mathbf{x}_i - \mathbf{x}_j|/h)]\; \mathbf{n}\, \mathrm{d}S \\ - \int\limits_{\Omega} u(\mathbf{x}_j,\, t) \cdot \nabla W(|\mathbf{x}_i - \mathbf{x}_j|/h)\; \mathrm{d}\mathbf{x}_j \tag{5.163}$$

Given that the support domain of the interpolating kernel is fully inside the problem domain Ω, the surface integral in (5.163) vanishes and we can write:

$$\langle \nabla \cdot u(\mathbf{x}_i,\, t) \rangle = - \int\limits_{\Omega} u(\mathbf{x}_j,\, t) \cdot \nabla W(|\mathbf{x}_i - \mathbf{x}_j|/h)\; \mathrm{d}\mathbf{x}_j \tag{5.164}$$

Since the precondition for the validity of (5.164) is that problem boundaries are not inside the kernel support, boundaries require specific treatment as we will see later. The discretized form of the divergence (5.164) is estimated using the summation:

$$\langle \nabla u_i \rangle = - \sum_j u_j \, \nabla W(|\mathbf{x}_i - \mathbf{x}_j|/h)\, \frac{m_j}{\rho_j} \tag{5.165}$$

Kernel Functions

Kernel Requirements

Basically, the approximating kernel function should satisfy the following requirements:

- *Compact support*:

$$W(|\mathbf{x}_i - \mathbf{x}_j|/h) = 0 \quad \text{for} \quad |\mathbf{x}_i - \mathbf{x}_j| > 2\,h \tag{5.166}$$

- *Normality*:

$$\int W(|\mathbf{x}_i - \mathbf{x}_j|/h)\; \mathrm{d}\mathbf{x}_j = 1 \tag{5.167}$$

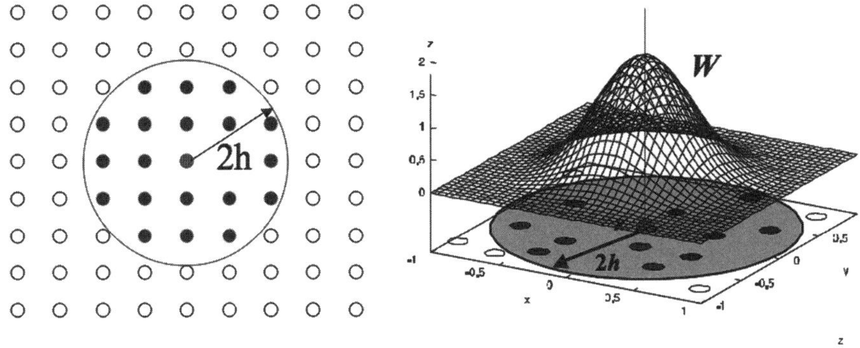

Fig. 5.26. Kernel approximation for functions $u(\mathbf{x}, t)$ at a particle i using neighbouring particles inside the $2\,h$ support domain of the kernel function $W(\mathbf{r}/h)$.

- *Symmetry*:

$$\int W(|\mathbf{x}_i - \mathbf{x}_j|/h) = \int W(|\mathbf{x}_j - \mathbf{x}_i|/h) \tag{5.168}$$

- and *Convergence* against the Dirac function:

$$\lim_{h \to 0} W(|\mathbf{x}_i - \mathbf{x}_j|/h) = \delta(\mathbf{x}_i - \mathbf{x}_j) \tag{5.169}$$

Cubic B-Spline

Various different kernel functions have been proposed since Gingold and Monaghan [135] proposed a Gaussian kernel while Lucy [255] preferred a third-order polynomial in bell-shape, respectively. A comprehensive overview on existing kernel functions and their properties is given in Liu and Liu [253]. The most often applied kernel estimate is probably the cubic B-spline, illustrated with its first and second derivative in Figure 5.30, which was proposed by Monaghan and Lattanzio [287]:

$$W\left(\frac{r}{h}\right) = \begin{cases} \dfrac{C}{h^\nu}\left\{1 - \dfrac{3}{2}\left(\dfrac{r}{h}\right)^2 + \dfrac{3}{4}\left(\dfrac{r}{h}\right)^3\right\} & \text{if} \quad 0 \leqslant \dfrac{r}{h} < 1 \\[2ex] \dfrac{C}{h^\nu}\left(2 - \dfrac{r}{h}\right)^3 & \text{if} \quad 1 \leqslant \dfrac{r}{h} \leqslant 2 \\[2ex] 0 & \text{if} \quad \dfrac{r}{h} > 2 \end{cases} \tag{5.170}$$

where the problem dimension is denoted by ν and C is a dimension dependent constant:

$$C = \begin{cases} \dfrac{2}{3} & \text{if} \quad \nu = 1 \\[2ex] \dfrac{10}{7\pi} & \text{if} \quad \nu = 2 \\[2ex] \dfrac{1}{\pi} & \text{if} \quad \nu = 3 \end{cases} \tag{5.171}$$

Compared to the former Gaussian kernels, the B-spline (5.170) has the advantage of automatically satisfying compact support (5.166). In addition it possesses a continuous second derivative. Liu and Liu [253] invented a new quartic kernel function and reported good results in simulating the Riemann problem applying it.

Smoothing Length

In the course of dynamic deformation processes as present in hypervelocity impact, changes in density can be comparable to thous of gas dynamical processes and, hence, reach order of magnitudes. The related change in inter particle distances can lead to either extremely high numbers of neighbours or to a complete loss of neighbours. To avoid such disadvantageous situations *variable smoothing lengths* are often used. A simple way of adjusting the smoothing length is realized in coupling it to its density by:

$$h = h_0 \left(\frac{\rho_0}{\rho} \right)^{\frac{1}{\nu}} \tag{5.172}$$

where density and smoothing length in the initial configuration are given by ρ_0 and h_0, respectively. The dimension of the problem is denoted by ν.

Alternatively, Benz [54] suggested to relate the smoothing length to the continuity equation via the divergence of the velocity vector. The time rate of change of the smoothing length is then expressed as:

$$\dot{h} = \frac{1}{\nu} h \nabla \cdot \mathbf{v} \tag{5.173}$$

which can be expressed by an SPH discretization:

$$\dot{h}_i = \frac{1}{\nu} h_i - \sum_j v_j \nabla W(|\mathbf{x}_i - \mathbf{x}_j|/h) \frac{m_j}{\rho_j} \tag{5.174}$$

Some forms of implementing variable smoothing length into code have been discussed in Hiermaier [176]. Common to all of these is the problem that anisotropic particle displacements lead to either overly big numbers of neighbours in one direction and related inaccurate approximations[17]. On the other

[17] Unphysical dispersion effects due to inadequate smoothing lengths have been investigated by Balsara [31] resulting in suggestions for optimal choices.

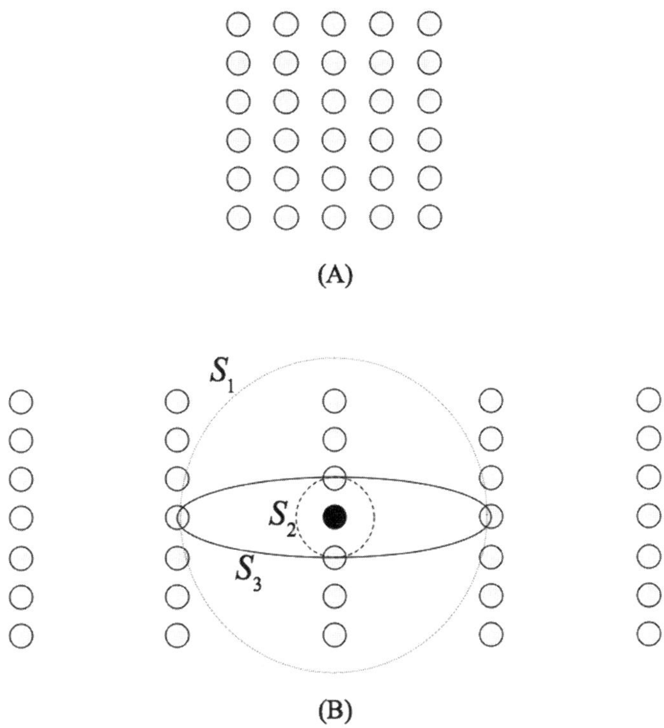

Fig. 5.27. Anisotropic displacement of SPH particles: (A) Original particle distribution. (B): Isotropic h-adjustment with support radii S_1 (dotted) and S_2 (dashed) leading to inaccurate estimates and numerical failure, respectively. An adequate anisotropic h-adjustment is illustrated by the S_3 support (solid line).

hand, the anisotropic displacement can cause a loss of neighbours and the related *numerical failure* in another direction (see Figure 5.27 for illustration).

Since the very dedication of SPH is its application to massive and not necessarily homogeneous deformations, an alternative basis for neighbour identification was needed. Libersky and Randles [250], [311] provided a method that utilizes *convex hull* and *interior hull* techniques (Figure 5.28) to identify nearest neighbours under arbitrary degrees of complexity regarding the deformation status. The basic idea is to map the coordinates of particles j which are candidate neighbours of particle i into a so called prime space by:

$$\mathbf{x}'_j = \frac{\mathbf{x}_{ij}}{\mathbf{x}_{ij} \cdot \mathbf{x}_{ij}} \tag{5.175}$$

As a result of (5.175), the position of particle i is mapped to infinity and infinity is mapped to the origin of the prime space. What is gained in terms of neighbour search is that the out-most particles in prime space forming a shell around all other particles can easily be identified as the neighbours of i. Figure 5.28 illustrates the procedure with a two-dimensional example.

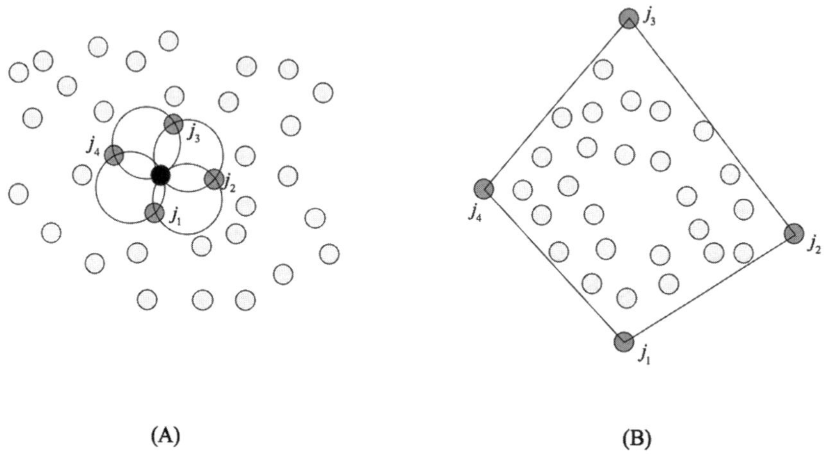

Fig. 5.28. Convex hull methodology to identify neighbours: (A) Original space with possible neighbours around a particle i and selected neighbours j_1-j_4 determined by a convex hull operation (B): $(0 \Rightarrow \infty)$, $(\infty \Rightarrow 0)$. (After Randles and Libersky [311]).

Symmetrization of the Kernel Estimate

Since the smoothing lengths of two interacting particles are not necessarily identical, the symmetry requirement (5.168) cannot be guaranteed without some kind of adjustment. That the requirement should be satisfied becomes apparent when a particle j is within the support domain of another particle i but not vice versa. In order to avoid that kind of violation of Newton's third law, various methods symmetrize the interactions between particles have been proposed. Benz [54] suggested an arithmetic mean of the smoothing lengths h_i and h_j:

$$h_{ij} = \frac{h_i + h_j}{2} \tag{5.176}$$

Liu and Liu [253] mention also geometric means as well as the maximal or minimal value as common smoothing lengths. Any of these symmetrizations of the smoothing lengths is finally used to construct a mutually used kernel function for particles i and j:

$$W_{ij} = W(|\mathbf{x}_j - \mathbf{x}_i|/h_{ij}) \tag{5.177}$$

As the only existing alternative a arithmetic mean of the kernel function values $W(|\mathbf{x}_j - \mathbf{x}_i|/h_i)$ and $W(|\mathbf{x}_j - \mathbf{x}_i|/h_j)$ which was proposed by Hernquist and Katz [174].

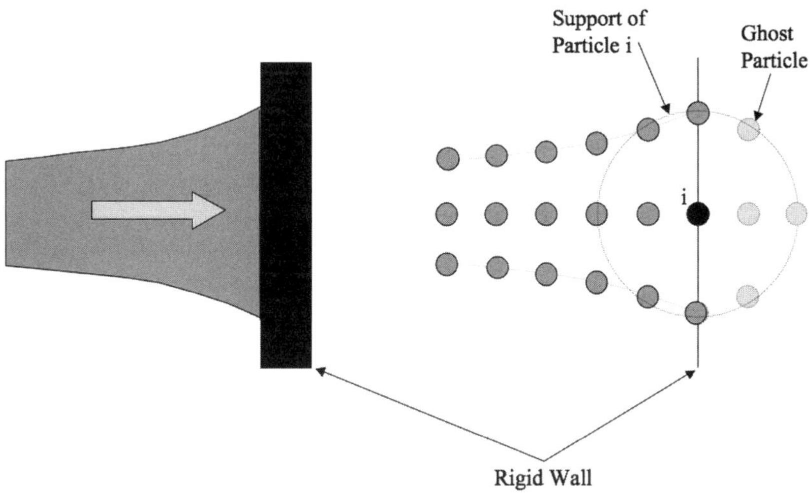

Fig. 5.29. Ghost particles employed to model a rigid boundary impacted by a cylindrical rod. The virtual particles are located at each time step individually for each particle reflecting all neighbours across the boundary.

Stability and Consistency Issues in SPH

Characteristic Properties of the SPH Approximation

It is essential for the understanding of the whole SPH methodology, as well as of its capabilities and limitations, to realize the character of the approximation (5.158):

- First of all, it is an approximation identical to FEM approximations but not an interpolation between particles since in contrast to the finite element shape functions the particle values are not identical to the approximations, i.e. $u_j \neq u(\mathbf{x}_j)$. As pointed out by Rabczuk et al. [316] this property leads to problems at boundaries and will need special attention for example provided in Dilts [110] or Belytschko et al. [44].

- Secondly, the Eulerian kernel estimate using current coordinates leads to a distortion of the stable domain and therewith to tensile instabilities as shown by Belytschko et al. [43].
- Thirdly, boundaries are problematic in the derivative approximation as well as in the function approximation. As long as boundary conditions, e.g. normal stress free surface conditions, the first problem type can be ignored. The second type of boundary problems results from the lack of supporting particles and the therefore truncated integral.

Directly related to these characteristic properties of approximation (5.158) are stability and consistency issues which turned out to be an inherent problem to the original SPH methodology. For the following discussion of the SPH stability problems, equations in index form will be used with the following conventions:

- Directional indices of vectors and tensors will be represented by upper Greek indices, e.g. the velocity component in α-direction is v^α.
- Lower Latin indices will denote the particle identity number. Therefore, the β-velocity at particle i is expressed by v_i^β.
- Double lower Latin indices represent the difference between values at two particle.

Hence, the expression

$$\sigma_{ij}^{\alpha\beta} = \sigma_j^{\alpha\beta} - \sigma_i^{\alpha\beta} \tag{5.178}$$

indicates the difference in a stress component between two particles j and i.

Completeness and Consistency

With the term *completeness* of nth order, the ability of an approximation to reproduce polynomials of order n is understood. Accordingly, the requirements for zeroth and first order completeness of a kernel approximation are:

$$\sum_j W_{ij} = 1 \quad \sum_j \nabla W_{ij} = 0 \tag{5.179}$$

and

$$\sum_j \mathbf{x}_j \, W_{ij} = \mathbf{x} \tag{5.180}$$

respectively.

For an exact reproduction of partial differential equations of order $n+1$, a completeness of order n must be required. Thus, the conservation equations, being of second order, demand linear or *first order completeness.*

The continuous approximation (5.158) fails for linear completeness but satisfies zeroth order completeness given a uniform particle distribution is provided and no boundaries are within th $2h$ support domain of a particle. Even worse, the discrete version of the SPH approximation does not satisfy zeroth order completeness on the boundary. This is due to the derivative term (5.165) and holds even true for uniform particle distributions.

These fundamental deficiencies of the original approximation lead to several improvement techniques in order to provide the needed first order completeness:

- Monaghan [284] suggested a *symmetrization* of the derivative approximation:
$$\langle \nabla \cdot u(\mathbf{x}_i, t) \rangle = -\sum_j (u_j - u_i)\, \nabla W_{ij}\, \frac{m_j}{\rho_j} \tag{5.181}$$
which bases on an assumption that is per se only valid for uniform particle distributions and in the absence of boundaries. Still, the resulting modified approximation (5.181) provides zeroth order completeness even for non-uniform particle spacings.

- Johnson and Beissel [213] introduced a *normalization* of the kernel approximation of a function which applied to density yields:
$$\rho_i = \frac{\sum_j m_j\, W_{ij}}{\sum_j \frac{m_j}{\rho_j}\, W_{ij}} \tag{5.182}$$

- Randles and Libersky [312] as well as Krongauz and Belytschko [233] formulated a *normalization* for derivatives that yields exact representations of the derivatives of constant and linear fields. Consequent application of kernel correction and normalization turned out to yield identical solutions as achieved with moving least squares (MLS) interpolants. The latter were invented by Lancaster and Salkauskas [242] and applied to particle methods by Belytschko et al. [48] and Dilts [108]. It replaces the original SPH approximation by
$$\langle u(\mathbf{x}_i, t) \rangle = \sum_j u_j\, \Phi_j(x_j) \tag{5.183}$$
where
$$\Phi_j(x_j) = \mathbf{p}(x_j)^T \cdot \mathbf{A}(x_j)^{-1}\, \mathbf{p}(x_j)\, W_{ij} \tag{5.184}$$
$$\mathbf{A}(x_j) = \sum_j p_j\, p_j\, W_{ij} \tag{5.185}$$
and an arbitrary given set of functions $\mathbf{p}$. If $\mathbf{p}^T = [1]$, then the MLS interpolant Φ is equal to the Shepard[18] functions:

[18] Shepard [356].

$$\Phi_i(x_i) = \frac{W_{ij}}{\sum\limits_j W_{ij}} \tag{5.186}$$

Dilts [108] gives a detailed derivation of the MLS type SPH methodology. Its equivalence in the case of a linear basis with corrected normalized SPH is documented in Randles and Libersky [311] as well as in Randles et al. [314]. Using the terminology $\sigma_{ij}^{\alpha\beta} = \sigma_j^{\alpha\beta} - \sigma_j^{\alpha\beta}$ and $r_{ij} = |\mathbf{x}_j - \mathbf{x}_i|$ as well as a MLS weighting function

$$w_{ij} = -\frac{m_j^2}{\rho_j} \frac{\nabla W_{ij}}{h_j\, r_{ij}} \tag{5.187}$$

the resulting kernel approximations for the divergence of the stress tensor and the velocity gradient are:

$$\left\langle \frac{\partial\, \sigma^{\alpha\beta}}{\partial\, x^{\beta}} \right\rangle_i = \frac{\sum\limits_j \sigma_{ij}^{\alpha\beta}\, x_{ij}^{\eta}\, w_{ij}}{\sum\limits_j x_{ij}^{\zeta}\, x_{ij}^{\eta}\, w_{ij}} \tag{5.188}$$

and

$$\left\langle \frac{\partial\, v^{\alpha}}{\partial\, x^{\beta}} \right\rangle_i = \frac{\sum\limits_j v_{ij}^{\alpha}\, x_{ij}^{\zeta}\, w_{ij}}{\sum\limits_j x_{ij}^{\zeta}\, x_{ij}^{\beta}\, w_{ij}} \tag{5.189}$$

, respectively.

- Vignjevic et al. [411] discussed the normalization with respect to preservation of spatial isotropy and established respective conditions.

Instabilities

Swegle et al. [381] and [382] were the first to investigated a specific numerical instability of SPH. The so called *tensile instability* manifests itself fist by unphysical oscillations in internal energy of individual particles and subsequently by clumping of many particles that leads to structural disintegration. In the course of a von Neumann stability analysis, Swegle, Hicks and Attaway [382] showed that small perturbations in a particle can lead to these instabilities not only under tension but also under compression. More precisely, the condition for the numerical problem is a combination of particle separation, kernel shape and sign of the stress tensor. Consequently, the found condition for instable behavior was found to be:

$$W''\, \sigma > 0 \tag{5.190}$$

Condition (5.190) separates the kernel support as illustrated in Figure 5.30 σ in two zones I and II at $|\mathbf{x}_j - \mathbf{x}_i| = 2/3\, h$, the location of the minimum of the kernel's first derivative. Under tensile stress states ($\sigma > 0$) instabilities can arise from interactions with neighbours in zone II, whereas compressive stress states are critical with particles in zone I. The resulting change in the character of the basic partial differential equations amplifies perturbations in the system.

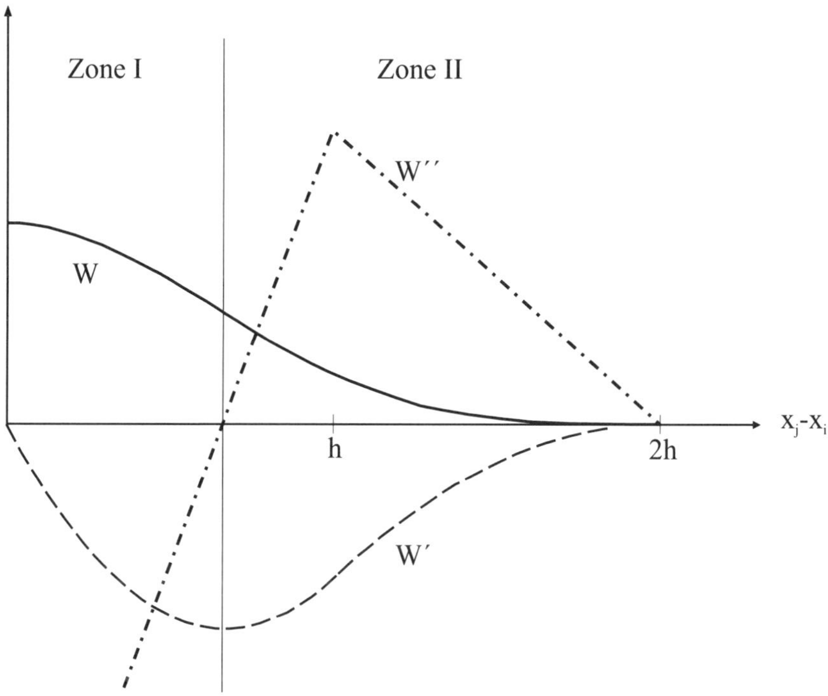

Fig. 5.30. Kernel function $W(|\mathbf{x}_j - \mathbf{x}_i|)$ (5.170) with its first and second derivatives and support zones I and II for which the instability criterion (5.190) predicts stable or instable behavior depending on the sign of the stress tensor σ.

Mainly based on the stability analyses performed by Swegle et al. [382], Belytschko et al. [43] as well as Belytschko and Xiao [49], the mathematical reasons for instabilities using SPH turned out to be of twofold nature:

- first it is due to distortion of the material, which is equal to the so called tensile instability
- and secondly it results from a rank deficiency of the discrete equations (see Section 8 in Belytschko et al. [47] for rank deficiency details).

Various improvements of the SPH methods have been developed reflecting the findings from the mentioned reviews. The most important ones are:

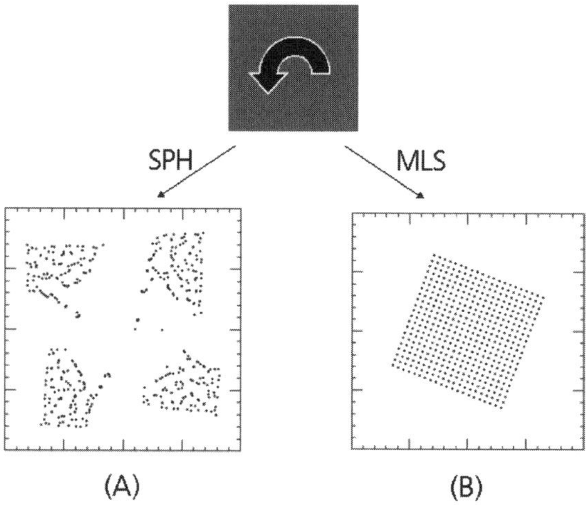

Fig. 5.31. Simulation of a counter-clock-wise rotating elastic bar with rectangular cross-section. Tensile instabilities in the original SPH methodology lead to virtual fragmentation. Application of MLS instead of the kernel approximation stabilizes the result partly as reported in Dilts [108] and Scheffer and Hiermaier [345] where the Dilts methodology was implemented in the FE-SPH coupled code SOPHIA.

- An early attempt to resolve this still as "tensile" denoted instability was suggested Guenther, Hicks and Swegle [154] by the *conservative smoothing* method. However, the related averaging of the approximations over neighbouring particles resulted in diffusive solutions and did not really solve the mathematical problem.

- A promising direction towards resolving the rank deficiency problem was opened by Dyka and Ingel [119] when they started to develop a *non-collocation SPH* version. What used to be a standard in finite element methods, i.e. the evaluation of the momentum equation to derive velocities and the calculation of strains and hence stresses at Gauss integration points, became now possible with particle methods. Necessarily, two types of particles had to be invented. One set, the *velocity points*, at which velocities are calculated from the stress divergence based on data coming from the second set of particles, called *stress points*. At the stress points, in turn, stresses are calculated from data gained at velocity points. Dyka and Ingel's first approach proofed to avoid the rank deficiency instability but was of one-dimensional type. At that time, it seemed quite impossible to expand the method to two and three dimensions due to limitations in computational capacities and the efficiency of the overall method.

- However, Randles and Libersky [311], Dyka, Randles and Ingel [120] as well as Vignjevic, Campbell and Libersky [411] successfully expanded the method to higher dimensions. Hence, a remedy for the rank deficiency caused instability was found. The *Dual Particle Dynamics* (DPD) called method is based on exactly that philosophy and was introduced by Libersky and Randles [252] as well as Randles et al. [314]. Still, the non-collocation SPH method is leaving the tensile instability unresolved (see Belytschko [43]).

- Rabczuk et al. [317] introduced the missing link to a stable particle methods: A total Lagrangean formulation of the kernel estimates allowed for additional stabilization of the "tensile" stability problem. The related total Lagrangean kernel uses the material coordinates $\mathbf{X}$:
$$W_{ij} = W\left(|\mathbf{X}_j - \mathbf{X}_i|/h\right) \tag{5.191}$$
Furthermore, a proper total Lagrangean SPH approximation uses finite strain measures and a related stress measure for the constitutive law and the conservation equations. An example for the resulting discrete formulations following this approach are given below in 5.8.3.

Boundary Treatment

To avoid inaccuracies from boundary influences, Campbell [77] discussed an inclusion of particular boundary terms into the approximation of derivatives (5.165).

Ghost particles are a concept introduced by Libersky and Petschek to model free surfaces. These virtual particles are located at the opposite side of the boundary edge. In case of free surfaces they only represent a mirror which reflects the velocity conditions of particles to the other side of a boundary. Ghost particles are also used to represent rigid boundaries. Figure 5.29 illustrates the use of ghost particles to model a rigid wall impacted by a rod.

A more sophisticated treatment of boundary zones including the application of ghost particles was introduced by Randles and Libersky [312]. It bases on a strict one-per-h particle distribution. Contributions in the density approximation of boundary particles come from interior, boundary and ghost particles, where the kernel summation for the boundary deficiency is corrected by normalization similar to the normalization used for derivative estimates. A boundary condition is applied to a field variable by assigning the same boundary value of the variable to all ghost particles. Interpolation of a constraint between specified boundary particle value and the calculated value on the interior particle imposes the boundary condition. As Vignjevic et al. [411] point out, the difficulties of the method are identified in defining the surface normals at the vertices and communication of the boundary values of a dependent variable from the boundary to interior particles.

Stable SPH Discretization for Hydrocodes

With the particle based kernel approximations (5.159) and (5.165) the continuous integral representation of a function u and its derivative $\nabla \cdot u$, respectively, were converted into discrete summations. As discussed in the foregoing paragraphs, these original approximations lead to instability problems and need to be corrected.

To avoid problems from both rank deficiencies and tensile instabilities, we will use the following descriptions:

- Following Rabczuk et al. [317] a *total Lagrangean* kinematic formulation with material coordinates $\mathbf{X}$ will be chosen. This implies the use of finite Green-Lagrangean strains (2.43) and Green stain rates (2.81):

$$\mathbf{E}(\mathbf{X},t) = \frac{1}{2}\left(\mathbf{F}^T\,\mathbf{F} - \mathbf{1}\right) \qquad \dot{\mathbf{E}} = \frac{1}{2}\left(\dot{\mathbf{F}}^T\mathbf{F} + \mathbf{F}^T\dot{\mathbf{F}}\right) \tag{5.192}$$

  as well as the related nominal stresses (2.102):

$$\mathbf{P} = (det\,\mathbf{F})\,\mathbf{F}^{-1}\,\sigma \tag{5.193}$$

- Secondly, the *MLS approach* derived in equation (5.183) now in total Lagrangean form is used which is equal to a corrected normalized SPH formulation:

$$\langle u(\mathbf{X}_i,\,t)\rangle = \sum_j u_j\,\hat{W}_{ij}\left(|\mathbf{X}_j - \mathbf{X}_i|/h\right) \tag{5.194}$$

  with the Shepard functions as MLS interpolant:

$$\hat{W}_{ij} = \frac{W_{ij}}{\sum\limits_j \frac{m_j}{\rho_j}\,W_{ij}} \tag{5.195}$$

  and its derivative:

$$\nabla\hat{W}_{ij} = \frac{\nabla W_{ij}}{\sum\limits_j \frac{m_j}{\rho_j}\,W_{ij}} \tag{5.196}$$

  With the total Lagrangean formulation of (5.187)

$$w_{ij} = -\frac{m_j^2}{\rho_j}\,\frac{\nabla\hat{W}_{ij}}{h_j\,r_{ij}} \tag{5.197}$$

the divergence of the nominal stresses $\nabla_0\cdot\mathbf{P}$ and the material velocity gradient $\dot{F}_i^{\alpha\beta}$ are approximated by:

$$\left\langle\frac{\partial\,P^{\alpha\beta}}{\partial\,X^{\beta}}\right\rangle_i = \frac{\sum\limits_j P_{ij}^{\alpha\beta}\,X_{ij}^{\eta}\,w_{ij}}{\sum\limits_j X_{ij}^{\zeta}\,X_{ij}^{\eta}\,w_{ij}} \tag{5.198}$$

and

$$\left\langle \dot{F}_i^{\alpha\beta} \right\rangle_i = \left\langle \frac{\partial v^\alpha}{\partial X^\beta} \right\rangle_i = \frac{\sum\limits_j v_{ij}^\alpha \, X_{ij}^\zeta \, w_{ij}}{\sum\limits_j X_{ij}^\zeta \, X_{ij}^\beta \, w_{ij}} \tag{5.199}$$

This allows us to approximate the conservation equations (ignoring body forces $\mathcal{B}$) for mass, momentum and energy in their total Lagrangean form (2.226), (2.228) and (2.227):

$$\rho = \frac{1}{J} \rho_0 \tag{5.200}$$

$$\dot{\mathbf{v}} = \frac{1}{\rho_0} \nabla_0 \cdot \mathbf{P} \tag{5.201}$$

$$\dot{e} = \frac{1}{\rho_0} \mathbf{P}^T : \dot{\mathbf{F}} \tag{5.202}$$

by a particle discretization.

Mass Conservation

To calculate the current density, the Jacobian determinant $J = det(\mathbf{F})$ is needed. The involved deformation gradient $\mathbf{F} = \nabla_0 \, \mathbf{u} + \mathbf{1}$ is approximated from the displacements $\mathbf{u}_i$ by:

$$\left\langle F_i^{\alpha\beta} \right\rangle = \frac{\sum\limits_j u_{ij}^\alpha \, X_{ij}^\zeta \, w_{ij}}{\sum\limits_j X_{ij}^\zeta \, X_{ij}^\beta \, w_{ij}} + \delta_{\alpha\beta} \tag{5.203}$$

where the indices of the Kronnecker delta $\delta_{\alpha\beta}$ are kept in the lower position.

Momentum Conservation

Momentum conservation is approximated using the stress divergence (5.198):

$$\langle \dot{v}_i^\alpha \rangle = \frac{1}{\rho_0} \frac{\sum\limits_j P_{ij}^{\alpha\beta} \, X_{ij}^\eta \, w_{ij}}{\sum\limits_j X_{ij}^\zeta \, X_{ij}^\eta \, w_{ij}} \tag{5.204}$$

Energy Conservation

With the MLS approximation of the material velocity gradient (5.199), total Lagrangean energy conservation is approximated by:

$$\langle \dot{e} \rangle = \frac{P_j^{\alpha\beta}}{\rho_0} \frac{\sum\limits_j v_{ij}^\alpha \, X_{ij}^\zeta \, w_{ij}}{\sum\limits_j X_{ij}^\zeta \, X_{ij}^\beta \, w_{ij}} \tag{5.205}$$

Typical SPH Applications

Fragmentation of Brittle Materials

Ceramic materials under impact loads are a perfect example for brittle failure processes that require highly flexible discretization if it shall be simulated. An experimental set-up that is often used to investigate crack propagation speeds in brittle materials is the so called *edge-on impact test*[19] illustrated also in Figure 7.21. The results in terms of pictures taken with 24 spark cameras are an excellent basis for the validation of numerical simulation, specifically of failure models.

The multiple fragmentation observed in a ceramic plate tested in the edge-on impact experiment makes meshfree methods an attractive discretization option. Although SPH seems to be the tool of choice for that kind of deformation, some specific adjustment is still to be made in order to allow for the simulation of fine crack propagation and crack opening as illustrated in Figure 5.32. Belytschko et al. [45] invented a *visibility criterion* that limits the support of a particle in the neighbourhood of a crack to its free surface. Hiermaier and Riedel [180] implemented the same in SOPHIA along with a reduced interpolation length for failed particles and a local particle refinement, i.e. location of additional, new particles. The applied statistical failure model used for the simulation of impacted aluminum-oxide Al_2O_3 illustrated in Figure 5.32 is explained in section 6.2.4.

Hypervelocity Impact on Space Vehicle Structures

In low earth orbits, i.e. at altitude of 250 to 400 km, impact of space debris or micro-meteoroids on space ship structures typically happens in a velocity regime of 3 to 20 $\mathrm{km\,s^{-1}}$. The use of Whipple shield technologies to protect the vehicles leads to massive fragmentation of the impacting objects as well as of parts of the impacted structures. Experimental investigation of the structural damage and the protective quality of the shield technologies is still limited to velocities around 10 $\mathrm{km\,s^{-1}}$. A predictive numerical simulation of the hypervelocity impact is of course reliant on a constitutive modelling including material characterization at strain rates in the regime of $10^6\,s^1$. Of equal importance for a predictive simulation is an adequate discretization. Starting with publications like Hayhurst and Clegg [170] and many more over the last ten years particle methods, and specifically SPH, has proven to be the methodology of choice for hypervelocity impact applications.

A comparably simple application of SPH to simulate hypervelocity impact induced fragment clouds is shown in Figure 5.25. The normal impact situation evaluated there allows for a two-dimensional axisymmetric discretization.

[19] Details to the experimental set-up and results derived with the edge-on impact test can be found in Straßburger and Senf [375].

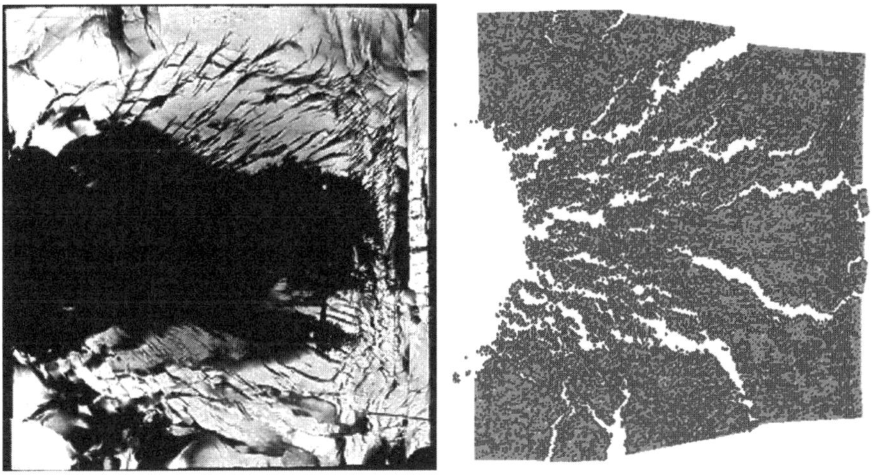

(A) (B)

Fig. 5.32. Multiple fragmentation in the course of an edge-on impact test on aluminum-oxide Al_2O_3: (A) Picture taken with an ultrahigh speed Cranz-Schardin camera (courtesy Straßburger and Senf [375]) and (B) Numerical simulation using the SPH code SOPHIA (see Hiermaier and Riedel [180] for details).

Discretization is restricted to the half-circular projectile and the target plate.

Still of two dimensional spatial character, but far more complex in its physical components is an impact on fluid filled vessels. Specifically when the fluid inside the vessel is pressurized and a impact driven fragment cloud is to propagate into this high pressure zone. The increase in complexity arises from the need for a spatial discretization over a wide area filled with a medium of far lower density and impedance.

An example for a pressurized vessel impacted by a spherical projectile is given in Figure 5.33. Hiermaier and Schäfer [182] investigated the shape and propagation speed of fragment clouds in high pressure containers usually attached to space vehicles. Two important aspects of fragment clouds in pressurized vessels can be observed from Figure 5.33:

- After an initial phase similar to the formation of fragment clouds in vacuum, an unusual formation of a spiky shape happens. The reason is that the friction induced deceleration of the individual particles depends on their kinetic energy and thus on the fragment mass. Smaller fragments are therefore decelerated much faster than bigger ones. This leads to the found spiky shape of the fragment cloud.

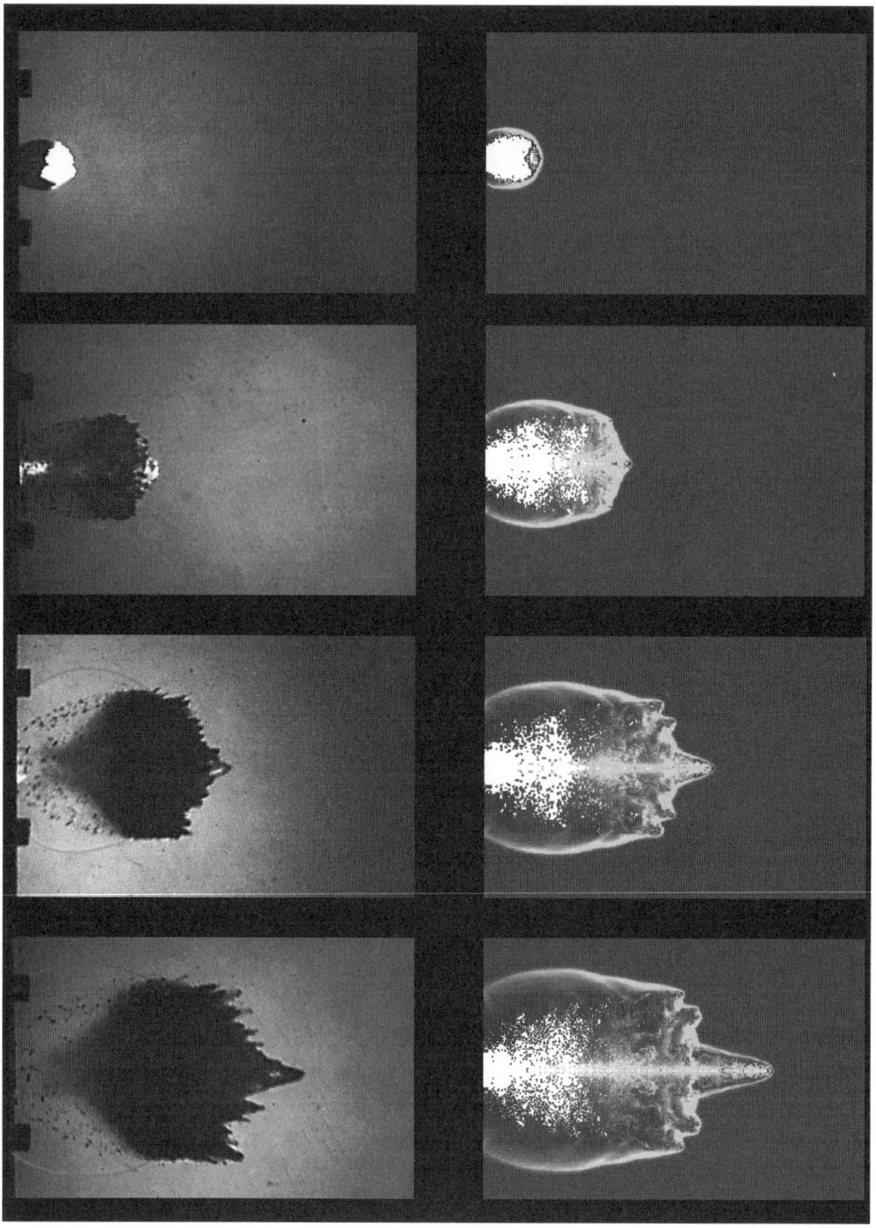

Fig. 5.33. Fragment clouds propagating into high pressure gas as observed in hypervelocity impact experiments (left column) and predicted by numerical simulation with Smooth Particle Hydrodynamics solver implemented in AUTODYN (right column) at 5, 15, , 25 and 35 [μs]. For details see Hiermaier and Schäfer [182].

- In the gaseous atmosphere around the fragment cloud, shock waves are initiated due to the high fragment velocities. These shocks propagate radially outward and may influence the structural behaviour of the pressure vessel. In the case of liquid fluid filled containers, the same effect is observed and called *hydraulic ram*. Numerical investigations on hydraulic ram effects are typically engaged with the problem of how to discretize the coupled fluid-structural problem.

Comparison of the experimental and numerical results provided in Figure 5.33 shows the potential of SPH discretizations for that kind of fluid-structure interaction.

5.9 Coupling and Adaptive Change of Discretizations

The high degree of specification in the various methods of numerical discretization allows for a both efficient and adequate discretization of individual loading conditions as well as material status. Structures undergoing large deformations and multiple fragmentation demand for an adequate representation in terms of discretization. The most complex and detailed material model looses its potential if the elementation is unable to account for the formation of new free surfaces or the partial loss in load carrying capabilities. Fluid-structure interaction, one of the most challenging processes for numerical simulation, by its multi-disciplinary nature, can neither be adequately described by pure Lagrangean methods nor by exclusively Eulerian ones.

Therefore, combinations of discretization methods as well as transitions from one method to another one upon criteria like failure or phase changes are reasonable. Adaptivity of finite methods was introduced to improve the accuracy of an approximation by error estimates and related changes in terms of mesh densities or shape functions. New ways of adaptivity include transitions from finite elements to meshfree particles in order to take account for basic changes in the state of the discretized material.

5.9.1 Meshfree - Finite Element Coupling

Coupling Classes

Combined application of different types of discretization requires a channel of communication and a definition of which information is to be communicated with which implications on the mutual sides. Basically, three types of coupling between meshfree methods and grid based finite element methods exist:

i) Master Slave coupling, which is more a contact algorithm then a coupling.

ii) Transition elements with boundary nodes that are at the same time SPH particles.
iii) Hybrid coupling with an extension of the kernel approximation by components from nodes inside the kernel support. Nodal force components are derived from contributions in the SPH summations of coupling particles.

iv) Adaptive coupling that can start with a certain initial discretization but also convert parts of the discretization into a more adequate one upon certain criteria. These criteria include situations where no failure in the material sense has occurred. Therefore, the partly changed discretization needs to be able to also represent an intact structure. That includes the ability to carry tensile and shear stresses across the evolved interface.

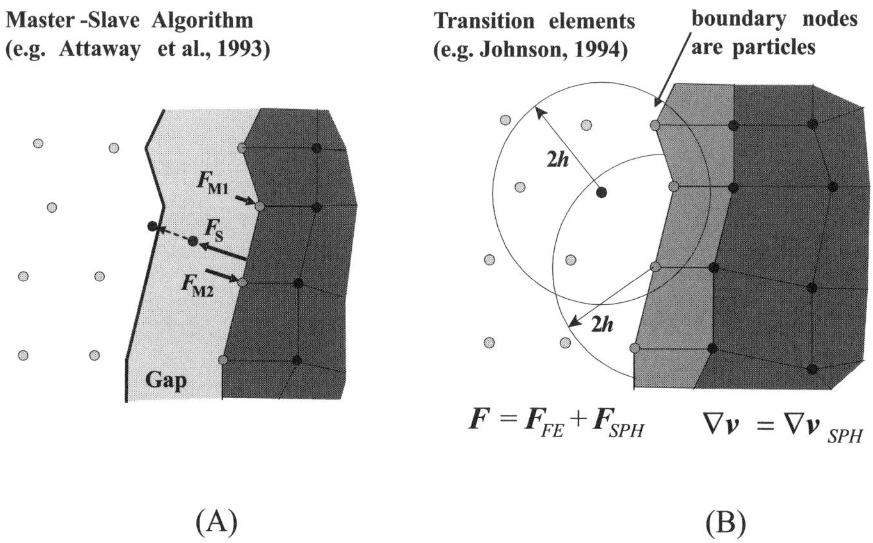

Fig. 5.34. Master slave (A) and transition element (B) coupling for combined SPH-FE discretizations. (Reprint with permission from Sauer [341])

The potential, but also the complexity rises with options i) - iv). An example for option i) was introduced in Attaway, Heinstein and Swegle [23]. As in many contact formulations for elements, the type i) coupling uses a gap space that defines the separation size between master and slave entities, in our case particles and elements. Figure 5.34 (A) illustrates a situation where a particle penetrating the gap space is repelled by a directed force F_S. Balancing the gap force nodal forces are applied to the surrounding element nodes.

Johnson [212] and Johnson et al. [218] proposed the use of transition elements with boundary nodes in the element regime that are at the same time particles. Figure 5.34 (B) shows the kernel supports of an original particle and of an element boundary particle. The deformation rates of boundary surface particles are calculated from all particles in its support but without nodal contribution from pure nodes. Forces, on the other hand, are calculated from both neighbour particles and transition elements.

Sauer Hybrid Coupling

Sauer [341] introduced a so called *hybrid coupling* and compared existing schemes of coupling meshfree methods with finite elements quantitatively. By the term *hybrid* the mixed contributions in evaluating the discrete balance laws is understood. If a particle has element nodes in its kernel support, then the element properties in terms of velocity and stress are used to calculate additional contributions to the standard meshfree formulations.

Sauer derived the new coupling method as a first step towards the ability of adaptively coupled particles and elements. An *adaptive change* means the change of discretization upon certain criteria, like for example element distortion, damage or failure.

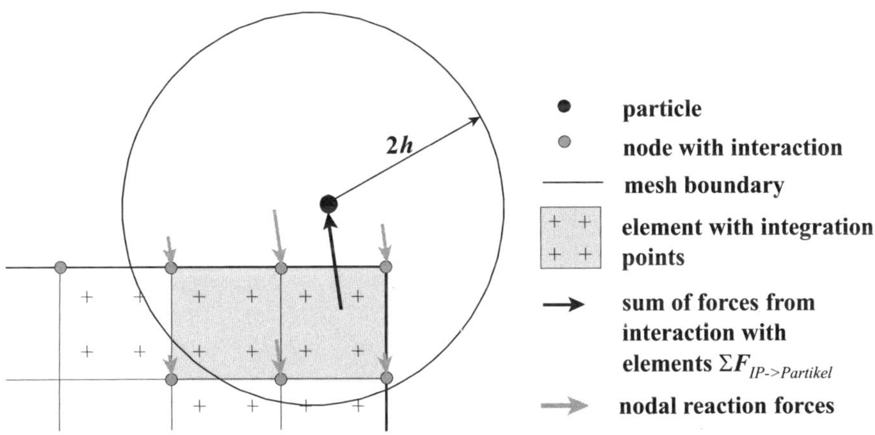

Fig. 5.35. Hybrid coupling as introduced by Sauer [341] uses variable properties at neighbouring elements and nodes to estimate the discrete SPH approximations of the balance laws. (Reprint with permission from Sauer [341])

The assessment of coupling methods provided by Sauer is performed with respect to the transmission quality of variables across a coupling interface. Investigated variables were

- velocities
- pressure
- momentum
- and energy.

The listed variables were calculated along the longitudinal direction of a bar to which a velocity boundary condition was applied on one side. The model was to one half discretized by finite elements and to the other side by SPH particles. Criterion for the quality of the coupling technique is the ratio of transmitted to reflected quantities in the variables as the transient waves cross the coupling interface. The more reflections occur, the more noise and artificial diffusion is registered in the system. As a result, the master-slave coupling was found to be inadequate for an adaptively changing discretization since did not represent tensile nor shear stresses.

Comparison of transition-element and hybrid coupling showed superiority of the latter due to improved momentum transmission with drastically reduced reflections. Additional quality in the hybrid method is gained if the particle resolution in the changed regime is variable. Details of the coupling assessment and the hybrid methodology are found in Sauer [341].

Examples of Application

Crack Propagation Under Quasi-Static Loading

Rabczuk and Belytschko [316] used a FE-EFG coupled discretization to model reinforced concrete slabs under quasi-static loading conditions up to failure. With the rebars modelled by finite element beam elements and the concrete using their Element-Free-Galerkin scheme, often faced problems in pure finite element solutions could successfully be overcome. The achieved degree in predicting the evolution and and propagation of multiple cracks including manifold intersections of cracks is an obvious proof of the potential in the applied coupling of numerical methods.

Fundamental Deformation Processes

A very unusual result of an impact test and at the same time a true challenge for numerical simulation was found in Rohr [331]. Originally, the impact experiments were designed to validate a material model for a 35NiCrMoV109 steel that covered the whole strain rate regime from quasi-static to rates of 10^6 $[s^{-1}]$. The tests consisted of impact experiments with penetrators and targets made from the same material. Cylindrical long rods with a length of 150 $[mm]$ and a diameter of 7.5 $[mm]$ were machined to serve as projectiles. The target was designed as semi-infinite, i.e. of much larger depth then what the projectile could penetrate at the given impact velocity of 1977 $[ms^{-1}]$. Flash X-ray instrumentation of the impact experiments revealed an interesting and surprising effect: the penetrator returned back out of the target after almost full penetration. As documented by high speed video and post-experiment investigations, the usual erosion process at the tip of the penetrator happened in way that the whole penetrator was flipped inside out. A process that could only happen due to a significant increase in the yield stress since the final deformation state consisted of a contiguous unbroken cylindrical hull. The inner surface of that cylinder was the outer surface of the former penetrator rod.

The challenge for the numerical simulation is twofold:

- The material model had to represent a wide range of strain rates in material with strongly rate dependent properties.
- The observed deformation process showed massive displacements without total disintegration of the structure.

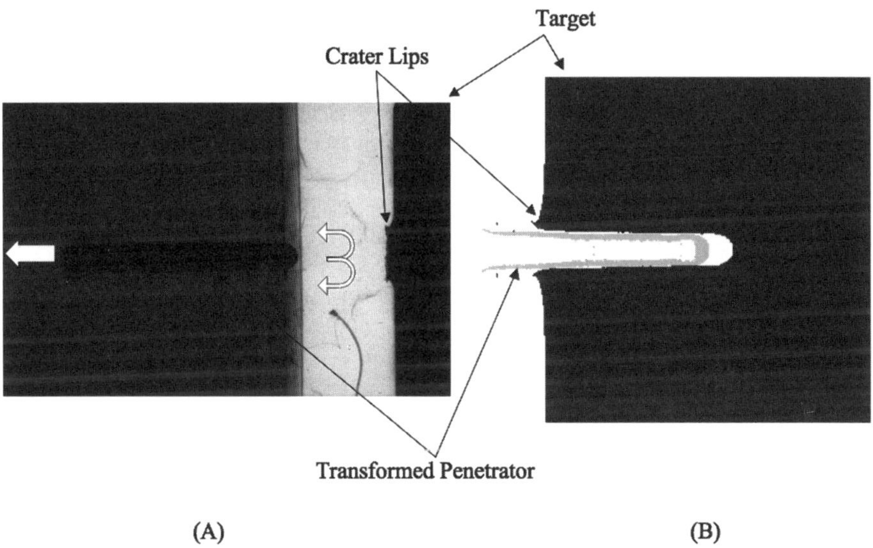

Fig. 5.36. Penetration process with $L = 150\,[mm]$, $D = 7.5\,[mm]$ 35NiCrMoV109 steel penetrator impacting a semi-infinite target at 1977 $[ms^{-1}]$. Experimental (A) and numerical (B) results showing an inside-out transformation of the penetrator and subsequent reverse motion at a velocity of -180 $[ms^{-1}]$ (experimental result) or -165 $[ms^{-1}]$ as calculated with AUTODYN. (Reprint with permission from Rohr [331])

Rohr choose a SPH discretization for the penetrator and finite differences for the target structure. The meshfree method was necessary since grid-based methods were not able to model the inside-out shape transformation of the penetrator. To describe the strain rate dependent plasticity, Rohr used his modified Johnson-Cook model (3.191) with the parameters given also in section 3.3.4. A Mie-Grüneisen shock equation of state (4.77) employing a quadratic $v_S - v_1$ relation:

$$v_S = 4622 + 1.35\, v_1 - 0.0007\, v_1^2 \tag{5.206}$$

was used. Finally, a stress-triaxiality dependent Johnson-Cook model (6.30) described the failure surface using the following parameters:

$$\begin{aligned} &D_1 = 0.48 \quad D_2 = 5 \quad D_3 = -3.7 \quad D_4 = 0 \\ &D_5 = \begin{cases} 0.0 & for\ T \leq 623\,[\mathrm{K}] \\ 5.5 & for\ T > 623\,[\mathrm{K}] \end{cases} \end{aligned} \tag{5.207}$$

The surprising experimental result is shown in Figure 5.36 (A) along with the solution calculated with the code AUTODYN, Figure 5.36 (B). As can be seen from the Flash X-ray picture, the penetrator is on his way out of the target visible at the right end of the picture with the impact shaped crater lips. The simulation result illustrates a moment briefly after the turn around point with a penetrator just about to move back out. Comparison of the experimental and numerical result for the crater depth shows 85 $[mm]$ and 82 $[mm]$, respectively. To measure the residual velocity of the penetrator, two pictures were taken at defined time instants. This revealed a residual velocity of -180 $[ms^{-1}]$, where the negative sign indicates the reverse flight direction with respect to the impact velocity. On the other hand, the numerical simulation predicted a very similar value of -165 $[ms^{-1}]$. In Rohr [331] more detailed results of the experiments can be found including cross sections of the penetration crater and pictures of the shape transformed projectile.

Fragmentation During Blast-Structure-Interaction

Blast-structure interaction on its own is a challenge for simulation since the two regimes of fluid dynamic and structural dynamic solutions need to be coupled in way that the coupling contains constraints for the respective other regime. Complexity rises if the structure is not only deformable but also likely to fail and to possibly break up into fragments. An application including brittle failure in structural components under blast loading was investigated by Hiermaier, Könke and Thoma [177] as feasibility study. Simulated was a blast wave in air interacting with a two-story building structure. The two-dimensional model contained a SPH based fluid solution for the air outside and inside the building, a finite difference discretization of the wall structures and a finite element shell representation of the windows. The applied coupling in AUTODYN utilized a master-slave type interaction between particles and elements similar to what was proposed by Attaway et al. [23] and Johnson [212].

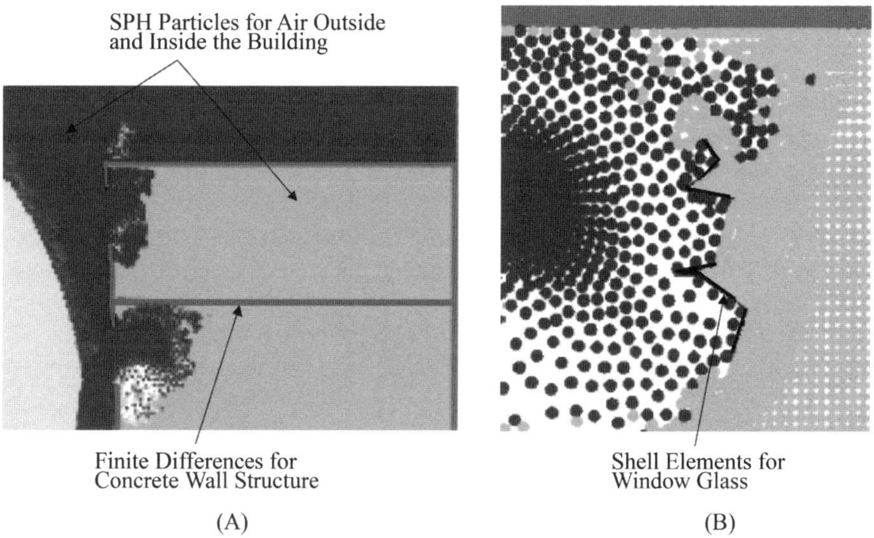

Fig. 5.37. Coupled FE-FD-SPH AUTODYN simulation of a blast-structure interaction involving fragmentation of glass windows.

Since window fragmentation is a major threat for people inside a building under blast loading, both the failure of the windows and the expected fragment sizes and velocities are of vital interest in a related numerical simulation. The described discretization was chosen to allow for a break-up of the window glass without loosing the interaction methodology. This is usually a problem if the standard Euler-Lagrange couplings are chosen instead. Here, the freedom of grids in particle methods provides a real alternative. At the same time a pure SPH discretization would miss the benefits of shell elements with respect to the achievable time step sizes. Several particle layers across the glass thickness direction would have drastically decreased the time steps.

Figure 5.37 (A) illustrates the position of the contact front, i.e. the interface between the expanding high pressure air and the initially unloaded ambient air, in front of the building. At that time, the blast wave has already hit the structure and shattered the windows. The exact position of window fragments originating from the lower window are shown in part (B) of Figure

5.37. Both the interfaces between fragments and air, as well as between originally outside air and originally inside air of the building are well recognizable.

Delamination of Composites under Crush Load

Composite failure can have many forms and mechanisms. Section 6.3 will introduce that variety to some extent. Crushing composite structures are an example for energy absorption processes in automotive crash. From experimental results it can be seen that failed fragments and specifically their interaction with neighbouring structures may influence the total deformation behaviour. Therefore, crushing composite structures are a good example for the application for an adaptive change in discretization. A SOPHIA implementation of Sauer's [341] hybrid coupling and related adaptive change from finite elements to particles was presented in Hiermaier et al. [179]. An example application was the vertical crushing of CFRP coupon specimen as illustrated in Figure 5.38. Delamination was modelled by element separation including the formation of new free contact surfaces. Failed or overly distorted elements were transformed into SPH particles to preserve the total structural mass including all fragments. A three-dimensional version of the same problem with a specific CFRP material model developed by Peter [305] will be given in the course of section 6.3.

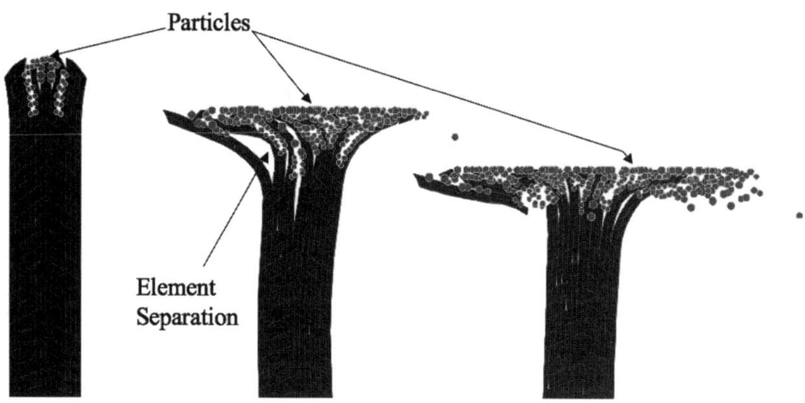

Fig. 5.38. Two-dimensional simulation of the vertical crushing of a carbon fibre reinforced plaastic coupon specimen. Depending on failure and distortion criteria, element separation and adaptive change of finite elements into SPH particles is performed. (For the related experimental set-up and a 3D simulation see Figures 6.15 and 6.16, respectively.)

Hypervelocity Impact on Sandwich Structures

Examples for the simulation of fragment clouds resulting from hypervelocity impact on thin protective Whipple shields were already given with the two calculations illustrated in Figures 5.25 and 5.33. A simulation of a more complex structure under hypervelocity impact is shown in Figure 5.39. The illustrated shielding system is a sandwich structure consisting of an aluminum honeycomb structure between two carbon fibre reinforced (CFRP) face sheets. Hypervelocity impact occurred, in the view of Figure 5.39, top down hitting the outer face sheet first. The resulting fragment clouds penetrates the honeycomb structure parallel to the foil planes. Due to its radial extension the fragment cloud perforates the honeycomb structure also radially.

The combination of impact induced fragmentation and thin walled structures behind the impacted face sheets makes a coupled discretization extremely desirable. Riedel et al. [324] applied the SPH-FE coupling of AUTODYN to discretize the aluminum projectile and the central impact zone of the outer CFRP face with SPH particles. This allowed for the simulation of fragment cloud. The outer zone of the impacted CFRP face sheet as well as the rear face sheet were modelled with eight node finite elements. The thin walled aluminum honeycomb structure was discretized by shell elements. With the coupling methodologies available in the code a representation of momentum transfer of the fragmented projectile and CFRP structure onto the honeycomb core became possible.

5.9.2 Coupling of Static and Dynamic Solvers

Sometimes it is desirable to formulate initial boundary value problems to calculate the initial conditions for a dynamic problem. Impact or blast loads on pre-stressed structures are a typical situation for a coupled approach of that type. Similarly, static solutions of a previously impact loaded structure can be of interest to predict the residual load carrying capability of a structure. In that case, an initial value problem would be formulated and solved to compute the initial and boundary conditions for the subsequent static solution. Some codes already provide capabilities of that type. Examples are ADINA, AUTODYN, LS-DYNA and PAM-Crash.

5.10 Shock Wave Simulation with Hydrocodes

A major field of application for hydrocodes is the simulation of shock waves in fluids and solids. With a decoupled evaluation of deviatoric and hydrostatic stresses along with the formulation of nonlinear equations of state the evolution and propagation of shock waves can be described mathematically. The

Fig. 5.39. Honeycomb structure between two CFRP face sheets impacted top down by a 3 [mm] diameter sphere at 4.49 $\mathrm{km\,s^{-1}}$. The AUTODYN simulation was performed using SPH particles coupled with both volume and shell elements for the face and honeycomb structure. (Reprint with permission from EMI report I-75/03)

quasi-discontinuous character of shock waves, however, makes the related numerical simulation at discrete positions in space and time a non-trivial task. Typical mesh densities in most real applications do not allow for a spatial resolution that could actually resolve typical shock wave dimensions of 10^{-5} [m] in solid metals to 10^{-7} [m] in gases. A course discretization smeares the shock out and produces less steep pressure jumps over longer rise times. Moreover, numerical schemes be it finite differences, elements or particle methods tend to predict unphysical oscillations right behind the shock jump. Therefore, to simulate shock waves with a reasonable spatial resolution specific adjustments in the schemes need to be made.

Two very common numerical treatments for shock waves are

- *Godunov methods*, a particular version of finite volumes that solves the Riemann problem over the volume boundaries (see section 5.6 for a brief introduction and links to further literature.)
- and secondly, the so called *artificial viscosity* which is used to damp the post-peak oscillations.

Applications of both options to model shock waves will be briefly discussed in the following. First, the basics of artificial viscosity will be introduced and finally used to compare shock propagation in solids using varying discretization and artificial viscosity. Then, the simulation of a shock loaded rigid structure due to a blast wave will be compared with experimental results. And finally an example for a full fluid-structure coupled blast simulation will be demonstrated.

5.10.1 Artificial Viscosity

More or less strong pressure oscillations right behind a shock front are a well know problem in all numerical schemes that do not explicitly solve the Riemann problem as the Godunov method does. To describe the problem and the effect of artificial viscosity, a one-dimensional case of shock wave propagation shall be discussed.

Unphysical Oscillations Upon Shock Wave Formation

As illustrated in Figure 5.40, a bar of 1 [cm] length shall be loaded at his left end by a triangular pressure boundary condition. The load shall be applied over a duration of 1 [μ s] with its peak pressure of 0.1 [M bar] at 0.2 [μ s]. A one-dimensional finite difference discretization with an element length of $2\,10^{-4}$ [cm] is used to resolve the physical dimensions of a shock wave over several elements.

Figure 5.40 also shows the formation of a shock wave in the one-dimensional simulation. The pressure and velocity distributions in the bar are illustrated at two instants of time: the first at $t_1 = 0.66$ [μ s] and the second at $t_2 = 1.5$ [μ s]. At t_1 the induced pressure wave already has a steep front but the shock wave has not yet formed. At t_2 the situation is different. Now a shock wave is propagating from left to right. At the same time, the peak pressure is followed by an unphysical oscillation. The influence of grid resolution is shown in Figure 5.41. A modified element size of $2\,10^{-3}$ [cm], i.e. ten times larger elements,

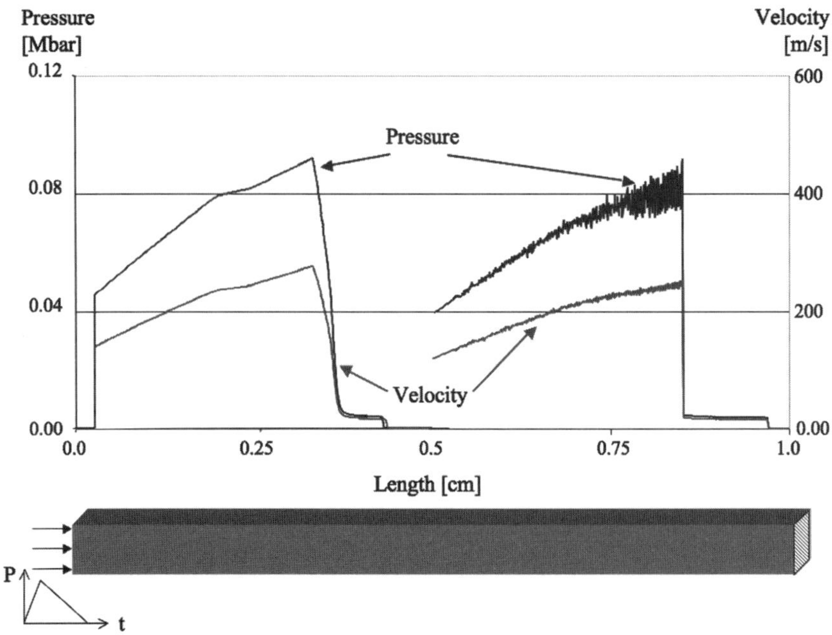

Fig. 5.40. Pressure loaded bar and calculated pressure and velocity distributions at $t_1 = 0.66$ [μ s] (left two curves) and $t_2 = 1.5$ [μ s] (right two curves), respectively.

results in a less steeper, step-wise pressure rise to the same peak pressure and very similar amplitudes in the subsequent oscillations. The oscillation is what methods like artificial viscosity shall change.

Though the name *artificial viscosity* sounds like an unphysical term is introduced to the method, there is a physical explanation for the idea. The conservation equations as well as the equation of state or constitutive equations do not reflect the dissipative aspects of shock waves. There is however always a certain amount of kinetic energy transformed into heat. The stronger the shock the more. Accordingly, a viscous pressure term is calculated from velocity gradients and added to the equation of motion and energy conservation.

Formulation of Artificial Viscosity

About ten years before the invention of the Godunov method, von Neumann and Richtmyer [295] proposed the use of an additional pressure term in order to damp these oscillations. From the Hugoniot equations (4.45) to (4.47) the pressure behind the shock can be derived as:

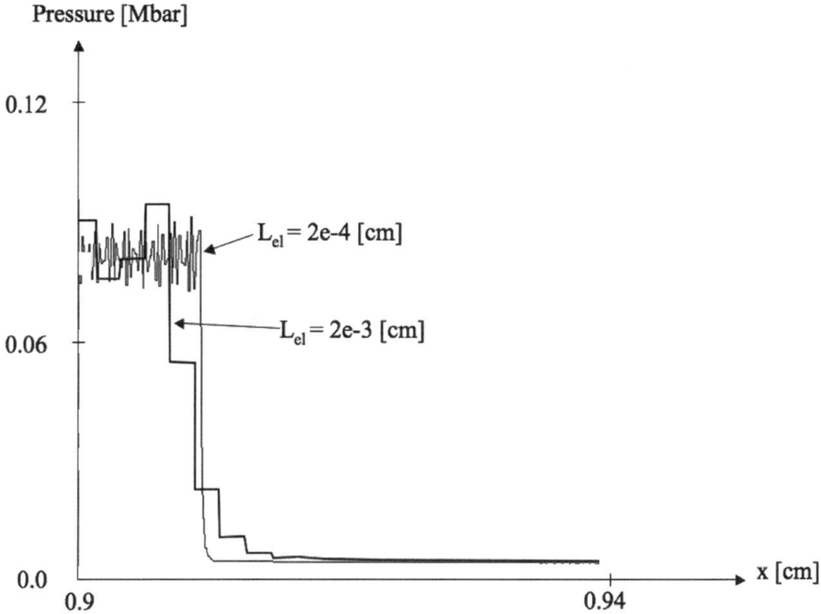

Fig. 5.41. Pressure and velocity distributions using element sizes of $2\,10^{-4}$ [cm] and $2\,10^{-3}$ [cm], respectively.

$$p_1 = p_0 + \frac{\rho_0}{\frac{\rho_1}{\rho_0} - 1}\,(v_1 - v_0)^2 - \rho_0\,v_0\,(v_1 - v_0) \tag{5.208}$$

containing a quadratic and a linear term in the velocity difference $(v_1 - v_0)$. In its original form, the von Neumann and Richtmyer viscosity was a nonlinear, quadratic term in the velocity difference Δv between two discretizing nodes:

$$q_{nl} = \alpha\,\rho\,(\Delta v)^2 \tag{5.209}$$

with a parameter α of dimension unity and a condition defining that q_{nl} should only be non-zero for compressive situations. Its primary task was to avoid zone inversion problems in shock simulations by smearing out the shock over a few cells or nodes[20]. Though with the nonlinear form (5.209) modelling of strong shocks was facilitated, it still produced the mentioned oscillations behind the shock. Therefore, Landshoff [245] invented a linear term

$$q_{lin} = \beta\,\rho\,c_0\,|\Delta v| \tag{5.210}$$

[20] An illustrative example of grid distortion due to strong shocks simulations without artificial viscosity is demonstrated in Wilkins [425].

with the sound speed c_0 was added. Thus, a methodology was found to spread the shock over a few cells and to damp the oscillations. A basic form of artificial viscosity q, applicable to both fluids and solids, that contains the quadratic and the linear velocity difference term is then:

$$q = \alpha \, \rho \, (v_1 - v_0)^2 + \beta \, \rho \, c_0 \, (v_1 - v_0) \tag{5.211}$$

where the parameters α and β define the degree of dissipation that is added to the system. The linear term, controlled by β, dissipates the noisy oscillations, whereas the α controlled quadratic contribution smeares out the shock front over a few discretizing nodes. Hence, a reduced oscillation is traded in against a less sharp shock front.

An implementation of the artificial viscosity term q into the conservation equations can be performed by adding q as a pressure term. Hence, the total stress tensor becomes:

$$\sigma_{ij} = S_{ij} - P \, \delta_{ij} - q \, \delta_{ij} \tag{5.212}$$

Artificial Viscosity in Finite Difference Schemes

Applied to a finite difference scheme[21], the artificial viscosity is calculated at nodes using a volumetric strain rate $\dot{\varepsilon}_v$ instead of the velocity difference $(v_1 - v_0)$ to get the nodal viscosity:

$$q_i = \alpha \, \rho \, \dot{\varepsilon}_v^2 + \beta \, \rho \, c_0 \, |\dot{\varepsilon}_v| \tag{5.213}$$

Figure 5.42 gives an impression of both the reduced oscillations and the reduced steepness of the shock front if an artificial viscosity is used with $\alpha = 1.0$ and $\beta = 0.2$ for the afore discussed bar problem. Apparently, the simulation with the lower spatial resolution is stronger affected by the artificial viscosity whereas in the high resolution case a pressure signal with reasonable rise time and eliminated oscillation results.

A detailed investigation of artificial viscosity formulations for multi-dimensional shock wave computations is for example provided in Caramana et al. [78] as well as in Wilkins [425].

Artificial Viscosity for SPH

Monaghan and Gingold [286] introduced an artificial viscosity for SPH. The so called *bond viscosity* is expressed as:

$$q_{ij} = C_Q \, \rho_i \, \mu_{ij}^2 + C_L \, \rho_i \, c_i \, |\mu_{ij}| \tag{5.214}$$

for negative relative velocities $\mu_{ij} < 0$, where

[21] See for example Walsh [415].

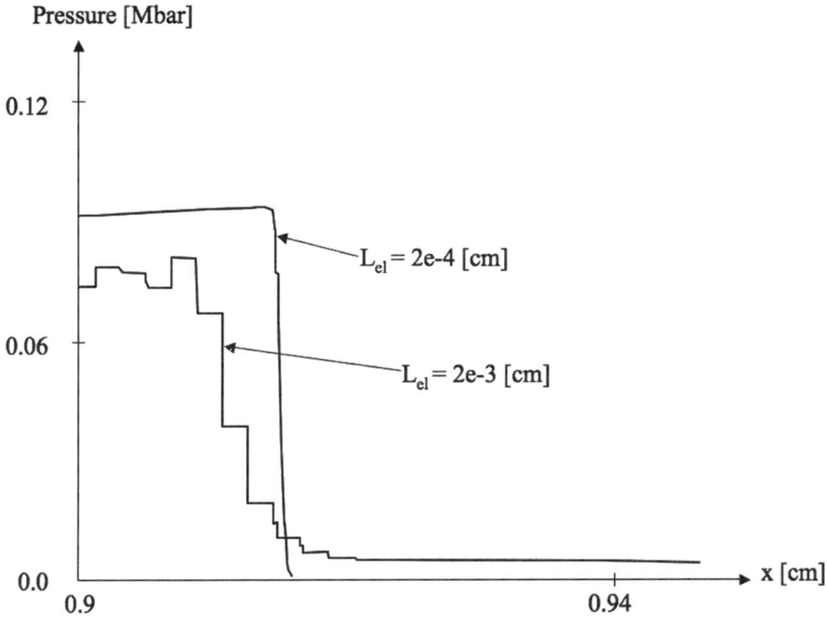

Fig. 5.42. Pressure and velocity distributions using element sizes of $2\,10^{-4}$ [cm] and $2\,10^{-3}$ [cm], respectively, and an artificial viscosity according to (5.211) with parameters $\alpha = 1.0$ and $\beta = 0.2$.

$$\mu_{ij} = \frac{h_{ij}\, v_{ij}^{\alpha}\, x_{ij}^{\alpha}}{r_{ij}^2 + \epsilon_0\, h_{ij}^2} \tag{5.215}$$

with the particle distance $r_{ij} = \sqrt{\mathbf{x}_{ij}}$ and a small number ϵ_0, usually set to a value around $\epsilon_0 = 0.01$, that avoids singularities of μ_{ij} for small particle distances. Double lower indices denote differences in values between two particles and upper indices denote directional components, both according to the SPH index formalism introduced in 5.8.

Balsara [31] points out deficiencies of (5.214) in the case of shear stresses as well as shear flows. To improve the particle viscosity term, Balsara suggests a correction factor to the relative velocities μ_{ij}:

$$f_i = \frac{|\nabla \cdot \mathbf{v}_i|}{|\nabla \cdot \mathbf{v}_i| + |\nabla \times \mathbf{v}_i| + \epsilon_1\, c_i/h_{ij}} \tag{5.216}$$

where $\epsilon_1 = 0.0001$ keeps the denominator positive for small velocity gradients $|\nabla \cdot \mathbf{v}_i|$ or rotation rates $|\nabla \times \mathbf{v}_i|$.

Libersky and Randles [250] investigated the application of artificial viscosity in SPH and suggest the use of a *directional viscosity*, also called *tensor viscosity* term. Their tensorial viscosity formulations for the quadratic shock smearing term $q_{2i}^{\alpha\beta}$ and for the linear noise reducing $q_{1i}^{\alpha\beta}$, respectively, are:

$$q_{2i}^{\alpha\beta} = a_2 \frac{\rho_i \sum_j v_{ij}^{\alpha} v_{ij}^{\beta} w_{2ij}}{\sum_j w_{2ij}} \tag{5.217}$$

and

$$q_{1i}^{\alpha\beta} = a_1 \frac{\rho_i c_i \sum_j \left(n_{ij}^{\beta} v_{ij}^{\alpha} + n_{ij}^{\alpha} v_{ij}^{\beta} \right) w_{1ij}}{\sum_j w_{1ij}} \tag{5.218}$$

with dimensionless coefficients a_1 and a_2, sound speed at a particle c_i, a unit vector

$$n_{ij}^{\alpha} = \frac{x_{ij}^{\alpha}}{|x_{ij}|} \tag{5.219}$$

and a weight function of unity $w_{1ij} = 1$ for the linear term, and a weight function for the quadratic term that allows viscosity only for positive relative velocities between particles, i.e. approaching particles with $x_{ij}^{\alpha} v_{ij}^{\alpha} < 0$:

$$w_{2ij} = \begin{cases} 0 & \text{for} \quad x_{ij}^{\alpha} v_{ij}^{\alpha} \geqslant 0 \\ \frac{\left(x_{ij}^{\alpha} v_{ij}^{\alpha}\right)^4}{\left(x_{ij}^{\beta} x_{ij}^{\beta}\right)^2} & \text{for} \quad x_{ij}^{\alpha} v_{ij}^{\alpha} < 0 \end{cases} \tag{5.220}$$

Randles et al. [313] specifically report excellent results using this particle viscosity applied to strong shock problems. Provided examples include the Noh problem, where a cold ideal gas moves radially inward at unit speed, as well as the Taylor anvil test, a symmetric cylindrical rod impact simulated for Copper.

5.10.2 Air Blast Effects on Structures

As a first step towards solutions for blast-structure-interactions, the blast propagation in the fluid needs to be predicted. A coupling of fluid and structural discretization facilitates the interaction in terms of a pressure boundary for the structure and a velocity boundary for the fluid. In the case of moderate structural deformation the modelling of the interaction process can be performed as a split task where a blast simulation with rigid structures predicts the pressure loads over the whole loading time. In a second step, the calculated pressure history is used in a pure structural simulation to calculate the resulting deformations.

For large deformations, possibly including fragmentation, this first approach is not adequate and a fully coupled simulation is needed. An example that used coupling of meshfree particles, finite differences and finite elements to simulate window failure under blast load was given in the last section.

Blast Interaction with Rigid Walls

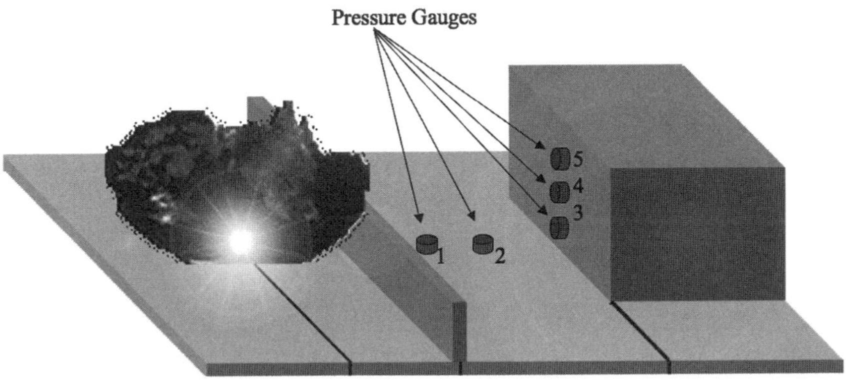

Fig. 5.43. Small scale blast wall experiment performed by Rice and Neuwald [321] with a 0.347 [g] Nitropenta charge in front of a blast wall. The ground between blast wall and building as well as the building front were instrumented with pressure gauges.

An example for the application of a Godunov scheme, here restricted to calculating pressures on a rigid structure, shall be given now. The example is intended to illustrate the quality in the simulation of blast waves with Godunov solvers.

To investigate the effectiveness of blast walls in front of buildings, Rice and Neuwald [321] performed small scale detonation tests and numerical simulation. The basic set-up is shown in Figure 5.43: A 0.374 [g] Nitropenta load was initiated. The blast propagation across the blast wall towards the building front was investigated by five pressure gauges. Figure 5.44 shows schlieren pictures from a 24-spark camera instrumentation at times 27, 67, 107 and 247 [μs] after initiation.

Rice and Neuwald performed two- and three-dimensional numerical simulations with the SHARC code. A comparison of numerical and experimental

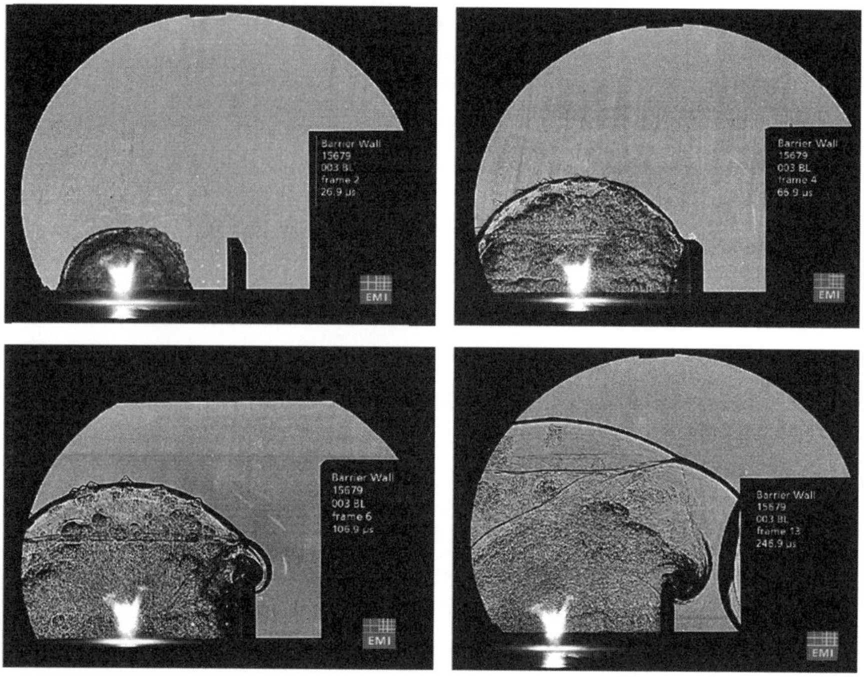

Fig. 5.44. Schlieren-optical pictures of the blast propagation at 27, 67, 107 and 247 [μs] after initiation.

results, illustrated in Figure 5.44, with respect to pressure and impulse histories at gauge 5 shows excellent agreement.

Blast Interaction with Deformable Structures

Deformable structures subject to blast effects are in many ways a real challenge for numerical simulation. Both, the fluid dynamical and the structural dynamics regime need to be modelled carefully to cover all relevant effects. In addition, the interaction process requires a specific code module. An example for calculating both regimes with individual codes coupled via a technique that is usually applied for parallel solutions on multi-processors computers is given next. The codes used were the Godunov solver APOLLO for the fluid domain and SOPHIA for the deformable structure part.

Figure 5.46 outlines the investigated situation where a shock tube is used to generate a plane shock wave that hits a deformable structure, in this case a simple pin made from aluminum. Instrumented with a 24-spark camera, the wave propagation and deformation process could be observed experimentally.

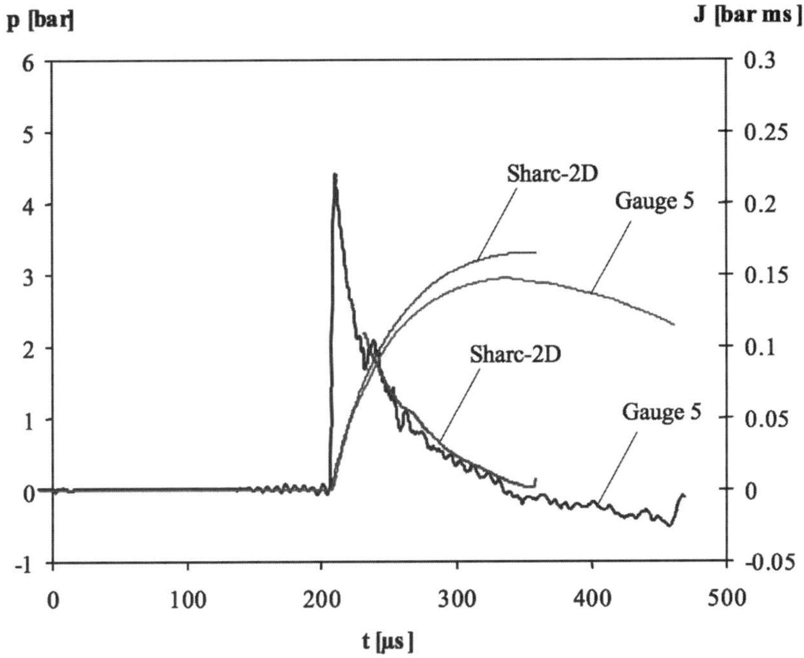

Fig. 5.45. Pressure and impulse histories at gauge 5 (Figure 5.43) as recorded in the experiment and calculated with SHARC-2D. (Reprint with permission from EMI, Annual Report 1998. For details see also Rice and Neuwald [321].)

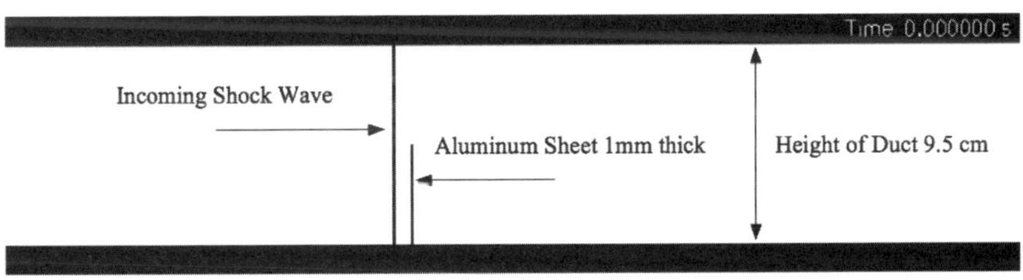

Fig. 5.46. Shock tube with a 1 [mm] aluminum pin.

Figure 5.47 compares the shock reflections and the growing deformation in the shock loaded pin as observed in experiment and numerical simulation.

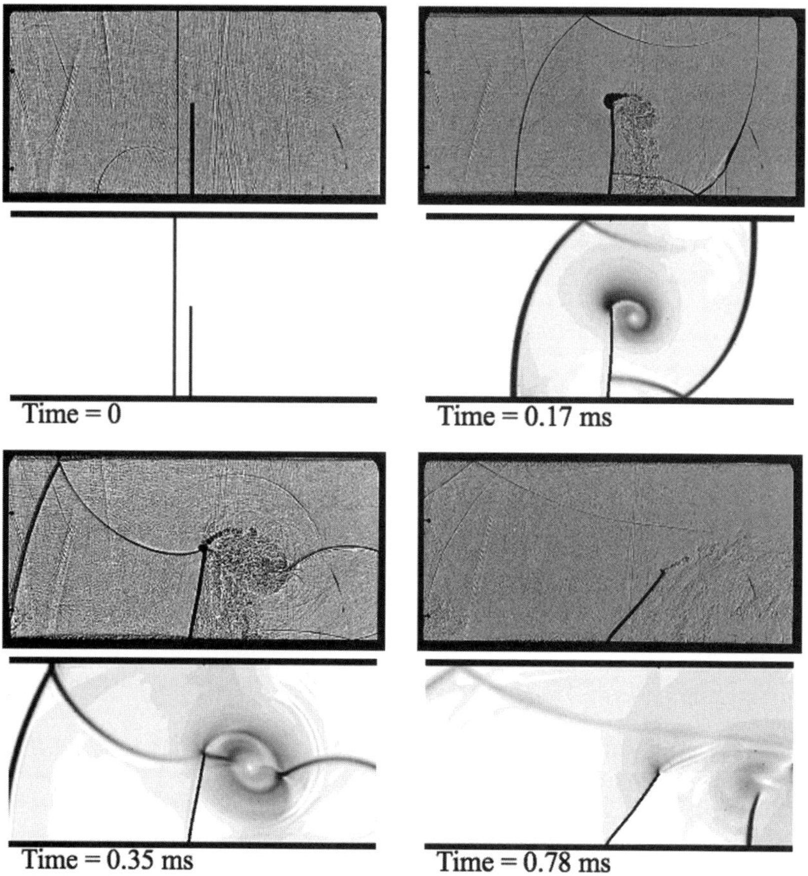

Fig. 5.47. Shock wave reflections and pin deformation as observed by Schlieren optics and numerical simulation with the EMI codes APOLLO and SOPHIA in a coupled mode.

6 Failure Models for Dynamic Loading Conditions

Failure, is an ultimate condition at which structures or materials loose the ability to carry loads, has become a broad field of theories and application in material science. The spectrum covers classic elastic and plastic fracture mechanics[1] for ductile and brittle materials under static, creep, fatigue and dynamic loading conditions observing individual or multiple cracks. In brittle materials, e.g. glasses or ceramics, failure under tensile loading typically occurs at strain state magnitudes of much less than one percent. In contrast, ductile materials show several tens (ductile metals) up to hundreds (polymers) of percent in plastic strain upon failing. Nucleation, coalescence and growth of voids in the materials are the driving mechanisms of large plastic strain in ductile failure.

Similar to plastic flow, failure thresholds may for some materials be rate dependent and for others not. Ultimate strain to failure can vary extremely. Ductile behaviour at quasi-static loads can turn into brittle failure at elevated rates (metals) and vice versa (concrete). A basic publication on nucleation and growth as well as rate dependent fracture mechanics in solids is written by Curran, Seaman and Shockey [100]. Experimental methods to investigate the high dynamic fracture mechanisms including spallation under shock loading and modes to describe them in codes are collected in Antoun et al. [10].

Fracture mechanics, formulated in the elastic-plastic regime, describes the local discontinuous propagation of cracks upon given crack shapes and loading conditions. Continuous formulations of initial versus damage induced degraded strengths of materials are utilized when the discretization of individual crack formations shall be avoided for the sake efficiency of solving problems involving multiple fragmentation. Nevertheless, continuous descriptions

[1] Historically, Griffith [146], Irwin [206] and Orowan [302] can be called the initiators of fracture mechanics. Standard textbooks on the topic are Gross and Seelig [147] or Broek [68]

of failure need to reflect the whole complexity of the related micro-mechanical processes, albeit at the comparably course resolution of a chosen macroscopic discretization, e.g. when energy dissipation under crash and impact shall be predicted. Therefore, the predictive capability of a material model including failure components is always a function of both the constitutive description and the quality of the discretization.

A chosen constitutive formulation may call for additional efforts on the discretization side. A typical example is the continuum damage approach leading to strain-softening. Strain softening leads to a change in the type of the governing equations, i.e a loss in ellipticity for quasi-static and a loss in hyperbolicity in dynamic problems. The result is a strong mesh dependency and an infinite number of solutions. In that case, additional methods like non-local or gradient formulations are needed to make physically meaningful solutions possible.

Multiple fragmentation, on the other hand, is a phenomenon that, once described by constitutive equations, demands for an adequate type of representation in the numerical discretization. A finite element discretized structure fragmenting under dynamic loads demands for additional functionalities like the discussed X-FEM method, grid opening or even transition to alternative discretizations, e.g. mesh-free methods. The more important local failure effects are for the overall energy dissipation behaviour of a structure, the higher the effort that must be taken in terms of formulating micro-mechanical influences on the constitutive model as well as in terms of its discretization.

Still, pure micro-mechanical descriptions as well as scale bridging so called *multi-scale material models* may be out of range for industrial applicability. However, the fast development in computational speed and available techniques in parallel processing along with advanced methods in adaptive discretizations of time and space are the basis for future refined descriptions of material behaviour including failure for standard industrial applications. Therefore, it is worthwhile directing more focus on replacing macroscopic empirical models by profound micro-mechanical based or on microscopic scale formulated relations.

In the course of this chapter, some failure characteristics of different kind will be explained through representative materials. After reviewing some models for dynamic failure in isotropic materials often used for metals, more complex failure mechanisms in polymer matrix fiber composites will be discussed. To reflect the evolution of failure and the related changes in mechanical properties before failure, the concept of continuum damage mechanics will be introduced first.

6.1 Continuum Damage Mechanics

Continuum damage approaches describe the degradation in strength of a material by the observation and modelling of microscopic effects. In contrast to empirical failure descriptions where macroscopic observations of failure initiation upon specific stress-strain states are modelled, damage mechanics account for the initial existence, deformation induced nucleation, growth and coalescence of micro-cracks or voids at grain boundaries, between crystals or components of composites. Individual cracks and crack tip motion through intact material are modelled within the micro-mechanical branch of fracture mechanics called *discrete damage mechanics*. Application of fracture theories on a microscopic scale allows for investigations of the mechanism at a high spatial resolution. A phenomenological approach to characterize the overall structural behaviour due to failure mechanisms is performed by the so called *continuum damage mechanics*. Here, in contrast, empirical findings define a set of internal state variables - as introduced in section 2.6.3) - in order to formulate and solve equations of irreversible thermodynamics. Of course, the precise location and distribution of micro-defects can be described by statistical methods. Continuum damage mechanics, however, tries to avoid statistics. Introductions to continuum damage mechanics and in depth discussions of the related theories are collected in Krajcinovic [231], Lemaitre [247] and in Allix and Hild [1]. A consistent framework on elastoplasticity coupled with damage is provided by Hansen and Schreyer [160].

6.1.1 Effective Stress and Strain Equivalence Concepts

Along with the formation of micro-cracks the macroscopic load carrying surface in a structure is reduced gradually. At the same time, there is a more or less gradual reduction of stiffness and or strength after damage initiation instead of a sudden loss of structural integrity. The idea originally formulated for creep processes by Kachanov and Rabotnov introduces a scalar damage variable d to describe material degradation. Kachanov [221] picked out a Representative Volume Element (RVE) from a damaged body. In that RVE, he considered an intersecting plane of total cross sectional area S and a damaged sub-area covered by internal voids S_d. The *surface damage scalar* d is then defined as ratio between the damaged and the total area:

$$d = \frac{S_d}{S} \tag{6.1}$$

Rabotnov's [309] *effective stress concept* relates the uni-axial tensile stress through the change of an effective area to the damage scalar:

$$\tilde{\sigma} = \frac{\sigma}{1-d} \tag{6.2}$$

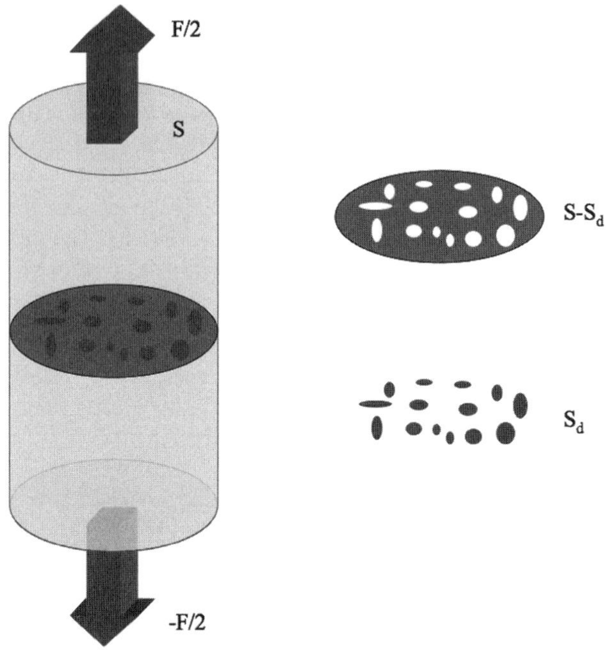

Fig. 6.1. Distribution of damaged area of total size S_d in cross section of size S in a representative volume loaded by a force F.

Following the idea of a surface damage scalar, a quantification of the damage parameter is achieved by

- either a micro-mechanics based modelling of the micro-voids including a calculation of the resulting effective stress-strain relations and the effective strength state or
- an approach that allows for direct or indirect experimental measurment of the amount of damage.

The various damage mechanisms under conditions of real application make a micro-analysis complex and computationally expensive. Specifically a whole spectrum of possible stress and strain states needs to be regarded. Therefore, a way of measuring damage is often preferred. Again, large variations in the size and distribution of micro-cracks make a direct method of counting the void area in a specific surface and accumulating the respective surface contributions almost impossible. With the *strain equivalence principle* and a related assignment of damage to strain softening and changes in stiffness, respectively, Lemaitre [246] proposed non-destructive indirect measurements of damage scalars. The first step of this method contains the assumption of a strain equivalence, i.e. the hypothesis that components of a strain tensor in a homogeneously damaged material state can be calculated the same way as for a virgin, undamaged material. The strain calculation is then facilitated by the effective stress $\tilde{\sigma}$ noted in equation (6.2) :

$$\varepsilon = \frac{\tilde{\sigma}}{E} = \frac{\sigma}{E(1-d)} \tag{6.3}$$

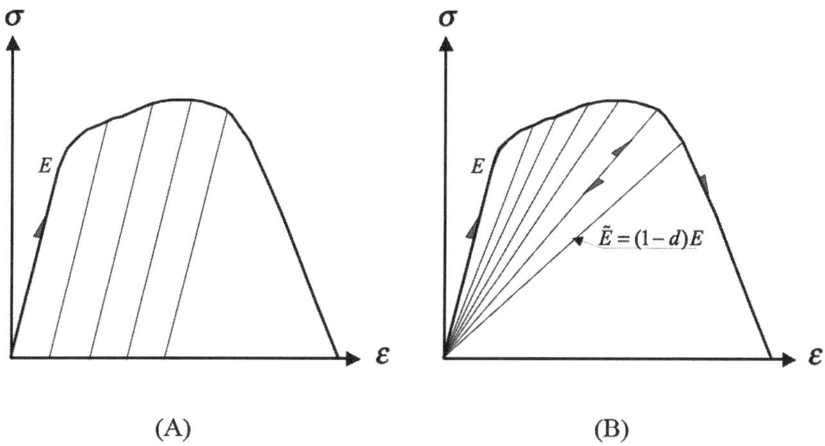

Fig. 6.2. Model representation of strain softening by (A) pure elasto-plasticity or (B) damage accumulation.

Hence, an indirect non-destructive way to characterize damage becomes possible by cyclic loading and unloading of specimen in material tests at well defined stress and strain states. Changes of elastic moduli, e.g. a reduction of the Young's modulus from E in the undamaged state to $\tilde{E} = E(1-d)$ after deformation induced damage initiation, are then a possible measure for the accumulated damage d:

$$d = 1 - \frac{\tilde{E}}{E} \tag{6.4}$$

Figure 6.2 (B) illustrates how damage accumulation and the related stiffness degradation can cause strain softening. Whereas an elasto-plastic behaviour shown in Figure 6.2 (A) leads to plastic strain components at constant elastic behaviour, the pure damage accumulation model changes the elastic modulus but implies no irreversible strain components. Realistic material behaviour is to be expected somewhere in between these two models, as for instance assumed in Figure 6.3.

Alternative approaches to measure damage are by observing the variations in wave propagation speeds, micro-hardnesses or the reduced strength under strain softening. The reader may be referred to Lemaitre [247] to find examples on the application of standard tension tests as well as ultrasonic wave propagation tests or experiments to evaluate the damage-induced variation of micro-hardness in order to derive further measures for a damage scalar

The strain state at which damage evolution initiates may be described by an equivalent strain value $\bar{\varepsilon}_i$ and the ultimate strain state at rupture of the material can be denoted by $\bar{\varepsilon}_u$. The amount of deformation and, thus, the critical accumulated equivalent strain under damage $\bar{\varepsilon}_c = \bar{\varepsilon}_u - \bar{\varepsilon}_i$ rupture of the material occurs varies between $\bar{\varepsilon}_c \approx 0$ for brittle materials and very high values for ductile ones.

A rather simple approximation to calculate a damage scalar in numerical simulation is therefore to track the accumulated effective strain $\bar{\varepsilon}$, to relate the damage parameter to it as $d = d(\bar{\varepsilon})$ and to describe the degradation in elastic moduli or strength accordingly. Depending on how the equivalent strain is calculated from the strain components, that approach describes a certain evolution of damage in the strain space and forms a critical surface to rupture. Application of a scalar measure to describe damage assumes isotropic damage behaviour.

6.1.2 Degradation and Damage Accumulation Functions

Elastic Degradation

Progressive degradation of the material stiffness under cyclic loading and unloading conditions, illustrated for example in Figure 6.3, starts when the load amplitudes exceed certain thresholds in the stress or strain space. The subsequent material response typically shows both a reduced elastic moduli and irreversible plastic strain components. To quantify the latter effect is the task of plasticity models. A simple concept to describe degradation was given in the last paragraph with the "$(1 - d)$"-factor and the related effective Young's modulus $\tilde{E}$. To meet the realistic material behaviour as close as possible, a combined degradation, damage and plasticity model is needed.

Changing material stiffness can be represented by a rank-four secant stiffness tensor E_{ijkl} in the stress space through:

$$\sigma_{ij} = E_{ijkl}\, \varepsilon_{kl} \tag{6.5}$$

As long as the loading conditions remain in the elastic regime, there is no change in the stiffness tensor, thus $\mathrm{d}E_{ijkl} = 0$. With the definition of a loading

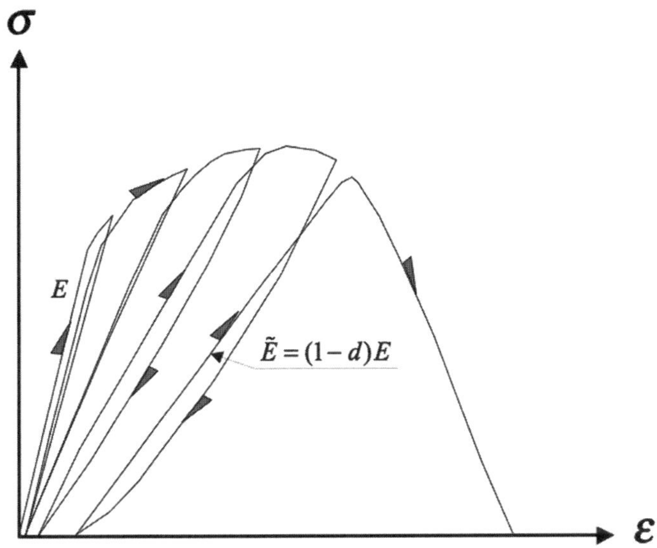

Fig. 6.3. Schematic combination of plasticity and damage induced strain softening under cyclic loading and unloading.

surface $F(\sigma)$ beginning degradation can be identified in analogy to plasticity models by a condition $F = 0$. Under this condition, the secant stiffness tensor changes ($\mathrm{d}F \neq 0$) and strain increments $\mathrm{d}\varepsilon$ can be decomposed into an elastic part $\mathrm{d}\varepsilon^{el}$ and a degrading excess strain increment $\mathrm{d}\varepsilon^{dg}$:

$$\mathrm{d}\varepsilon_{ij} = \mathrm{d}\varepsilon_{ij}^{el} + \mathrm{d}\varepsilon_{ij}^{dg} \tag{6.6}$$

giving a related stress increment

$$\mathrm{d}\sigma_{ij} = E_{ijkl}\left(\mathrm{d}\varepsilon_{ij}^{el} + \mathrm{d}\varepsilon_{ij}^{dg}\right) \tag{6.7}$$

In section 2.6.5 we identified damage as an internal state variable with a strain energy density release rate Y as conjugate thermodynamic tension. Therefore, the Helmholtz free energy Ψ of a damaged element is formulated first in order to derive a damage evolution model that is consistent with thermodynamics.

For elastic isothermal deformation processes the specific Helmholtz free energy given by equation (2.286) reduces to:

$$\psi = \frac{1}{2\rho}\varepsilon_{ij}\, E_{ijkl}^{el}\,(1-d)\,\varepsilon_{kl} \tag{6.8}$$

leading to the strain energy density release rate Y

$$Y = -\rho\,\frac{\partial\psi}{\partial d} = \frac{1}{2}\varepsilon_{ij}E^{el}_{ijkl}\varepsilon_{kl} \quad . \tag{6.9}$$

With the maximum value of Y over the load history $\tau \leq t$ at time t

$$\tilde{Y}(t) = \max_{\tau \leq t}\left(\sqrt{Y_{ij}(\tau)}\right) \tag{6.10}$$

and an experimentally derived critical value Y^c, the accumulated damage can be described as:

$$d = \frac{Y(t)}{Y^c} \tag{6.11}$$

An often found description of damage accumulation is directly related to increments in the equivalent plastic strain $\Delta\,\bar{\varepsilon}^{pl}$ normalized to the equivalent plastic strain to failure $\bar{\varepsilon}^{pl}_f$:

$$d = \frac{\sum \Delta\,\bar{\varepsilon}^{pl}}{\bar{\varepsilon}^{pl}_f} \tag{6.12}$$

6.2 Isotropic Failure Models

As pointed out in the introduction to this chapter, it cannot be the intention of this book to provide the reader with a complete failure theory but with a selection of models relevant to dynamic deformation processes and their implementation in hydrocodes.

6.2.1 Maximum Stress or Strain Criteria

Since failure is a threshold above which materials loose their ability to carry loads, stress or strain criteria are used to quantify these thresholds. As in the case of plasticity the criteria are identified by surfaces in the stress or strain space. The most simple failure models are maximum stress or maximum strain criteria. A maximum stress model, for example, compares the current components of the stress tensor σ_{ij} with experimental results for ultimate stresses to failure under tension, compression or shear:

$$max\left\{\left|\frac{\sigma_{xx}}{X_{T,C}}\right|, \left|\frac{\sigma_{yy}}{Y_{T,C}}\right|, \left|\frac{\sigma_{zz}}{Z_{T,C}}\right|, \left|\frac{\sigma_{xy}}{S_{xy}}\right|, \left|\frac{\sigma_{xz}}{S_{xz}}\right|, \left|\frac{\sigma_{yz}}{S_{yz}}\right|, \right\} = 1 \tag{6.13}$$

The comparison of individual stress components σ_{ij} with its related ultimate values does not regard effects from combined loading conditions. A straight forward approach to compare a multi-axial stress state with ultimate stresses derived from material tests under uni-axial stress is the use of equivalent stresses. An equivalent stress is a scalar derived from the components of the stress tensor according to a specific hypothesis. Common hypotheses and related equivalent stresses expressed in terms of principal stresses are:

- Maximum Principal Stress Criterion

$$\bar{\sigma}_P = \sigma_1 \tag{6.14}$$

- Maximum Shear Stress Criterion

$$\bar{\sigma}_S = \frac{1}{2}\,(\sigma_3 - \sigma_1) \tag{6.15}$$

- Maximum Distortional Strain Energy Criterion, also called von Mises equivalent stress

$$\bar{\sigma} = \frac{1}{\sqrt{2}}\,\sqrt{(\sigma_1 - \sigma_2)^2 + (\sigma_2 - \sigma_3)^2 + (\sigma_3 - \sigma_1)^2} \tag{6.16}$$

In the same way as for the stresses, measures for equivalent strains can be defined. Using the von Mises equivalent strain scalar

$$\bar{\varepsilon} = \frac{1}{\sqrt{2}}\,\sqrt{(\varepsilon_1 - \varepsilon_2)^2 + (\varepsilon_2 - \varepsilon_3)^2 + (\varepsilon_3 - \varepsilon_1)^2} \tag{6.17}$$

various empirical models containing a equivalent strain to failure $\bar{\varepsilon}_f$ have been proposed. Examples are:

- a constant Equivalent Strain criterion:

$$\bar{\varepsilon} = \bar{\varepsilon}_f \tag{6.18}$$

- the Cockcroft-Latham [94] maximum principal stress model:

$$\int_0^{\bar{\varepsilon}_f} \frac{\sigma_1}{\bar{\sigma}}\,\mathrm{d}\bar{\varepsilon} = C_{CL} \tag{6.19}$$

- the Brozzo et al. [69] hydrostatic pressure dependent model:

$$\int_0^{\bar{\varepsilon}_f} \frac{2\,\sigma_1}{3\,(\sigma_1 + p)}\,\mathrm{d}\bar{\varepsilon} = C_B \tag{6.20}$$

- or, finally, also using hydrostatic pressure, the LeRoy et al. [249] model developed to describe ductile fracture based on void nucleation and growth:

$$\int_0^{\bar{\varepsilon}_f} (\sigma_1 + p)\,\mathrm{d}\bar{\varepsilon} = C_{LR} \tag{6.21}$$

An inherent problem to all models mentioned so far is the instantaneous failure of an element at the moment of satisfying an individual criterion. A model that takes into regard a spatial dimension in terms of a critical fracture zone was proposed by Wilkins et al. [427]. According to the so called $D_C - R_C$ criterion failure occurs when the integral term D_C reaches a critical value over the fracture zone R_C:

$$D_C = \int_0^{\bar{\varepsilon}_f} \frac{1}{(1 + a\,p)^{\lambda}} \, (2 - A)^{\mu} \; \mathrm{d}\bar{\varepsilon} \tag{6.22}$$

6.2.2 Gurson Micro-mechanical Model for Ductile Fracture

A very popular model based on a micro-mechanical instead of purely phenomenological theory is the Gurson criterion. Its modified version by Needleman and Tvergaard [293] is implemented in several commercial and academic codes. The micro-mechanical aspect of the Gurson model describes the time rate of change of the void volume fraction f by the concept of nucleation and growth according to:

$$\dot{f} = (1 - f)\, \dot{\varepsilon}_V^{pl} + A\, \dot{\varepsilon}_M^{pl} \tag{6.23}$$

where $\dot{\varepsilon}_V^{pl}$ is the volumetric plastic strain rate and $\dot{\varepsilon}_M^{pl}$ denotes the equivalent plastic strain rate of the matrix material. The nine non-independent damage parameters ε_N, s_N, q_1, q_2, q_3, f_0, f_n, f_c and f_f need to be characterized, e.g. from tension tests with smooth and notched specimens, in order to calculate the factor A in (6.23):

$$A = \frac{f_n}{s_N\, \sqrt{2\,\pi}} \exp\left[-\frac{1}{2} \left(\frac{\varepsilon_M^{pl} - \varepsilon_N}{s_N} \right)^2 \right] \tag{6.24}$$

This leads to an effective void volume fraction ν described by:

$$\nu(f) = \begin{cases} f & for \quad f < f_c \\ f_c + \frac{1/q_1 - f_c}{f_f - f_c}\,(f - f_c) & for \quad f > f_c \end{cases} \tag{6.25}$$

and hence to the Gurson failure surface

$$\frac{\bar{\sigma}^2}{\sigma_Y^2} + 2\, q_1\, \nu \cosh\left(\frac{3\, q_2\, p}{2\, \sigma_Y} \right) - 1 - (q_1\, \nu)^2 = 0 \tag{6.26}$$

with the actual yield stress of the matrix material σ_Y. Tvergaard [403] and [404] identified the dimensionless parameters $q_1 = 1.5$, $q_2 = 1.0$ and $q_3 = 2.25$ based in investigations of a void under general loading conditions.

A detailed discussions of the Gurson model in the context of strain rate and temperature dependent materials was published by Hao and Brocks [161].

Recent examples for numerical studies of failure processes using the Gurson model for an aluminum motor carrier and for a high strength steel B-column, both under crash loads, can be found in Feucht et al. [125]. Ockewitz and Sun [298] successfully used the Gurson model for extruded aluminum profiles and aluminum cast alloy.

6.2.3 Phenomenological Stress Triaxiality Dependent Failure Models

An often used scalar parameter representing the amount of multi-axiality of the stress state is the so called *stress triaxility*

$$\tilde{\sigma} = \sigma_m / \bar{\sigma} = -p / \bar{\sigma} \tag{6.27}$$

which is the ratio of the hydrostatic or mean stress σ_m to the von Mises equivalent stress $\bar{\sigma}$. Models depending on $\tilde{\sigma}$ are described next.

The relevancy of stress triaxiality for failure initiation has been often investigated. A recent example is Hopperstad et al. [197] investigating the influence of stress triaxiality at varying strain rates for Weldox 460 E steel. A Split-Hopkinson tension bar was used to test axisymmetric notched specimens with initial notch radii R_0 of 0.4, 0.8 and 2.0 [mm]. The initial specimen radius in the center of the notch was $a_0 = 1.5$ [mm]. Optical strain measurement was used for precise evaluation of true-stress true-strain data including observation of the current strain rates ranging between 10^{-1} $[s^{-1}]$ and 10^4 $[s^{-1}]$.

Being aware of some reported problems with the Bridgman analysis to calculate the initial stress triaxiality[2]:

$$\tilde{\sigma}_0 = \frac{1}{3} + \ln\left(1 + \frac{a_0}{2\,R_0}\right) \tag{6.28}$$

numerical analyses were performed in a second study published by Børvik et al. [61]. The combination of experimental and numerical results allowed for a much more accurate determination of the actual stress triaxiality as well as of the related failure locus.

Bao [32] found that the second term in equation (6.28) should be multiplied by $\sqrt{2}$ for a better correlation.

[2] Specifically mentioned in Hopperstad et al. [197] are the findings of Alves and Jones [3].

Rice-Tracey Model

Investigating the ductile void enlargement under triaxial stress states Rice and Tracey [322] found the stress triaxiality $\tilde{\sigma}$ as a key model parameter:

$$\int_0^{\bar{\varepsilon}_f} \exp\left(\frac{3}{2}\tilde{\sigma}\right) \, \mathrm{d}\bar{\varepsilon} = C_{RT} \tag{6.29}$$

Johnson-Cook Model

An often implemented and used failure model postulating that a critical equivalent fracture strain $\bar{\varepsilon}^f$ is a function of the stress triaxiality $\tilde{\sigma}$ was published by Johnson and Cook [215]. Following an earlier approach by Hancock and Mackenzie [159], the equivalent strain to failure $\bar{\varepsilon}^f$ is modeled as decreasing with increasing hydrostatic tension σ_m. The postulated criterion for the failure strain is described via five materal constants D_1-D_5 as:

$$\bar{\varepsilon}^f = (D_1 + D_2 \exp(D_3 \tilde{\sigma})) \left(1 + D_4 \ln \dot{\varepsilon}^*_{pl}\right) (1 + D_5 T^*) \tag{6.30}$$

with the homologous temperature

$$T^* = \frac{T - T_A}{T_M - T_A} \tag{6.31}$$

where T_A is the ambient or reference temperature and T_M the melting temperature. With the second and third terms in (6.30) a strain rate and temperature dependency of the failure strain is accounted for. As in the strain rate and temperature dependent yield model (3.188), the equivalent plastic strain rate is normalized to $\dot{\varepsilon}_0 = 1\, s^{-1}$:

$$\dot{\bar{\varepsilon}}^*_p = \frac{\dot{\bar{\varepsilon}}_p}{\dot{\varepsilon}_0} \tag{6.32}$$

In order to account for a path dependency, the accumulation of a damage scalar with the increments in equivalent plastic strain $\Delta \varepsilon_{pl}$ according to

$$D = \sum \frac{\Delta \varepsilon_{pl}}{\bar{\varepsilon}_f} \tag{6.33}$$

is performed. The failure criterion (6.30) is first activated when $D = 1.0$.

CrachFEM Model

A criterion specifically developed for thin sheets and extrusions that distinguishes between two distinctive ductile fracture mechanisms, i.e. the ductile void growth/coalescence and shear failure, is the so called *CrachFEM* model. It was introduced by Gese et al. [134] and Hooputra et al. [195]. Separate stress triaxiality dependent descriptions of the equivalent strains to failure under the two types of mechanisms are:

$$\bar{\varepsilon}_f^{duct} = d_0 \exp\left(-3\,c\,\tilde{\sigma}\right) + d_1 \exp\left(3\,c\,\tilde{\sigma}\right) \tag{6.34}$$

and

$$\bar{\varepsilon}_f^{shear} = d_2 \exp\left(-f\,\theta\right) + d_3 \exp\left(f\,\theta\right) \tag{6.35}$$

where c, d_0, d_1, d_2, d_3 an f are material constants. The so called *shear fracture parameter* θ is defined as:

$$\theta = \frac{\bar{\sigma}}{\tau_{max}} \left(1 - 3\,k_s\,\tilde{\sigma}\right) \tag{6.36}$$

with an additional material parameter k_s and the maximum shear stress τ_{max}. The actual strain criterion to failure ε_f under a given stress state is identified in the CrachFEM model as the smaller value of either (6.34) or (6.35):

$$\varepsilon_f = \min\left(\bar{\varepsilon}_f^{duct},\, \bar{\varepsilon}_f^{shear}\right) \tag{6.37}$$

Bao-Wierzbicki Model

Bao and Wierzbicki [33] performed an extensive experimental campaign to investigate the stress triaxiality dependent failure behaviour of aluminum Al 2024-T351. The 15 individual types of material tests ranged from uni-axial compression tests between cone-shaped platens[3] over shear tests with various specimen shapes to uniaxial tension tests with straight and notched cylindrical specimens. From the collection of data illustrated in Figure 6.4 it is apparent that monotonically decreasing models cannot fit the whole range of failure strains. Therefore, Bao and Wierzbicki [33] developed a combined model that separately covers the regime of compressive-shear loads $-1/3 < \tilde{\sigma} < 0$, a transition regime of tensile-shear states $0 \leq \tilde{\sigma} < 1/3$ and the third regime of uniaxial to multiaxial tensile stresses $1/3 \leq \tilde{\sigma}$. The resulting model including the parameters derived in the model calibration is:

$$\bar{\varepsilon}^f\left(\tilde{\sigma}\right) = \begin{cases} \infty & \tilde{\sigma} \leqslant \frac{1}{3} \\ 0.1225\left(\tilde{\sigma} + \frac{1}{3}\right)^{-0.46} & -\frac{1}{3} < \tilde{\sigma} \leqslant 0 \\ 1.9\,\tilde{\sigma}^2 - 0.18\,\tilde{\sigma} + 0.21 & 0 < \tilde{\sigma} \leqslant 0.4 \\ 0.0846 + 1.0694 \exp\left(-2.7149\,\tilde{\sigma}\right) & 0.4 < \tilde{\sigma} \end{cases} \tag{6.38}$$

where an important aspect is also the cut-off value $\tilde{\sigma} = -1/3$, i.e. uni-axial compression, for the failure mechanisms. Figure 6.4 illustrates the derived model along with two sets of data for the Johnson-Cook model fitting either the first or the third regime.

---

[3] Cone shaped loading devices are used to avoid the barrel effect (see for example Siebel, Pomp [358]).

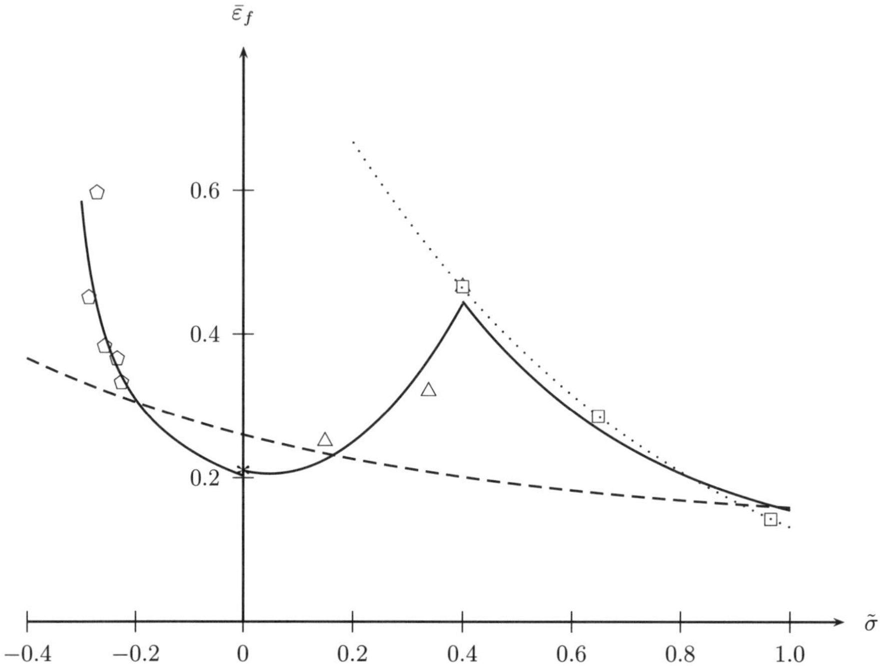

Fig. 6.4. Stress triaxiality $\tilde{\sigma}$ dependent failure locus for aluminum Al-2024-T351 as experimentally evaluated by Bao and Wierzbicki (□: Tension Tests, △: Shear-Tension, ∗ : Shear, ⬠: Compression). Approximations represent: Johnson-Cook model (6.30) with $D_1 = -0.07$, $D_2 = 1.02$, $D_3 = -1.62$ (dotted line) and with $D_1 = 0.13$, $D_2 = 0.13$, $D_3 = -1.5$ (dashed line). Bao-Wierzbicki model (6.38) (solid line).

Xue-Wierzbicki Model

Recently Wierzbicki and Xue proposed a more general version of (6.38) including the influence of the third invariant of the stress deviator J_3 in combination with stress triaxiality. Using a so called deviatoric state parameter ξ:

$$\xi = \frac{27}{2} \frac{J_3}{\bar{\sigma}^3} \tag{6.39}$$

the postulated model for the yield locus is:

$$\bar{\varepsilon}^f(\tilde{\sigma}, \xi) = C_1\, e^{-C_2 \tilde{\sigma}} - \left(C_1\, e^{-C_2 \tilde{\sigma}} - C_3\, e^{-C_4 \tilde{\sigma}} - \right) \left(1 - \xi^{\frac{1}{n}}\right)^n \tag{6.40}$$

with parameters C_1 to C_4 and a hardening exponent n.

A critical assessment of seven different fracture models was performed by Wierzbicki et al. [421] including calibration of all models for aluminum Al 2024-T351. The study covers the constant equivalent strain criterion (6.18), a fracture forming limit diagram model, a maximum shear stress criterion (6.15), an isothermal constant strain rate version of the Johnson-Cook model (6.30), the Xue-Wierzbicki model (6.39), the Cockcroft-Latham model (6.19) as well as the CrachFEM (6.37) and Wilkins models (6.22). The comparative study was performed under the assumption of plane stress. Comparing the results of all modes used in the course of the evaluation, the maximum shear stress criterion was found as the best model since it not only predicts correctly the experimental results but is also the one with the minimum amount of calibration effort. A comparable quality in the results were achieved using the physically based CrachFEM model and the Xue-Wierzbicki model. The Wilkins as well as the Johnson-Cook models are either predictive in the low or in the high stress triaxiality regime, but not in both. The same applies in even more narrow bands of $\tilde{\sigma}$ for the forming limit and constant equivalent strain criteria.

6.2.4 Brittle Failure

Typical dynamic problems with brittle materials involved are glasses, concrete or ceramic materials under impact. However, also materials which are showing ductile behaviour under moderate loading dynamics, may fail in a quasi-brittle manner as will be shown by the failure mode called *spallation*. From the variety of existing constitutive models to describe brittle failure of materials under impact loads, only two examples will be briefly mentioned in the present section. The first one is a phenomenological model and the second one a statistical model for brittle failure, respectively.

Johnson-Holmquist Phenomenological Brittle Failure Model

Johnson and Holmquist [216] introduced a brittle failure model that found implementation in many hydrocodes. Its main features are pressure dependent damage, strength and fracture. After some concerns about the first version, gradual softening and a more thorough description of parameter derivation were published as an improved version in [217].

Accumulation of damage to provide gradual softening with increasing plastic strain is modelled via the equivalent plastic strain $\Delta\,\bar{\varepsilon}^{pl}$ and the equivalent plastic strain to failure $\bar{\varepsilon}_f^{pl}$ at constant pressure by:

$$d = \frac{\sum \Delta\,\bar{\varepsilon}^{pl}}{\bar{\varepsilon}_f^{pl}} \tag{6.41}$$

where $\bar{\varepsilon}_f^{pl}$ is calculated as:

$$\bar{\varepsilon}_f^{pl} = D_1 \left(p^* + p^*_{min}\right)^{D_2} \tag{6.42}$$

with the constant parameters D_1 and D_2 as well as a pressure p^* and a minimum tensile hydrostatic pressure p^*_{min} to failure, both normalized to the Hugoniot elastic limit σ_{HEL} (see section 3.3.5):

$$p^* = \frac{p}{\sigma_{HEL}} \qquad p^*_{min} = \frac{p_{min}}{\sigma_{HEL}} \tag{6.43}$$

Hence, under constant pressure and subjected to plastic strain increments, damage begins to accumulate.

Using the damage scalar (6.41), softening of the material is described by a change in the normalized equivalent stress $\sigma^* = \sigma/\sigma_{HEL}$ according to:

$$\sigma^* = \sigma_0^* - d\left(\sigma_{0^*} - \sigma_f^*\right) \tag{6.44}$$

where σ_{0^*} and σ_{f^*} are the normalized intact equivalent stress and the normalized equivalent stress to failure, respectively, defined by:

$$\sigma_0^* = A\left(p^* + p^*_{min}\right)^N \left(1 + C\,\ln\dot{\varepsilon}^*\right) \tag{6.45}$$

and

$$\sigma_f^* = B\left(p^*\right)^M \left(1 + C\,\ln\dot{\varepsilon}^*\right) \tag{6.46}$$

with the normalized strain rate $\dot{\varepsilon}^* = \dot{\varepsilon}/\dot{\varepsilon}_0$ and $\dot{\varepsilon}_0 = 1.0\,[\mathrm{s}^{-1}]$.
Thus, equations (6.45) and (6.45) describe the failure surfaces for the intact and failed material, respectively. The actual failure surface after a loading history containing plastic strain increments is smoothly changed by the damage variable and described by (6.44).

With the softening (6.44) internal energy is dissipated. To account for this dissipation in the energy conservation, a pressure change Δp was suggested in Johnson and Holmquist [217]. Using a polynomial equation of state (4.84), they suggest a pressure change in a time step $t \rightarrow t + \Delta t$ of:

$$\Delta p_{t+\Delta t} = -K_1\,\mu_{t+\Delta t} + \sqrt{\left(K_1 \mu_{t+\Delta t} + \Delta p_t\right)^2 + 2\,\beta\,K_1\,\Delta e} \tag{6.47}$$

where $\Delta e = e_{t+\Delta t} - e_t$ is the change in elastic internal energy evaluated from the von Mises equivalent stresses $\bar{\sigma}$ and the elastic shear modulus G by

$$e = \frac{\bar{\sigma}}{6\,G} \tag{6.48}$$

SPHSF - Statistical Failure Models

Statistical models for brittle materials usually describe a certain random distribution of flaws in material per unit volume. Evolution equations describe the growth and coalescence of these flaws and the associated local damage. The damage state, in turn is used again to degrade material strength in terms of both deviatoric and hydrostatic stress components. Meshfree methods turned out to be a useful discretization since the evolving macroscopic cracks can be represented in a robust and comparably simple way.

A statistical failure model for brittle materials was described by Benz and Asphaug [55] and used with SPH by the same authors in [56]. A fundamentally modified version of the Benz-Asphaug model was developed by Mandell, Wingate and Swegle [264] and implemented in the SPHINX SPH code[4]. The model was applied to impact and penetration simulations in ceramic targets. Hiermaier and Riedel implemented the same model into SOPHIA to predict fracture patterns in various ceramic materials subjected to the edge-on impact test. This modified version, by the authors Mandell et al. [264] called smooth particle hydrodynamics statistical fracture (SPHSF) model shall be introduced here.

Its most important difference to the original Benz-Asphaug model is that SPHSF starts with a Poisson distribution of flaws whereas the former assumes the flaws as randomly assigned to particles. With the Poisson distribution, a number of N_f flaws in the neighbourhood of the average, empirical number of flaws per volume N_a is assigned to each particle according to:

$$P(N_f = N) = \frac{N_a^{N_f}}{N_f!} \exp(-N_a) \tag{6.49}$$

Each flaw is related to a critical stress σ_m representing its strength. In other words, a flaw activates damage as soon as the stress in its particle exceeds the critical value σ_m. A Weibull distribution defines the strength of each particle according to:

$$P(\sigma) = 1 - \exp\left[-\frac{V}{V_0}\left(\frac{\sigma - \sigma_u}{\sigma_o}\right)^m\right] \tag{6.50}$$

where m, σ_u and σ_o are material parameter, V describes the volume of the sample and V_0 is the average sample size per flaw.

Thus, with growing cracks in a volume discretized by a particle i, damage starts to accumulate. The time rate of change in the related damage scalar D is defined using the number of flaws in the particle n_i, the crack propagation

[4] See Wingate and Stellingwerf [428]

speed c_g and the characteristic length r_s (usually set equal to the cube root of the particle volume) as:

$$\frac{\mathrm{d}D}{\mathrm{d}t} = \left(n_i \frac{c_g}{r_s}\right)^3 \tag{6.51}$$

For the crack propagation speed Mandell et al. [264] recommend a value of 0.4 times the material's bulk sound speed. From edge-on impact tests it is possible to measure the impact velocity dependent crack speeds. An example is given in Figure 7.22 of section 7.4.3.

6.2.5 Spallation Modelling

The failure mode called *spallation* can happen for example under high or hypervelocity impact. An illustrative example of spallation is shown in Figure 6.5 where the impact of a 10 [mm] aluminum sphere into a massive aluminum target at an impact velocity of 7.0 [$\mathrm{km\,s^{-1}}$] evoked a strong shock wave. Propagated to the edges of the target, the shock was reflected at the free surface boundaries as rarefaction wave with extremely short rise times and high amplitude. Rarefaction waves resulting from reflected shock waves may represent a hydrostatic tension that is strong enough to instantaneously rupture the material. This was the case along the failure line at the bottom of the target shown in Figure 6.5.

The spallation strength is often modelled with a very simple criterion which is a minimum hydrostatic tension value p_{min}. Elements, cells or particles fulfilling the criterion

$$p \leq p_{min} \tag{6.52}$$

fail with p_{min} negative pressure.

Numerical simulation of the hypervelocity impact into the massive target of Figure 6.5 shows the evolution of the spall plane through wave superposition. As illustrated in Figure 6.6, the impact induces a spherical compressive wave travelling into the target. At the same time a compressive wave propagates into the projectile and is reflected at the rear end of it. The reflected rarefaction waves from the projectile and the rear end of the target follow the compressive wave. After reflection of the compressive wave at the target bottom the resulting rarefaction wave hits the second first rarefaction wave. Superposition of both waves leads to a negative hydrostatic pressure of 6 to 9 [GPa] in an interaction zone parallel to the bottom surface. This is the locus of *hydrodynamic tensile failure*, called *spallation*.

To predict the both the spallation locus and the subsequent local crack opening and related deformation the p_{min}-criterion (6.52) was used in combination with an element erosion criterion that eliminates hydrodynamic tensile failed cells. Even the chosen course discretization delivers a good prediction

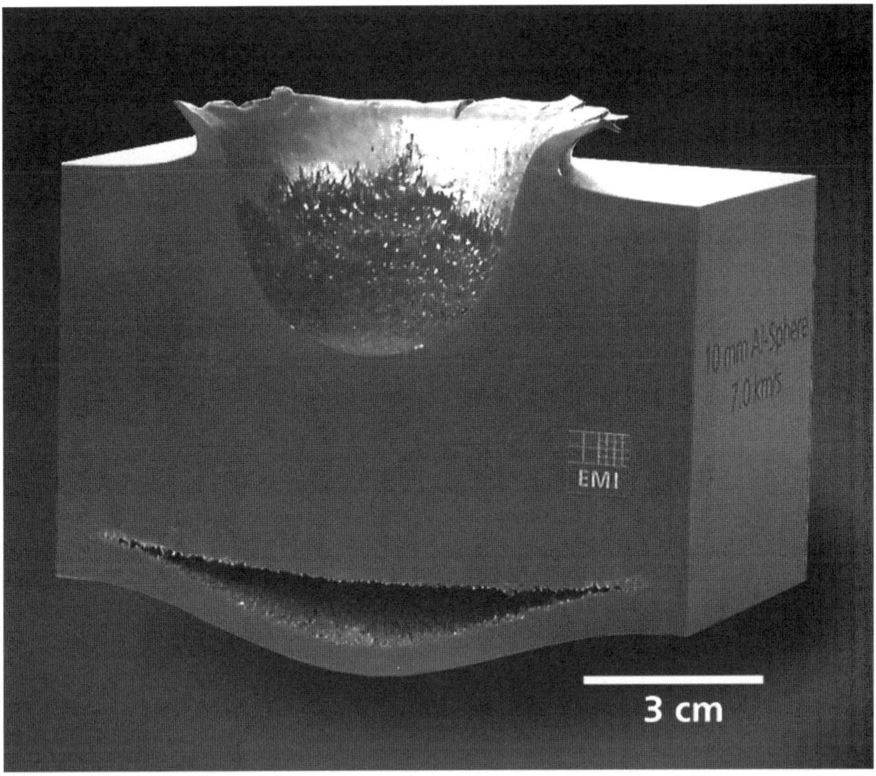

Fig. 6.5. Impact crater and spallation zone in an aluminum target impacted by a 10 mm aluminum sphere. Originally, the target was a massiv cylinder of dimensions 180 [mm] in diameter and 75 [mm] in length. After the impact test the target was cut to the illustrated cubic shape.

of the spall plane and the crater shape. Other than in the experiment close observation of the simulated spall plane reveals a smaller second plane separated by a single line of cells (see Figure 6.7).

More complex than the mathematical criterion is the experimental procedure to determine the spallation strength of a material. A reliable method to derive p_{min} is the flyer plate test (See section 7.4.2). From the free surface velocity measured by VISAR-technique in the plane strain wave experiment, spall signals and the related stress state can be precisely derived.

A detailed investigation of flyer plate specimens impacted at different velocities and with varying flyers to achieve loads below and beyond spall fracture shows that spallation can also be described by void nucleation, growth and coalescence models. Curran, Seaman and Shockey ([100], [99], [349], [357])

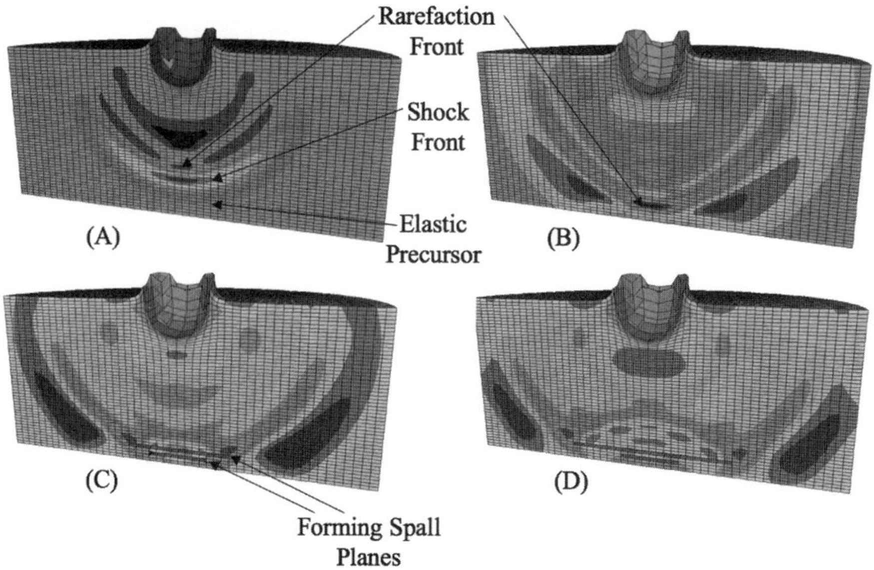

Fig. 6.6. Numerical simulation (AUTODYN) of the hypervelocity impact scenario illustrated in Figure 6.5. Superposition of rarefaction waves leads to the spallation effect. The illustrated instants of the impact process represent pressure distributions at (A):10.0, (B):12.5, (C):13.5 and (D):15.0 [μs], respectively.

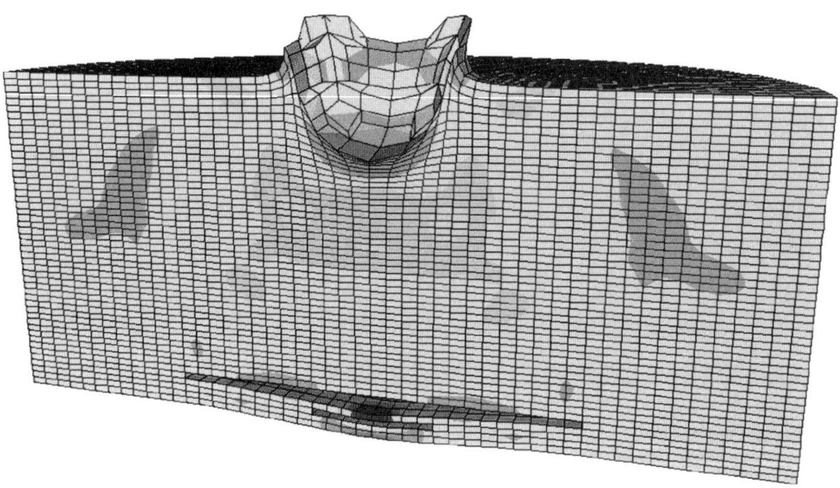

Fig. 6.7. Deformation state of the impacted aluminum target after 500 [μs].

provided pioneering work in the field of void growth measurement and modelling. A derivation of the basic equations used for the brittle fracture model BFRACT, as well as detailed investigations on the spallation effect can be found in Antoun et al. [10]. Meyers [276] gives a detailed introduction to the phenomenology and modelling of spallation as well as to dynamic fracture in general.

6.3 Failure Models for Composites

Inhomogeneity of materials poses additional complexity to the derivation of constitutive models. Distribution, size and orientation of filling material in the matrix of a composite influence the overall behaviour. Matrix and filling reinforcement may have independent failure mechanisms and thresholds. In addition, bonding limits between the individual components along their interfaces represent new failure criteria under various loading conditions. Damage initiation and propagation may therefore completely differ from comparable homogeneous materials.

In order to characterize composite failure on the level of continuum models, the micro mechanics of failure in the individual components and along their interfaces needs to be understood. Damage in composite materials usually occurs in various ways and influences the stress-strain history of composite structures. Experimental techniques need to be identified to measure that influence of damage development on the "elastic" deformation under cyclic loading, on the changes in the composite stiffness and finally on the manifold failure thresholds.

As mentioned above, there are several principal locations and mechanisms in a composite material where failure occurs under various loading conditions. A typical deformed state of a box beam made from carbon fiber reinforced polymer (CFRP) material after axial crushing in a drop tower experiment is shown in Fig.6.8. Concerning the material failure locus in laminated composites, one can distinguish between failure

- in the matrix,
- in the fibers,
- between the matrix and the fibers and
- between lamina.

Whereas the first three are *intra-laminar* failure mechanisms, the latter is an *inter-laminar* one, called *delamination*. At the respective loci, failure occurs differently under various loading conditions. The predominance of tensile, compressive or shear stresses as well as the stress-triaxiality or, in case of delamination, the crack opening mode can initiate different failure phenomena

Fig. 6.8. Carbon fiber reinforced box beam after longitudinal crushing in a drop tower experiment.

as well as varying strain rates an temperatures. Finally, the failure mode occurs in a brittle manner or gradually by accumulation of a damage state in the material. An experimental setup to observe various failure mechanism under compression is the so called *coupon test*. Fig.6.9 shows a coupon specimen made from carbon fiber reinforced polymer material before, during after the test. The drop weight induced crushing of the sample leads to different failure mechanisms depending on the orientation of the fibers relative to the crushing direction and the laminate setup.

All the mentioned aspects need to be addressed by a mathematical material model in order to simulate the structural behaviour including energy absorption of composite structures. Currently available failure descriptions are formulated

- either as macroscopic empirical models
- or as meso-scale continuum damage formulations referring to micro-mechanical observations

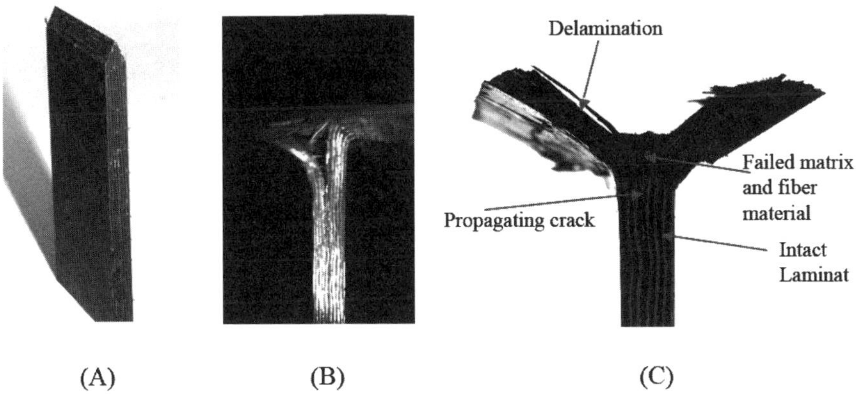

Fig. 6.9. CFRP coupon specimen (A) before, (B) during and (C) after crush loading in a drop tower test.

- or, finally, as atomistic and molecular dynamics models.

Though both the empirical and the meso-scale attempts are formulated for applications on a homogeneous macroscopic level, empirical models usually observe failure processes as seen in the force-displacement-curves of tests with large scale samples. Micro-mechanically based formulations, on the other hand, identify the processes leading to rupture on microscopical scales to establish thermo-mechanical relations describing them. Of course, these models also need parameters which are again derived from macroscopic tests.

In the course of the following sections on composite failure, the nomenclature for fiber orientations and vectors will follow the coordinate system given for an elementary ply in Fig. 6.10.

6.3.1 Analytical Models for Intra-Laminar Failure

Uncoupled Maximum Stress or Strain Criteria

Existing models in formerly available numerical codes and limited experimental capabilities defined the character and limitations of the first failure models for composite lamina. As in the case of metals, they were formulated as *maximum stress* or *maximum strain criteria* for a homogeneous model representing average properties of the components in an actually much more complex composite material. Thus, in the case of a maximum stress criterion, the six independent components of the stress tensor σ_{ij} are individually compared

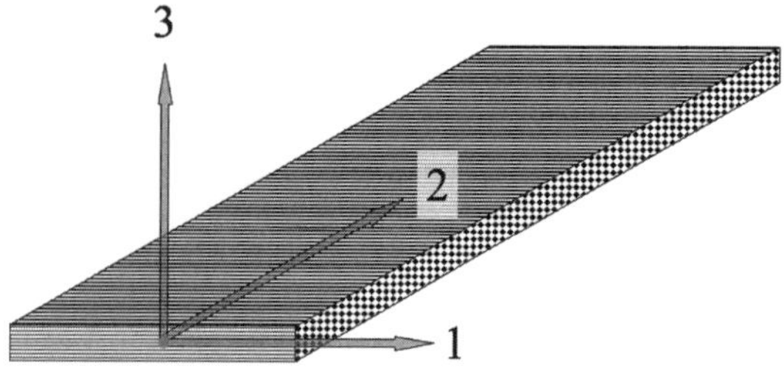

Fig. 6.10. Elementary ply coordinates.

to corresponding critical values defining failure. Taking into account different thresholds for the normal stress components under tension and compression, respectively, nine equations define a corresponding maximum stress failure model:

$$max \left\{ \left| \frac{\sigma_{11}}{X_{T,C}} \right|, \left| \frac{\sigma_{22}}{Y_{T,C}} \right|, \left| \frac{\sigma_{33}}{Z_{T,C}} \right|, \left| \frac{\sigma_{12}}{S_{12}} \right|, \left| \frac{\sigma_{13}}{S_{13}} \right|, \left| \frac{\sigma_{23}}{S_{23}} \right|, \right\} = 1 \qquad (6.53)$$

where the material dependent thresholds X_T, Y_T, Z_T, X_C, Y_C and Z_C are derived from uni-axial tests in the 1-, 2- or 3-direction under tension and compression, respectively. The remaining threshold values S_{ij} are accordingly defined by shear tests. Basically, maximum strain models can be formulated accordingly. However, concerning the derivation of the critical strain values it is important to note that uni-axial stress tests imply shear strain components in combination with the normal strains.

Another mathematical formulation for the model described in (6.53) is the simplest version of the so called *polynomial criteria*:

$$\begin{aligned}&(\sigma_{11} - X_T)(\sigma_{11} - X_C)(\sigma_{22} - Y_T)(\sigma_{22} - Y_C)(\sigma_{33} - Z_T)(\sigma_{33} - Z_C)\\&(\sigma_{12} - S_{12})(\sigma_{13} - S_{13})(\sigma_{23} - S_{23}) = 0\end{aligned} \qquad (6.54)$$

Given a precise measurement of the limit values and a mature code implementation, a maximum stress or strain criterion will reproduce the material tests performed for parameter identification and similar loading conditions in other structures very well. General loading conditions, specifically in the case of crash and impact situations, will, however, include multi-axial stress states reducing the predictive capabilities of models with non-interacting stress functions dramatically.

Smooth quadratic failure criteria

Hill criterion

A first model that allows to describe failure with interactive stress components was introduced by Hill [183]. The proposed polynomial model was originally developed to describe anisotropic yielding of metals. Formulated as generalized von Mises criterion it was first applied to the orthotropic behaviour of sheet metals. As already shown in equation (3.174), the *Hill criterion* assumes a quadratic relation between the stress components and does not distinguish between tension and compression:

$$F\left(\sigma_{22}-\sigma_{33}\right)^2+G\left(\sigma_{33}-\sigma_{11}\right)^2+H\left(\sigma_{11}-\sigma_{22}\right)^2+L\sigma_{23}^2+M\sigma_{31}^2+N\sigma_{12}^2=1 \tag{6.55}$$

To identify the six anisotropy parameters F,G,H,L,M and N critical stresses X, Y, Z and S_{ij} according to (6.53) need to be derived from uni-axial tension or compression tests and from shear tests. The correlation between these individual yield stresses and the Hill parameters is given by:

$$\begin{aligned}
2L &= \tfrac{1}{S_{23}^2} & G+H &= \tfrac{1}{X^2}\\
2M &= \tfrac{1}{S_{31}^2} & H+F &= \tfrac{1}{Y^2}\\
2N &= \tfrac{1}{S_{12}^2} & F+G &= \tfrac{1}{Z^2}\\
2F &= \tfrac{1}{Y^2}+\tfrac{1}{Z^2}-\tfrac{1}{X^2} & &\\
2G &= \tfrac{1}{Z^2}+\tfrac{1}{X^2}-\tfrac{1}{Y^2} & &\\
2H &= \tfrac{1}{X^2}+\tfrac{1}{Y^2}-\tfrac{1}{Z^2} & &
\end{aligned} \tag{6.56}$$

Azzi-Tsai-Hill model

Its anisotropic character with coupled stress components was the main reason for Hill's yield criterion to become the baseline of a series of failure criteria for composites. Azzi and Tsai [26] assumed transverse isotropic behaviour in unidirectional fiber composites. Tensile and compressive loading conditions are still treated equally in the resulting *Azzi-Tsai-Hill model*:

$$\begin{aligned}
&\left(\tfrac{\sigma_{11}}{X}\right)^2+\left(\tfrac{\sigma_{22}}{Y}\right)^2+\left(\tfrac{\sigma_{33}}{Z}\right)^2\\
&\quad-\left(\tfrac{1}{X^2}+\tfrac{1}{Y^2}-\tfrac{1}{Z^2}\right)\sigma_{11}\sigma_{22}-\left(\tfrac{1}{Y^2}+\tfrac{1}{Z^2}-\tfrac{1}{X^2}\right)\sigma_{22}\sigma_{33}\\
&\quad\quad-\left(\tfrac{1}{X^2}+\tfrac{1}{Z^2}-\tfrac{1}{Y^2}\right)\sigma_{11}\sigma_{33}+\left(\frac{\sigma_{23}}{S_{23}}\right)^2+\left(\frac{\sigma_{31}}{S_{31}}\right)^2+\left(\frac{\sigma_{12}}{S_{12}}\right)^2=1
\end{aligned} \tag{6.57}$$

Hoffman's version of the Azzi-Tsai model

To overcome that deficiency, Hoffman [189] proposed a slight modification to achieve an orthotropic formulation with separate regard to tensile and compressive failure stresses:

$$\begin{aligned}\left(\tfrac{1}{X_T}-\tfrac{1}{X_C}\right)\sigma_{11}+\left(\tfrac{1}{Y_T}-\tfrac{1}{Y_C}\right)\sigma_{22}+\left(\tfrac{1}{Z_T}-\tfrac{1}{Z_C}\right)\sigma_{33}&\\
+\tfrac{1}{2}\left(\tfrac{1}{X_T X_C}+\tfrac{1}{Y_T Y_C}-\tfrac{1}{Z_T Z_C}\right)(\sigma_{11}-\sigma_{22})^2&\\
+\tfrac{1}{2}\left(\tfrac{1}{Y_T Y_C}+\tfrac{1}{Z_T Z_C}-\tfrac{1}{X_T X_C}\right)(\sigma_{22}-\sigma_{33})^2&\\
+\tfrac{1}{2}\left(\tfrac{1}{X_T X_C}+\tfrac{1}{Z_T Z_C}-\tfrac{1}{Y_T Y_C}\right)(\sigma_{33}-\sigma_{11})^2&\\
+\left(\tfrac{\sigma_{23}}{S_{23}}\right)^2+\left(\tfrac{\sigma_{31}}{S_{31}}\right)^2+\left(\tfrac{\sigma_{12}}{S_{12}}\right)^2&=1\end{aligned} \tag{6.58}$$

Tsai-Wu tensor polynomial

A further improvement aiming at the ability to cover not only orthotropic failure aspects but arbitrary anisotropies was achieved through the so called *general tensor polynomial* type of models. With a symmetric strength tensor **F** its most general form was described by Gol'denblat and Kopnov [137]. A simplified quadratic version is represented by the *Tsai-Wu model.* As Hashin [164] already emphasised, the quadratic form of describing a failure surface instead of cubic or higher order polynomials has no physical reason. It is fitting the experimental results by far better than linear approaches and is, thus, pure empirical finding.

Using the tensors F_α and $F_{\alpha\beta}$ the Tsai-Wu fully anisotropic model takes the form:

$$F_\alpha\sigma_\alpha + F_{\alpha\beta}\sigma_\alpha\sigma_\beta = 1 \qquad \alpha,\beta = 1,...,6 \tag{6.59}$$

with

$$\begin{aligned}\sigma_1 &= \sigma_{11} & \sigma_2 &= \sigma_{22} & \sigma_3 &= \sigma_{33}\\ \sigma_4 &= \sigma_{23} & \sigma_5 &= \sigma_{13} & \sigma_6 &= \sigma_{12}\end{aligned}$$

which in the expanded three dimensional form yields:

$$\begin{aligned}&F_1\sigma_{11}+F_2\sigma_{22}+F_3\sigma_{33}\\
&+F_{11}\sigma_{11}^2+2F_{12}\sigma_{11}\sigma_{22}+2F_{13}\sigma_{11}\sigma_{33}+2F_{14}\sigma_{11}\sigma_{23}+2F_{15}\sigma_{11}\sigma_{13}+2F_{16}\sigma_{11}\sigma_{12}\\
&+F_{22}\sigma_{22}^2+2F_{23}\sigma_{22}\sigma_{33}+2F_{24}\sigma_{22}\sigma_{23}+2F_{25}\sigma_{22}\sigma_{13}+2F_{26}\sigma_{22}\sigma_{12}\\
&+F_{33}\sigma_{33}^2+2F_{34}\sigma_{33}\sigma_{23}+2F_{35}\sigma_{33}\sigma_{13}+2F_{36}\sigma_{33}\sigma_{12}\\
&+F_{44}\sigma_{23}^2+2F_{45}\sigma_{23}\sigma_{13}+2F_{46}\sigma_{23}\sigma_{12}\\
&+F_{55}\sigma_{13}^2+2F_{56}\sigma_{13}\sigma_{12}\\
&+F_{66}\sigma_{12}^2=1\end{aligned} \tag{6.60}$$

Reduced to a plane stress state in the ply plane of a laminate, the Tsai-Wu model takes the following form:

$$F_1\sigma_{11} + F_2\sigma_{22} + F_3\sigma_{33} + F_{11}\sigma_{11}^2 + 2F_{12}\sigma_{11}\sigma_{22} + F_{22}\sigma_{22}^2 + F_{66}\sigma_{12}^2 = 1 \tag{6.61}$$

In this way, the anisotropic tensile and compressive behaviour along the principal directions is modeled through the linear terms with F_α. To complete the description of the failure surface in the stress space, the quadratic terms are adjusted via $F_{\alpha\beta}$. An experimental characterization of the strength tensors' components F_i and F_{ik} can be realized by tensile, compressive, shear and bi-axial tests. In case of an orthotropic lamina, the measured limit values $X_{T,C}$, $Y_{T,C}$ and S_{ij} are used to identify the needed parameter:

$$\begin{aligned} F_1 &= \tfrac{1}{X_T} - \tfrac{1}{X_C} & F_{11} &= \tfrac{1}{X_T X_C} \\ F_2 &= \tfrac{1}{Y_T} - \tfrac{1}{Y_C} & F_{22} &= \tfrac{1}{Y_T Y_C} \\ F_6 &= 0 & F_{66} &= \tfrac{1}{S_{12}^2} \end{aligned} \tag{6.62}$$

To determine the coupling parameter F_{12}, one option is to perform a biaxial test under equal stresses in 0°- and 90°-directions. Alternatively, a simpler approximate characterization can be achieved through a tensile test along the 45°-direction. In the first case, the biaxial test performed until rupture at a measured stress to failure value of $\sigma_{biax,fail}$ delivers together with the above introduced uni-axial values $X_{T,C}$ and $Y_{T,C}$:

$$F_{12} = \frac{1}{2\sigma_{biax,fail}^2}\left\{1 - \left(\frac{1}{X_T} - \frac{1}{X_C} + \frac{1}{Y_T} - \frac{1}{Y_C}\right)\sigma_{biax,fail} + \left(\frac{1}{X_T X_C} + \frac{1}{Y_T Y_C}\right)\sigma_{biax,fail}^2\right\} \tag{6.63}$$

Applying a tension test along the 45°-direction until failure at $\sigma_{45,fail}$, the interaction parameter F_{12} is found as:

$$F_{12} = \frac{2}{\sigma_{45,fail}^2}\left\{1 - \left(\frac{1}{X_T} - \frac{1}{X_C} + \frac{1}{Y_T} - \frac{1}{Y_C}\right)\frac{\sigma_{45,fail}}{2} - \left(\frac{1}{X_T X_C} + \frac{1}{Y_T Y_C}\right)\frac{\sigma_{45,fail}^2}{4}\right\} \tag{6.64}$$

There is an extensive literature summarizing and discussing Hill type failure criteria and its various extensions. Examples are Berthelot [57], Vinson and Sierakowski [413] or Burk [71].

Direct Mode Criteria

A common feature to all the so far mentioned models is that failure of a composite does not distinguish between the composite's constituents. A fiber composite with transversely isotropic behaviour in the elastic regime does not necessarily follow a smooth Tsai-Wu failure surface (6.61) as a similarly transverse isotropic hexagonal crystal would. Individual failure of the components and the interfaces between them often necessitate to make changes in the model. Moreover, the definition of the interaction parameter F_{12} in (6.63) and (6.64) contains an implicit dependence of the biaxial tensile failure on uni-axial compressive failure parameters X_C and Y_C. Obviously, there is no physical foundation for such an assumption.

Hashin's principal failure modes

Hashin [164] addressed these problems and formulated a model with separate failure modes for matrix and fiber material under tensile, compressive and shear loads, respectively. Separating failure along the 0°- and the 90°-direction by the use of fiber and matrix criteria prevents from the application of an ambiguous F_{12}-term as needed in the Tsai-Wu model (6.61). In that context, Hashin also makes reference to the work of Narayanaswami and Adelman [291] where, on the basis of numerical investigations, an assumption $F_{12} = 0$ was estimated to cause errors of less than 10%.

The separation into principal mode failure types lead to a piecewise smooth surface in the Hashin model instead of a smooth failure surface. The following components for principal modes of failure characterize the *Hashin model*:
Fiber tensile failure:

$$\frac{\sigma_{11}^2}{X_T^2} + \frac{\sigma_{12}^2 + \sigma_{31}^2}{S_{12}^2} = 1 \tag{6.65}$$

for which Hashin optionally offers to neglect the influence of shear components and, thus, replacing (6.65) by the maximum normal stress criterion:

$$\sigma_{11} = X_T \tag{6.66}$$

Fiber compressive failure:

$$\sigma_{11} = X_C \tag{6.67}$$

Matrix tensile failure:

$$\frac{(\sigma_{22} + \sigma_{33})^2}{Y_T^2} + \frac{\sigma_{12}^2 + \sigma_{13}^2}{S_{12}^2} + \frac{\sigma_{23}^2 - \sigma_{22}\sigma_{33}}{S_{23}^2} = 1 \tag{6.68}$$

Matrix compressive failure:

$$\left(\frac{Y_C^2}{4S_{12}^2} - 1\right)\frac{\sigma_{22} + \sigma_{33}}{Y_C} + \frac{(\sigma_{22} + \sigma_{33})^2}{4S_{23}^2} + \frac{\sigma_{12}^2 + \sigma_{13}^2}{S_{12}^2} + \frac{\sigma_{23}^2 - \sigma_{22}\sigma_{33}}{S_{23}^2} = 1 \tag{6.69}$$

Explicit formulation of mode specific failure, as introduced by the Hashin model, also opens the possibility of observing the sequence of different failure mechanisms in structures during complex loading situations. Therefore, after initiation of one specific mode, additional ones can be distinguished as they successively follow and, thus, a structural design to avoid one or another mode or an optimization task in the context of energy dissipation can be performed by the help of corresponding models.

Another modification of the Tsai-Wu model, also intended to predict the precises failure modes in fiber reinforced structures under transient loading conditions, was suggested by Maier [262] and applied by Thoma et al. [388]. The failure modes of fibers in tension and compression as well as matrix material under tension, compression and shear in and between the lamina planes were taken into account. The Tsai-Wu approach was kept unchanged for the matrix components, whereas for the fiber failure a maximum stress formulation was chosen. Given a loading state fulfilling the failure criterion, each of the related stress components σ_{ij} was compared to its critical value $\sigma_{ij,c}$ and the actual failure mode identified as the one with the highest ratio of

$$\left| \frac{\sigma_{ij}}{\sigma_{ij,c}} \right| \tag{6.70}$$

The resulting model allows for discrimination of nine different modes and is comparably easy to supply by material parameters.

Chang-Chang failure and degradation model

Having achieved a modeling complexity that allows for separation of individual modes, a next issue is to describe the post-failure behaviour. Identification of a specific stress- and strain-condition in time that is leading to local failure automatically rises questions on how the affected area looses strength over the loading process. Is it an instantaneous way of reducing all strength components to zero, e.g. total disintegration and possible phase changes of material under hyper-velocity impact? Or is it only partial failure, e.g. with the tensile strength diminished whereas compressive load my still be carried? Is the reduction of one or all strength components a function of ongoing deformation and can it be modeled via energy release functions similarly as in fracture mechanics?

To answer these questions, Chang and co-workers ([83], [81],[82] investigated notched laminated composites pin-loaded to tension and specifically the damage progression due to the related stress concentrations. To model fiber and matrix tensile failure, they used a quadratic stress formulation similar to the respective components in the Hashin model, i.e. equations (6.65) and (6.68). Instead of shear stresses and shear strengths Chang and Chang integrate strain energy and take into account nonlinear shear stress - shear strain relations, e.g. a cubic approach with the new material parameter α:

$$\varepsilon_{12} = \frac{1}{2G_{12}}\sigma_{12} + \alpha\sigma_{12}^3 \tag{6.71}$$

With the ultimate shear strain at failure $\varepsilon_{12,c}$ the *Chang-Chang failure model* is then written as:
Fiber tensile failure:

$$\frac{\sigma_{11}^2}{X_T^2} + \frac{\int_0^{\varepsilon_{12}} \sigma_{12} \mathrm{d}\varepsilon_{12}}{\int_0^{\varepsilon_{12,c}} \sigma_{12} \mathrm{d}\varepsilon_{12}} = 1 \tag{6.72}$$

Matrix tensile failure:

$$\frac{\sigma_{22}^2}{Y_T^2} + \frac{\int_0^{\varepsilon_{12}} \sigma_{12} \mathrm{d}\varepsilon_{12}}{\int_0^{\varepsilon_{12,c}} \sigma_{12} \mathrm{d}\varepsilon_{12}} = 1 \tag{6.73}$$

Assuming a linear elastic behaviour, i.e. $\alpha = 0$, the two tensile components of the Chang-Chang model reduce to:

$$\frac{\sigma_{11}^2}{X_T^2} + \frac{\sigma_{12}^2}{S_{12}^2} = 1 \tag{6.74}$$

and

$$\frac{\sigma_{22}^2}{Y_T^2} + \frac{\sigma_{12}^2}{S_{12}^2} = 1 \quad , \tag{6.75}$$

respectively.

Concerning a post-failure behaviour, Chang and Chang suggested a property degradation model which, depending on the failure mode, describes the change in elastic properties once failure has occurred. Consequently, within the *Chang-Chang degradation model* elastic properties of a failed material are reduced differently, depending on the type of failure. In case of matrix cracking, the in-plane stiffnesses are set to zero except for the fiber direction. For fiber failure, it is postulated that both the longitudinal modulus E_{11} and the shear modulus G_{11} are reduced according to a Weibull distribution with the size of the predicted damage area as a parameter (see [82] for details). Strength degradation and continuum damage models to describe that effect will be the topic of the following chapter 6.3.2.

6.3.2 Continuum Damage Based Intra-Laminar Failure Models

Ladeveze-Le Dantec – Damage Kinematics

A model that on the ground of micro-mechanical observations accounts for matrix micro-cracking, fiber-matrix debonding and fiber breakage in transverse direction was originally developed by Ladeveze and co-workers. Following earlier publications, Ladeveze and Le Dantec [241] follow the approach of formulating failure through damage mechanics on the elementary ply scale. Assuming a two-dimensional plane stress condition in the ply plane, inter-laminar stresses are not accounted for and thus, in the original version, delamination is not addressed by this model. Later, the fiber longitudinal damage was added in a very similar way as shown in Greve and Pickett [145]. The same authors present a further extension to three-dimensional loading conditions in order to model the important influence of delamination effects on energy absorption under impact and crash loads. Figure 6.10 shows the addressed ply plane with its coordinate system and fiber orientation along the 1-axis. To define variable directions in that plane, e.g. for material testing, the fiber longitudinal direction is identified with the 0°-direction and the 2-axis with 90°.

Within the framework of the modified Ladeveze type damage kinematics for an elementary ply, three damage scalars d_{11}, d_{22} and d_{12}, constant in through-ply-thickness direction, are assumed to be accumulated over the loading history. With these scalars, stiffness reductions in the 0°-, 90°- and 45°-direction are formulated, respectively. In the longitudinal and transverse (0°, 90°) direction, fiber damage is described; the 45°-component represents in plane shear failure with fiber-matrix debonding.
Given an arbitrary damage state causing stiffness reduction, the damage dependent elastic relations are

$$\begin{aligned} \varepsilon_{11}^e &= \frac{\sigma_{11}}{E_{11}^0(1-d_{11})} - \frac{\nu_{12}^0}{E_{11}^0}\sigma_{22} \\ \varepsilon_{22}^e &= \frac{\langle\sigma_{22}\rangle_+}{E_{22}^0(1-d_{22})} + \frac{\langle\sigma_{22}\rangle_-}{E_{22}^0} - \frac{\nu_{12}^0}{E_{11}^0}\sigma_{11} \\ \varepsilon_{12}^e &= \frac{\sigma_{12}}{2G_{12}^0(1-d_{12})} \end{aligned} \tag{6.76}$$

where E_{ij}^0 and G_{12}^0 are initial undamaged elastic moduli. Variables in $\langle\rangle_+$ and $\langle\rangle_-$ brackets indicate

$$\langle a\rangle_+ = \begin{cases} a & if\; a \geq 0 \\ 0 & otherwise \end{cases}$$

$$\langle a\rangle_- = \begin{cases} a & if\; a \leq 0 \\ 0 & otherwise \end{cases}$$

Such a separation of positive and negative contributions allows for modelling closing effects of transverse micro-cracks under compressive loads.
With equations [6.76], the corresponding strain energy e_D of damaged material can be formulated as

$$e_D = \frac{1}{2}\left[\frac{\sigma_{11}^2}{E_{11}^0(1-d_{11})} - \frac{2\nu_{12}^0}{E_{11}^0}\sigma_{11}\sigma_{22} + \frac{\langle\sigma_{22}\rangle_+^2}{E_{22}^0(1-d_{22})} + \frac{\langle\sigma_{22}\rangle_-^2}{E_{22}^0} + \frac{\sigma_{12}^2}{2G_{12}^0(1-d_{12})}\right] \tag{6.77}$$

Damage accumulation model

Ongoing damage in a loaded composite material is caused by various micro-cracks as mentioned before and goes along with stiffness reductions in the corresponding loading directions, called *degradation.* As suggested before with equation (6.4) for isotropic damage scalars, also the components of a non-isotropic damage tensor d_{ij} can be observed as ratio between the current release modulus after a certain history of cyclic loads and the initial modulus, e.g. in case of the shear modulus $G_{12,i}$ at a specific cycle i:

$$d_{12,i} = 1 - \frac{G_{12,i}}{G_{12}^0} \tag{6.78}$$

Cyclic loading and unloading tests on the material can be performed to derive a relation between predefined strain conditions and resulting damage values according to (6.78). The *damage energy release rate* is calculated as accumulating function $Y_{ij}(t)$ over time:

$$\begin{aligned} Y_{11} &= \left.\frac{\partial e_D}{\partial d_{11}}\right|_{\sigma,d_{22},d_{12}} = \frac{1}{2}\frac{\sigma_{11}^2}{E_{11}^0(1-d_{11})^2} \\ Y_{22} &= \left.\frac{\partial e_D}{\partial d_{22}}\right|_{\bar{\sigma},d_{11},d_{12}} = \frac{1}{2}\frac{\langle\sigma_{22}\rangle_+^2}{E_{22}^0(1-d_{22})^2} \\ Y_{12} &= \left.\frac{\partial e_D}{\partial d_{12}}\right|_{\bar{\sigma},d_{11},d_{22}} = \frac{1}{2}\frac{\sigma_{12}^2}{G_{12}^0(1-d_{12})^2} \end{aligned} \tag{6.79}$$

Then, the square root of the maximum values of $Y_{ij}(t)$ at an individual time τ preceding the current time t are taken :

$$\underline{Y}_{ij}(t) = \max_{\tau \leq t}\left(\sqrt{Y_{ij}(\tau)}\right) \tag{6.80}$$

to accumulate the damage energy release rates $\underline{Y}_{ij}$ which, in a linear function, deliver the current damage scalar according to:

$$d_{ij} = \frac{\langle \underline{Y}_{ij}(t) - Y_{ij}^0\rangle_+}{Y_{ij}^c} \tag{6.81}$$

Material dependent damage controlling parameters in that procedure are Y_{ij}^0 and Y_{ij}^c. Ladeveze and LeDantec [241] validated this linear approach for unidirectional lamina. However, Johnson [210] found poor correlation with experimental results in the case of fabric reinforced composites and, therefore, proposed a nonlinear relation in the shear damage accumulation by $ln\,(\underline{Y}_{12}(t))$ which in the above used notation yields:

$$d_{12} = \frac{\langle ln\underline{Y}_{12}(t) - lnY_{12}^0\rangle_+}{Y_{12}^c} \tag{6.82}$$

Damage associated inelastic strain

Damage processes enable both elastic and inelastic deformation. To quantify the inelastic part of the strain tensor for unloading processes, a plasticity

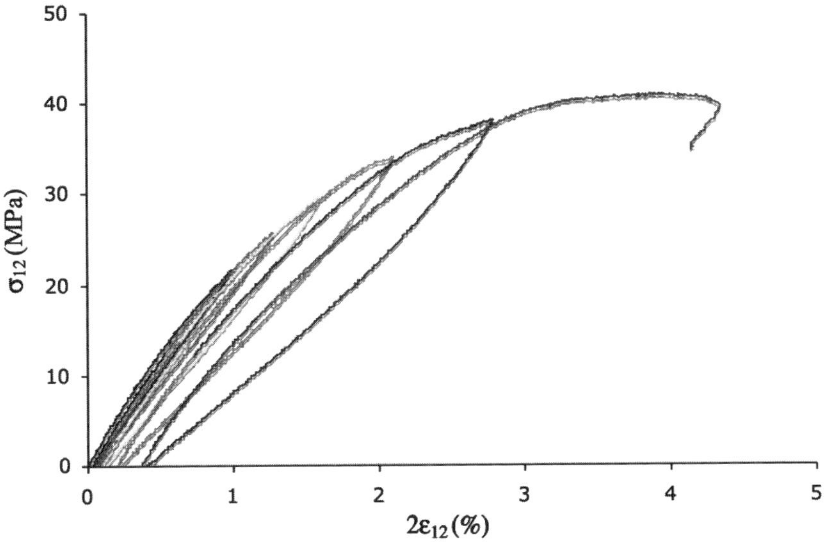

Fig. 6.11. Typical results for step cycle tests with CFRP

model accounting for damage induced irreversible deformation is needed. Assuming that in the ply layer only transverse strain $\varepsilon_{22,p}$ and shear strain components $\varepsilon_{12,p}$ contribute to plastic strain ($\varepsilon_{11,p} = 0$) a plasticity model with isotropic hardening was formulated in [241]. In that model, the elastic domain function f is established:

$$f = \sqrt{\frac{2\sigma_{12}^2}{(1-d_{12})^2} + a^2\frac{\langle\sigma_{22}\rangle_+^2}{(1-d_{22})^2} + \langle\sigma_{22}\rangle_-^2} - \sigma_H(\hat{\varepsilon}_p) \tag{6.83}$$

with the hardening function

$$\sigma_H(\hat{\varepsilon}_p) = \sigma_Y^0 + \beta\hat{\varepsilon}_p^m \tag{6.84}$$

and three material constants a^2, β and m. The accumulated effective plastic strain $\hat{\varepsilon}_p$ is integrated from the plastic shear strain $\varepsilon_{12,p}$ according to

$$\hat{\varepsilon}_p = \int_0^{\varepsilon_{12,p}} 2(1-d)\mathrm{d}\varepsilon \tag{6.85}$$

The model found various applications, validations and modifications. Concerning the plasticity model, Johnson et al. [211] neglect the transverse stress and strain components in addition to the longitudinal components.

6.3.3 Delamination models

Generally, delamination may occur under different modes of loading and under combinations of them, so called mixed mode delamination. In analogy to classic fracture mechanics, the three orthogonal modes of delamination of composite materials, as illustrated in 6.12, are called

- Mode I for crack opening normal to the inter laminar surface
- Mode II and Mode III for shear loading in the inter laminar surface along the crack opening direction and normal to it, respectively.

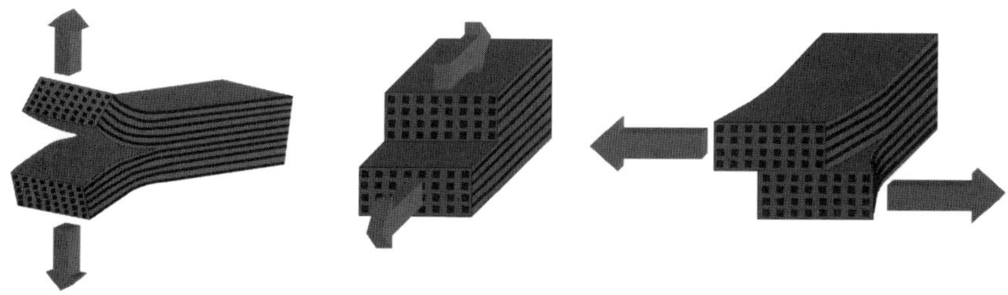

Fig. 6.12. Crack opening patterns in fracture mechanics typically defined as (A) mode I, (B) mode II and (C) mode III.

Examples for an experimental set-up to test composite specimens in mode-I and mixed mode-I and mode-II are illustrated in Figures 6.13 and 6.14.

Brewer-Lagace – Equivalent Strain Based Damage

In the stress space, the Brewer-Lagace delamination model uses an empirical criterion for the initiation of failure which is similar to the direct mode criteria discussed in chapter 6.3.1. Using the critical inter-laminar tensile stress to failure, i.e. the failure stress in the normal direction to the lamina Z_T, and the related critical intra-laminar shear stress S_{13} and S_{23} the criterion is formulated as:

$$\left(\frac{\sigma_{33,T}}{Z_T}\right)^2 + \left(\frac{\sigma_{13}}{S_{13}}\right)^2 + \left(\frac{\sigma_{23}}{S_{23}}\right)^2 = 1 \qquad (6.86)$$

The intended damage evolution based on the Brewer-Lagace criterion uses the equivalent strain formulations. Therefore, the criterion (6.86) must be

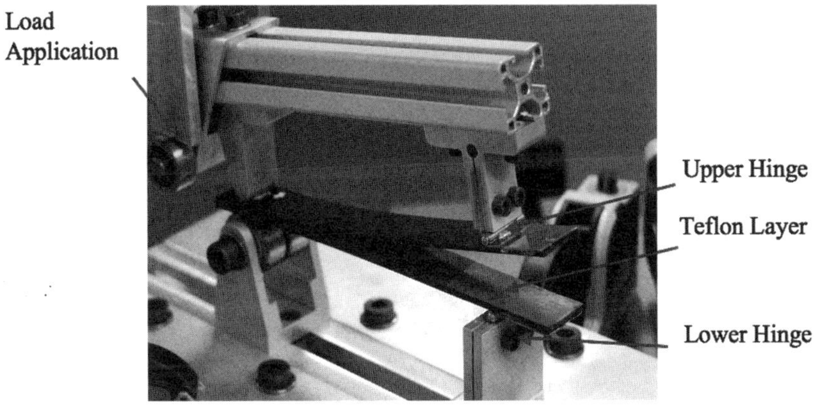

Fig. 6.13. Set-up of a double cantilever beam test. (Reprint with permission from Peter [305])

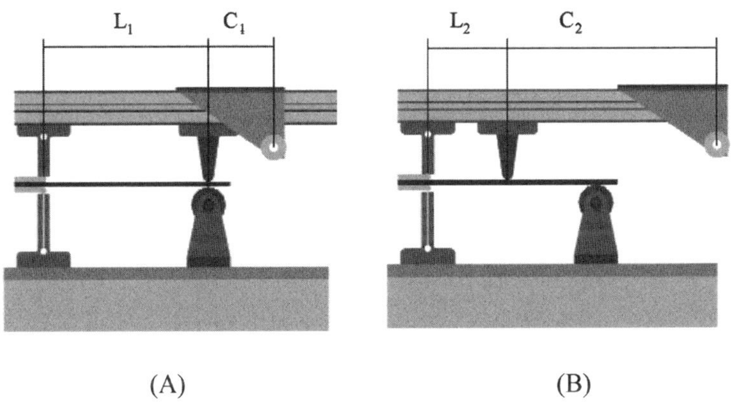

Fig. 6.14. Setup of an adjustable DCB designed by Peter [305] for variable application to mode I and II as well as mixed modes tests.

transformed to the strain space. To get there, Hörmann [193] re-arranged the criterion to

$$\sigma_{33,T}^2 + \sigma_{13}\frac{Z_T^2}{S_{13}^2}\sigma_{13} + \sigma_{23}\frac{Z_T^2}{S_{23}^2}\sigma_{23} = Z_T^2 \tag{6.87}$$

which, in tensorial form, can also be written as

$$\sqrt{\tilde{\sigma}^T \mathbf{P} \tilde{\sigma}} - Z_T = 0 \tag{6.88}$$

using the stress tensor of damaged material, i.e. the effective stresses $\tilde{\sigma}$:

$$\tilde{\sigma} = (1 - d)\sigma \tag{6.89}$$

and the coupling tensor P_{ijkl}, a fourth order tensor which in Voigt notation writes:

$$\mathbf{P} = diag\left\{0\;0\;1\;0\;\frac{Z_T^2}{S_{23}^2}\;\frac{Z_T^2}{S_{13}^2}\right\} \tag{6.90}$$

The coupling tensor is the key element for a transformation of the stress criterion into the strain space. For example the following equivalent strain measure can be introduced:

$$\bar{\varepsilon} = \frac{1}{E_{33}}\sqrt{\varepsilon^T\,\mathbf{E}^{elT}\,\mathbf{P}\,\mathbf{E}^{el}\,\varepsilon} \tag{6.91}$$

For a solution that is marching in time, the effective strain $\bar{\varepsilon}$ can be calculated at each time step. A tracking variable γ can be used to denote the maximum equivalent strain over time from the start of the calculation at $t = 0$ until a current time $t = \tau$:

$$\gamma = \max_{0<t\leq\tau}\{\bar{\varepsilon}(t)\} \tag{6.92}$$

To accumulate the damage scalar over the loading history, Hörmann [193] suggests to use a modified power law in the following equivalent strain based formulation:

$$d = \begin{cases} 0 & for \quad \gamma < \gamma_i \\ 1 - \left(\frac{\gamma_i}{\gamma}\right)^\beta \left(\frac{\gamma_u - \gamma}{\gamma_u - \gamma_i}\right)^\alpha & for \quad \gamma_i \leq \gamma < \gamma_u \\ 1 & for \quad \gamma_u \leq \gamma \end{cases} \tag{6.93}$$

where the initial equivalent strain γ_i is related to the critical stress to failure:

$$\gamma_i = \frac{Z_T}{C_{33}} = \varepsilon_{33,i} \tag{6.94}$$

and γ_u denotes the ultimate effective strain when the critical value of fracture toughness G_c is reached:

$$\gamma_u = \gamma_i - \frac{2G_c}{Z_T h_{PZ}} \tag{6.95}$$

where h_{PZ} is the thickness of the process zone between the lamina.

6.3.4 Discretization Aspects of Composite Failure

For the simulation of laminated composite materials under crash and impact loading both the inter- and the intra-laminar failure represent non-negligible contributions to the energy absorption process. Standard approaches are usually developed for two-dimensional shell element discretizations with restriction to model plane stress conditions. Basically these discretizations are not capable of inter-laminar effects. Greve and Pickett [145] solve that dilemma using tied shell elements as mechanical constraints to represent the three dimensional effects. Another approach was presented by Peter [305] who modeled

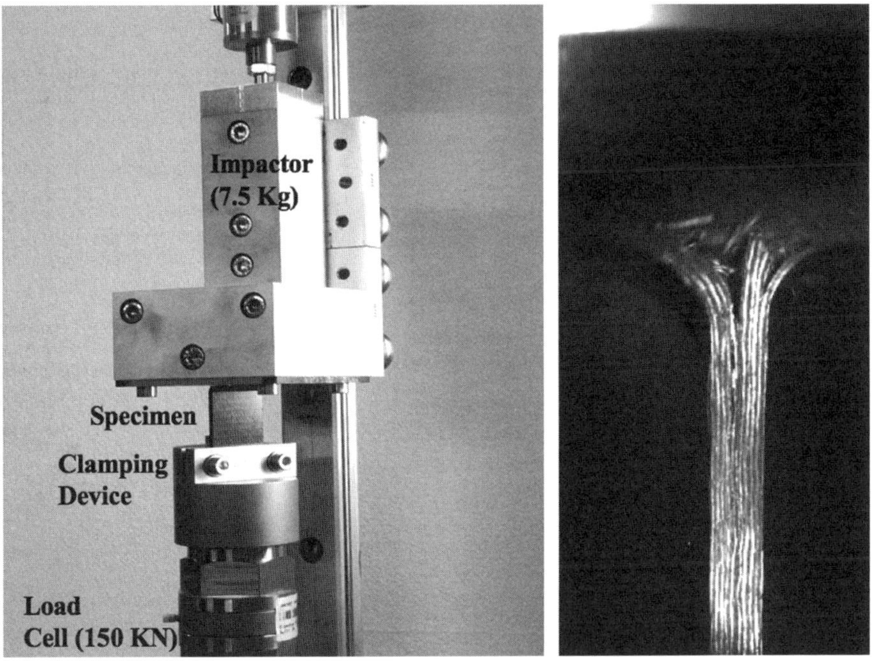

Fig. 6.15. Set-up of the coupon test and typical failure patterns of a uni-directional laminate sample during the dynamic test.

the individual lamina by three-dimensional elements - a solution that is computationally much more expensive. Applications of volume elements demand for new strategies in the time and space discretization like local sub-cycling and adaptive methods. But the achievable results in terms of predictive quality regarding the failure processes in composites are convincing.

Representative results of the simulations performed by Peter [305] are the 3D coupon crush illustrated in Figure 6.16 and the axial crushing of a quarter box-beam made from the same CFRP material shown in Figure 6.17. Applying his combined failure and discretization model to box beams with varying corner radii demonstrated the predictive capabilities of the Peter model.

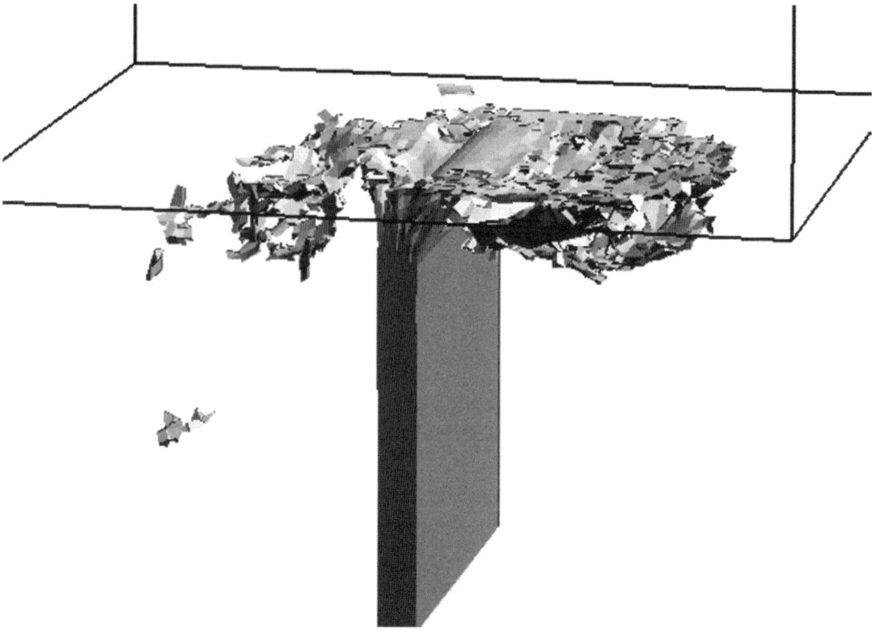

Fig. 6.16. SOPHIA simulation of the deformation and failure mechanisms in the dynamically crushed coupon sample applying the model developed by Peter [305].

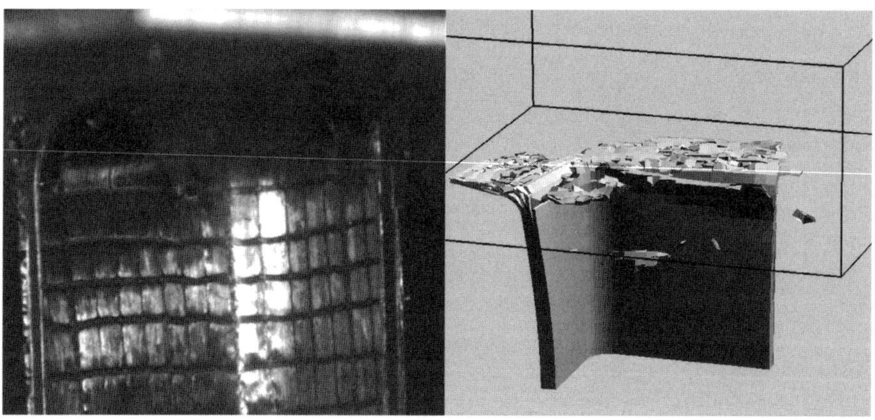

Fig. 6.17. High speed video image of a crushing quarter box and related simulation. (Reprint with permission from Peter [305])

7

Aspects of Advanced Dynamic Material Testing

7.1 Objectivity of Material Parameter Derivation

Predictive numerical simulation of any structural deformation process requires objective constitutive equations including parameters that are derived objectively for the material. Objective, in this context, means that the parameters are valid for the whole spectrum of loading conditions covered by the material model. This includes its applicability to arbitrary domains or geometries without restriction to specific structures. Experimental parameter derivation providing that kind of data can be called *material test.* The objectivity criterion to the parameters distinguishes the material test from a structural test used for verification or validation purposes.

A constitutive law is expected to describe a unique relation between stresses and strains valid for a specific material and independent from restrictions like structural geometries or boundary conditions. Therefore, the material specific parameters of the model need to be derived under conditions with well defined states of stress and strain. That means that the prevailing non-zero stress and strain components under a given loading situation in a chosen specimen shape need to be well known. When the dimensions actually measured are for example forces and deflections, the precise cross-sectional dimensions, relative lengths and cross-sectional distributions of forces need to be known throughout the testing time interval.

Material behaviour under dynamic loading conditions involves the strain rate dependency of material properties and, hence, experiments characterizing that influence. Therefore, in dynamic material tests also the actual strain rate needs to be known during the whole measurement period.

An additional, mostly non-negligible effect results from temperature changes. Constitutive equations including the option to describe temperature effects are given in the previous chapters. However, the actual separation of strain rate and temperature influences is often also a non-trivial challenge. Therefore, temperature effects are often measured implicitly and the results used in isothermal models. For applications of the material at significantly varying ambient temperatures, this means that the tests are to be performed at various ambient temperatures in order to then use the results for isothermal simulations at the individual temperatures.

Thus, in order to guarantee an objective derivation of strain rate dependent material parameters, a *material test* must at least provide:

- well defined states of stress and strain in the measurement of the specimen
- well known actual strain rates in the specimen
- both over the whole measurement duration.

Otherwise, the test cannot be called a material test. If material parameters are derived from tests that do not obey the requirements, the resulting material model is very likely to *reproduce* the process measured in this test but not to *predict* deformation processes in *generalized configurations* of structural shapes and loading conditions.

With increasing complexity of a test set-up the precise conditions of stress and strain may not always be easy to determine. A closely coupled numerical simulation of the material test can then be very useful.

In the course of this chapter a selection of set-ups and instrumentations for material tests at various strain rates will be given. This selection can of course not reflect all existing techniques that have been derived for the manifold classes of materials and loading conditions. Instead some representative methods in the regimes of low, moderate and high dynamic strain rates will be introduced along with recent developments in terms of instrumentation. The low dynamics are considered to be around $\dot{\varepsilon} = 10^2$ $[\mathrm{s}^{-1}]$, the moderate regime covers $\dot{\varepsilon} < 10^4$ $[\mathrm{s}^{-1}]$ and the high dynamic strain rates are related to shock wave propagation processes with strain rates up to $\dot{\varepsilon} \leq 10^7$ $[\mathrm{s}^{-1}]$.

7.2 Material Characterization in the Low Dynamic Regime

Material characterization in the strain rate regime up to few hundred s^1 is typically performed to supply modelling tasks engaged with crash or other accidental load case simulation. Common to these tasks is the need to investigate maximum loads and energy absorption characteristics of related structures. Concerning the material model it means that the stress-strain relations are known until ultimate loads, i.e. failure, including post-failure behavior.

Material characterization ranging from elastic over plastic to failure behavior poses significant additional problems and challenges to the experimental procedures and set-ups. One of them is to provide a measurement technique that allows for the measurement of the related large strains which may range to several hundred percent at. At the same time, the actual strain rate at the measuring locus needs to be observed and correlated with the stress and strain evaluation.

Implications of large strain failure characterization in the lower strain rate regime will be covered by the examples of uniaxial tension tests of thermoplastic materials and by the measurement of shear failure in general.

7.2.1 Uniaxial Tension to Failure with Optical Strain Measurement

For low dynamic strain rates the most common facilities are *servo-hydraulic systems* and *drop-weight towers*. Both types allow for tensile, compressive and shear loading. With increasing strain rates the influence of wave propagation effects resulting from the load application become more and more dominant. To keep the noise component in the measured signals as little as possible, eigen-frequency analyses of the set-up and related filter techniques are used. Standard instrumentations are force, strain and acceleration gauges.

Investigations of materials showing large strain values to failure, e.g. thermoplastics with 100 % failure strain or more, encouraged the development of new *strain measurement techniques*. Standard strain gauge instrumentation is limited in maximum strains around 20 %. Application of strain gauges using adhesives can significantly influence the material behaviour. Specifically in the case of many polymer materials. In addition, dynamic deformation of the specimen can lead to partial or total loss of contact between gauge and specimen surface.

These were the reasons why early investigations of failure in thermoplastic materials used a so called *Longitudinal Velocity Differential Transformer* (LVDT) technique, i.e. the machine signal to derive the strain from the motion of the clamping system. The resulting strain, however, is a measure of the global strain over the total specimen length. As already illustrated in section 3.3.7, Figure 3.23, many thermoplastic materials show a distinct strain localization that cannot be investigated by LVDT means. The simulation results shown in Figure 7.1 illustrate the difference in the strain measures that would result from either a global LVDT or a local strain measurement.

Fortunately, the rapid improvement of high speed cameras in terms of resolution quality and speed allowed for the observation of dynamic deformations with satisfying precision. Greyscale correlation of tracked markers on the

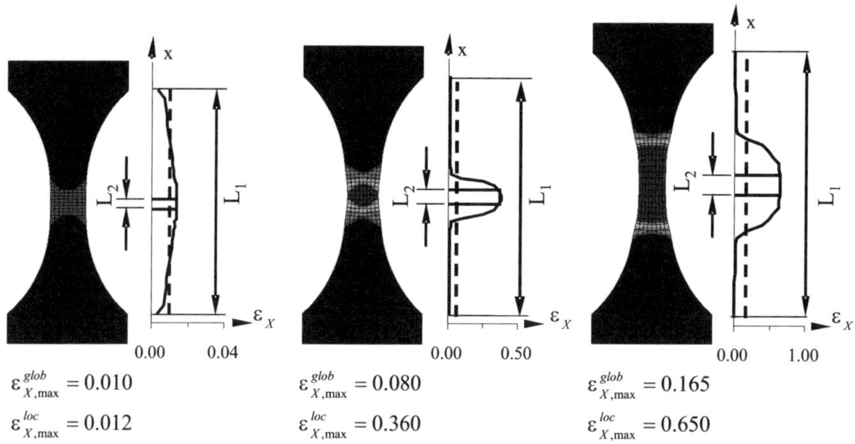

Fig. 7.1. Strain distribution for a uniaxially tensile loaded thermoplastic specimen showing strain localization in the necking zone. Global strain measures refer to a LVDT method with an measuring length of 35 [mm] whereas the local values refer to an initial length of 1 [mm]. Junginger [220] performed the simulations to show the necessity of a local, in his case optical, strain measuring technique. (Reprint with permission from Junginger [220])

specimen and the related derivation of deformation gradients for each picture taken over the deformation time enables the measurement of strain by *optical means*.

An early application of optical strain measurement technologies was presented by G'Sell et al. [150] and [151]. Junginger [220] describes the derivation of the two-dimensional strain tensor and its distribution on a specimen surface.

With Figures 3.24 and 3.25 the influence of the gauge length on the resulting strain and strain rate measures, respectively, were shown. Three representative frames of the optical measurement used for the derivation of local data are shown in Figure 7.2.

7.2.2 Shear Failure Characterization

Arcan Specimen - The Basic Problem

A well established method to characterize elastic material behavior under shear loads is the application of butterfly shaped specimens in universal testing facilities. The apparatus was developed by Arcan et al. [11] to test fiber reinforced materials under a uniform state of plane-stress. Its validity and usefulness has been proven in its application to many materials.

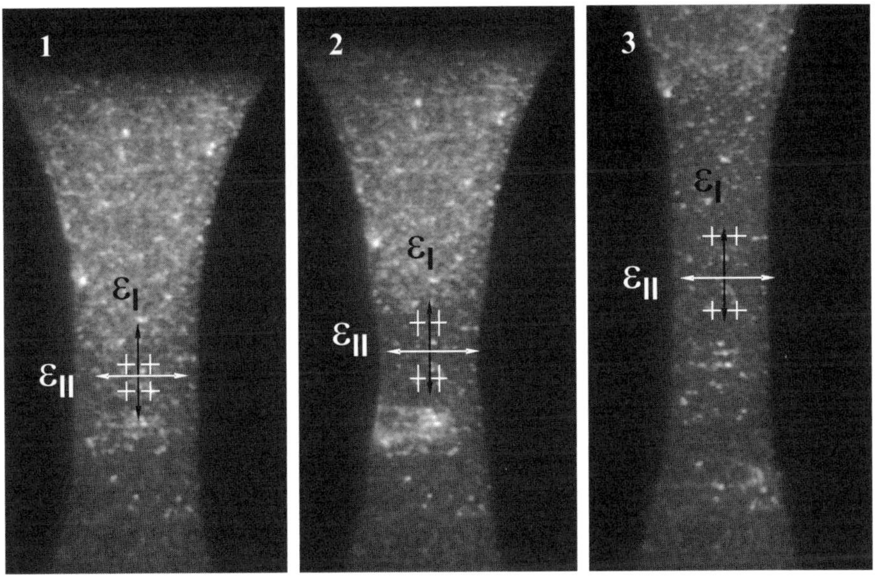

Fig. 7.2. Three significant instances in the uniaxial tension of localizing materials as observed with high speed cameras by Junginger [220]: 1) Initial homogeneous strain distribution, 2) Necking starts accompanied by strain localization and 3) Cold drawing phase with the necks stretching out. Tracking of individual pixel combinations on the specimen surface by means of grey-scale correlation provides displacement vectors, displacement gradients and, from these, the strain measure. (Reprint with permission from Junginger [220])

Usually application of the clamped version (see Figure 7.3) ignores measurement of the horizontal forces F_H. Mohr and Doyoyo [282] analyzed the set-up by experimental and numerical means and found significant importance for the horizontal force component to be measured. Application of the modified set-up enabled the identification of a corrected biaxial failure surface for a honeycomb material. Doyoyo and Wierzbicki [115] applied the modified Arcan test for the failure characterization of ductile and brittle aluminum foams.

A critical point in the application of the Arcan set-up arises, however, when the elastic and early plastic regime is left. Large deformation of the specimen leads to shear stresses in the central specimen zone more and more super-imposed by tensile stresses. Initially, the tensile stresses are restricted to zones reaching from the clamping boundaries to the free surfaces above and below the central area. With increasing deformation these tensile zones expand towards the central, for the measurement relevant zone. Moreover, as indicated in Figure 7.4, failure initiation does not occur first in the shear loading area, which would be in the middle of the narrow specimen center,

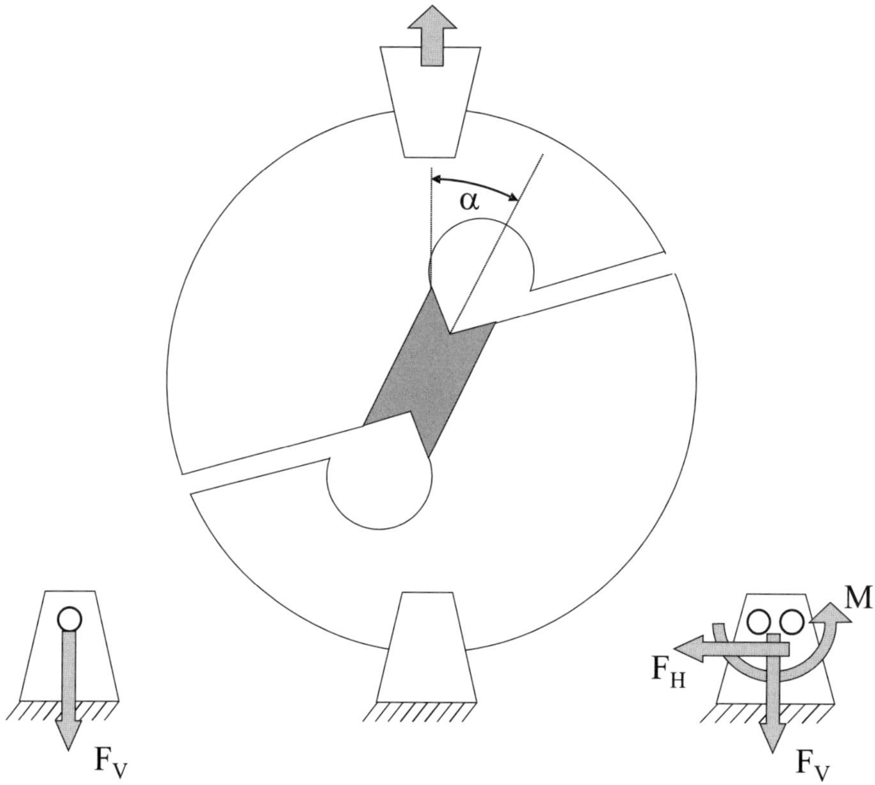

Fig. 7.3. Basic set-up of the test proposed by Arcan et al. [11] to investigate materials under uniform plane-stress conditions. Left: Moment-free clamping. Right: Fixed clamping.

but rather in the meanwhile totally tensile zone above.

For the characterization of the investigated material the observed stress states together with failure initiation means that the measured maximum stress does not identify failure under shear stress loading. Therefore, alternative configurations need to be used if *shear failure* shall be characterized.

Biaxially Combined Compression-Tension

An alternative that, based on the above results, was proposed by Jansen [208] is a combined compression-tension biaxial loading. A cross-shaped specimen compressively loaded in one direction and subjected to tensile loads in the perpendicular direction provides with a perfect shear stress state in the center of the specimen. An illustrative simulation of the actual stress state in the biaxially loaded cross-shaped sample is given in Figure 7.5 along with the

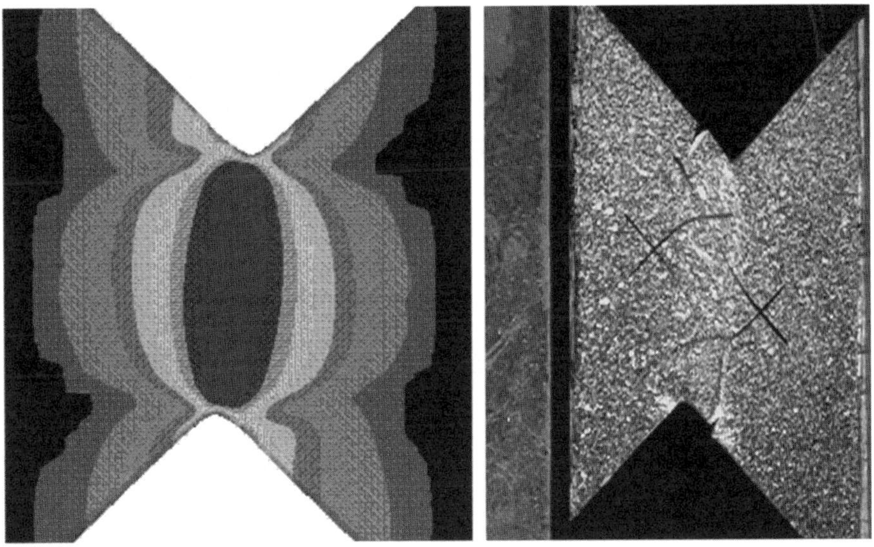

(A) (B)

Fig. 7.4. Shear stress states in an Arcan specimen calculated by numerical simulation (A) and failure initiation in the tensile free surface boundary zone as observed experimentally (B) for a cast aluminum sample. (Reprint with permission from Jansen [208])

experimental proof that failure actually starts in the middle of the specimen. Hence, the chosen set-up is valid to characterize shear failure.

A drawback of the cross-shaped sample is the stability problem associated with the compressive direction. The compressive arms need to be long enough to allow for shear loading up to failure. At the same time its length is limited by stability thresholds. The latter depend on the stiffness of the investigated material as well as on the thickness of the sample and, of course, the actual shear strain failure magnitude. For materials with high failure strains under shear load, the needed specimen dimensions are likely to be to large compared to the stability limit.

Tension-Torsion Cylinders

As a second alternative, a cylindrical specimen under mixed torsion and tension should be mentioned. If a tube sample can be fabricated from the material of interest, combinations of torsion and tension loads to the tube allow for a wide range of stress biaxiality. The reason for its limited application is probably related to the not always guaranteed precision or even general feasibility

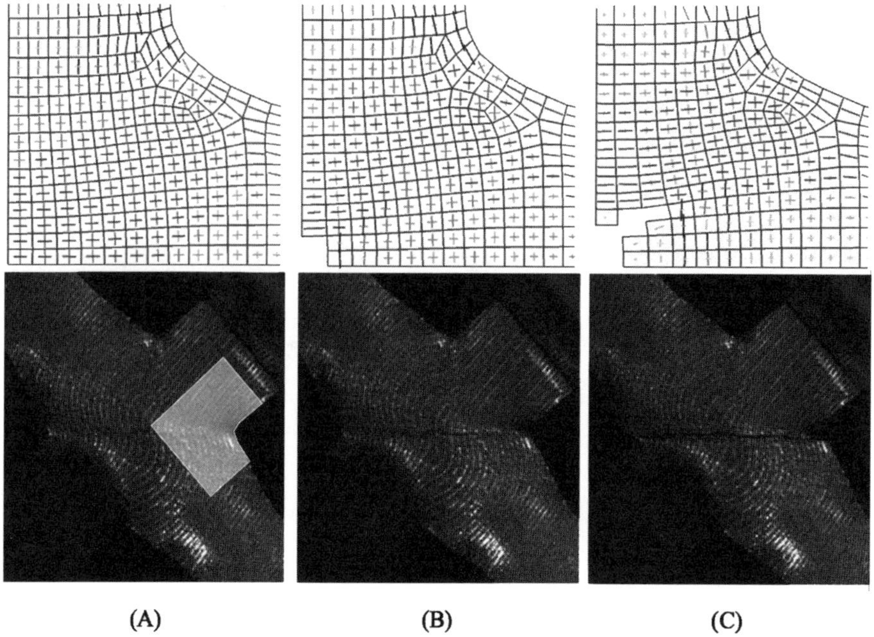

Fig. 7.5. Cross-shaped specimen under combined tensile-compressive load as investigated by numerical simulation (top row) and in the experiment (second row) with a light grey area indicating the simulated quarter of the symmetric geometry. Vectors in the simulation result represent principal stress directions indicating the pure shear stress state in the center. Illustrated are an intact state (A), failure initiation taken place in center (B) and crack growth towards the corner (C). (Reprint with permission from Jansen [208])

in the specimen production.

7.3 Material Tests at Moderate Dynamic Strain Rates

7.3.1 Hopkinson-Bar Facilities

A classical facility to test materials in the strain rate regime between a few hundred and several 10^3 [s^{-1}] is the *Split Hopkinson bar* (SHB). Originally developed by Hopkinson [196] as a single bar facility the second bar was added by Kolsky [230] leading to the Split Hopkinson Pressure Bar (SHPB), or now also called Kolsky bar. Meanwhile, there are various modes to operate Hopkinson-Kolsky bars including compressive, tensile and torsional loading to specimens. Comprehensive introductions and discussions can for example

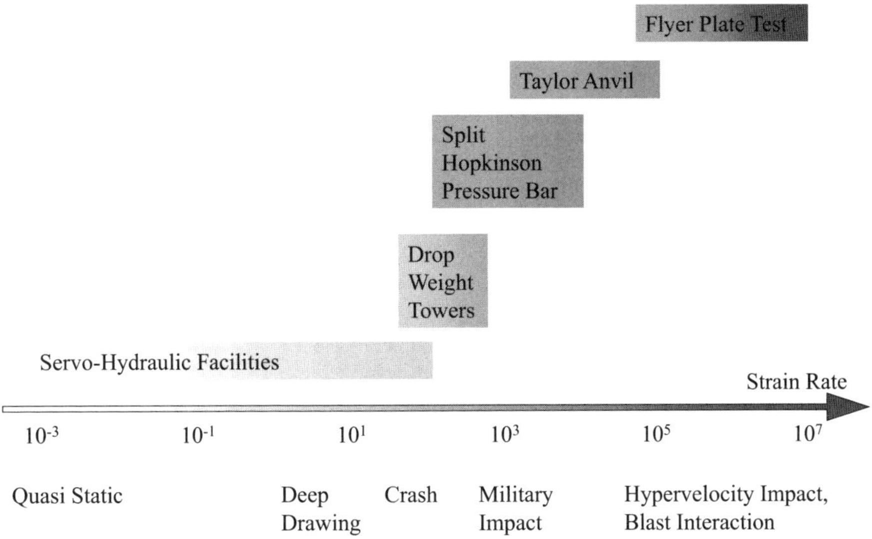

Fig. 7.6. Strain rate regimes for some classes of application and typical testing facilities to characterize the related material behavior.

be found in Follansbee [129] or Gray [141].

The basic principle of the compressive Split Hopkinson Pressure Bar (SHPB) is to position a specimen between two bars, an incident and an transmitter bar (see Figure 7.7). Impacting the incident bar with a striker introduces an acoustic wave propagating towards the specimen. At the interface between incident bar and specimen, the acoustic wave is partly transmitted into the specimen and partly reflected back into the incident bar. The transmitted component travels through the specimen. At the interface to the transmitter bar again a partly transmission and reflection takes place. An elastic wave of finally transmitted intensity travels into the transmitter bar. The derivation of stress-strain relations from the set-up uses strain signals measured on the incident bar and on the transmitter bar, respectively.

From the above description it is clear that the quality of the signal measured in the transmitter bar depends on the percentage of the transmitted wave intensities. From energy conservation it can be derived that the amount of transmitted wave energy W_T is described using the acoustic impedances $Z_i = \rho_i \, c_i$ of the materials according to

$$W_T = \frac{4 \, Z_2 \, Z_1}{(Z_2 + Z_1)^2} \tag{7.1}$$

For the case of two materials with differences in their acoustic impedances by orders of magnitude, equation (7.1) means that the transmitted signal is

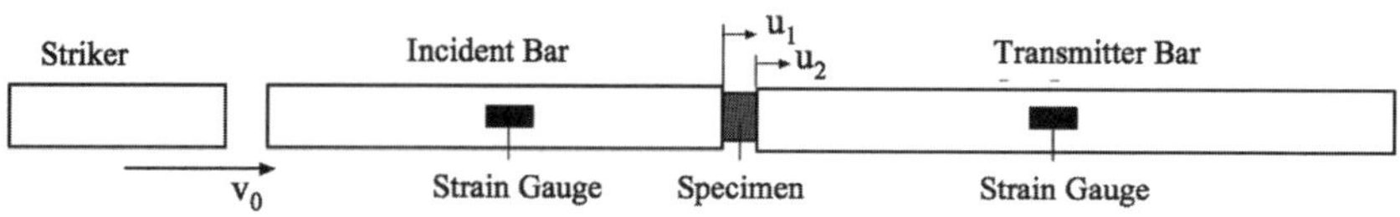

Fig. 7.7. Set-up of a classical Split-Hopkinson Pressure Bar.

very weak compared to the initial signal. In other words, a specimen material of low acoustic impedance as e.g. rubber or weak foam materials between metal bars leads to signals with a very weak signal to noise ratio.

Meenken [269] investigated the limitations of the SHPB with respect to the signal to noise ratio and calculated the transmitted portions for some combinations of media. Figure 7.8 illustrates these results. Clearly, specimen media like rubber materials or weak foams offer extremely low signal to noise ratios if metal bars in a SHPB are used. On the other hand, application of lower impedance bar materials like PMMA results in additional problems caused by dispersion effects. Application is only possible if a related a re-evaluation of the measured strain signals accounting for the dispersion effects is performed[1].

7.3.2 Direct-Impact Test for Low-Impedance Materials

For extremely low impedance materials like weak foams even PMMA bars represent an inappropriate impedance mismatch. The application of weak foams in automotive seats puts significant importance on their dynamic deformation behavior up to compression states of 80 to 90 percent. Both, the extremely low impedance and the needed large compression states rule out Hopkinson bars for the related characterization task.

An experimental set-up that attacks both weaknesses of the classical SHPB applied to weak foams was proposed by Meenken [269] via a direct impact facility. As illustrated in Figure 7.9 the specimen is loaded directly by the striker's impact rather than by wave transmission. Application of a piezoelectric polymer called polyvenylidenfluorid (PVDF) in thin stress gauge foils[2] attached to the interfacial surfaces delivers a clear and smooth signal for the

[1] Examples for the application of polymer bars are Sawas et al. [343] and Sogabe et al. [360]

[2] The piezoelectric nature of polyvenylidenfluorid (PVDF) was known since the late sixties (Kawai [222]). Bauer and Graham [38] were the first to successfully

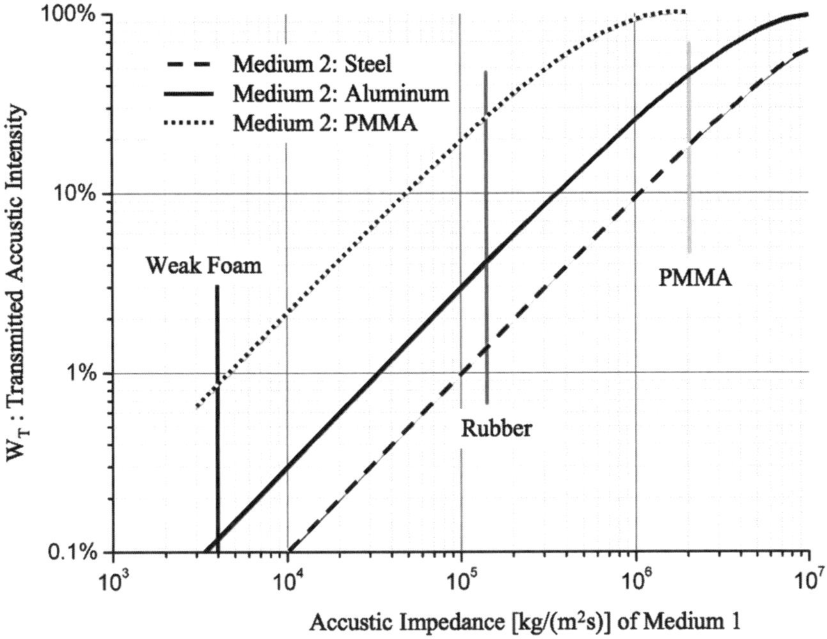

Fig. 7.8. Transmitted acoustic intensities at interfaces between two media. (Reprint with permission from Meenken [269])

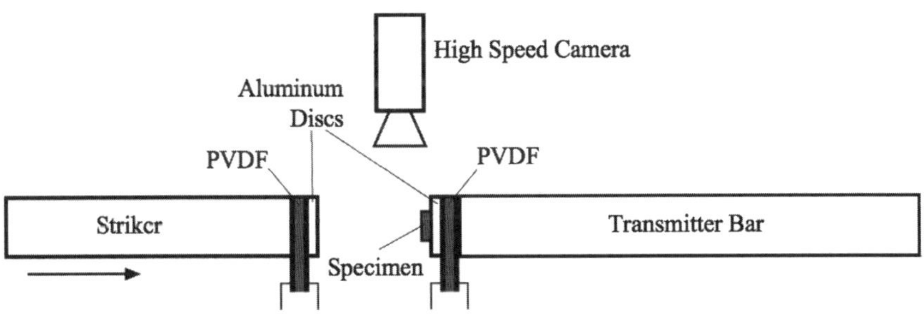

Fig. 7.9. Set-up of the direct impact facility as proposed by Meenken [269] with PVDF thin foil stress gauges and high speed camera instrumentation for stress and strain measurement, respectively.

stress measurement. The related compressive strain evolution is derived from optical instrumentation with high speed cameras. Figure 7.9 illustrates the set-up of the direct impact facility. An example application of the direct impact facility to derive stress-strain relations for Confor-Blue foam at average strain rates between 160 and 1115 [s^{-1}] as reported by Meenken is illustrated in Figure 7.10.

Two stress-strain curves were evaluated at each experiment. One by the stress-gauge applied to the striker, the second derived from the stress-gauge mounted at the transmitter bar. Comparing the two signals allows for an estimate of whether a homogeneous stress distribution is present in the specimen. Meenken [269] investigated the known problem of stress equilibration taken place through several wave reverberations.

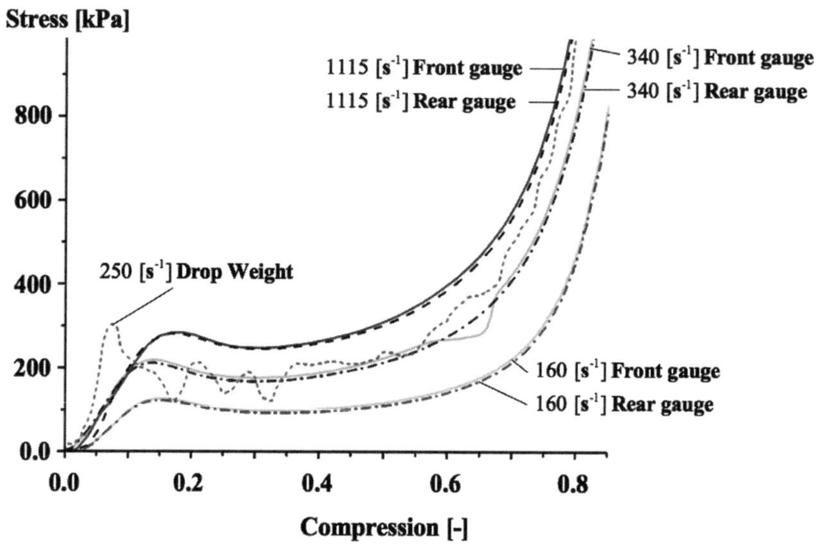

Fig. 7.10. Stress-strain curves measured with the instrumented direct impact facility for Confor-Blue foam at average strain rates of 160 [s^{-1}], 340 [s^{-1}] and 1115 [s^{-1}]. Front (solid lines) and rear gauge (dashed lines) signals are mostly overlapping. For comparison a result from a drop weight tower test at 250 [s^{-1}] is added. (Reprint with permission from Meenken [269])

use the effect for stress measurement with the then so called PVDF-sensors. See also Bauer [39].

7.4 Material Characterization at Extreme Strain Rates

Shock wave loaded materials undergo extremely high strain rates due to the particular short rise times of shock pulses. Starting from impacts with a few hundred meters per second, the spectrum of application ranges over blast loaded structures to hypervelocity impact and contact detonations. To simulate the exposure of structures to such extreme dynamic conditions, experimental facilities are required that reproduce the related processes in terms of strain rates. And since it shall be material tests, as elaborated above, these tests have to be performed under well defined stress and strain states.

As representative experiments for that strain rate domain the Taylor anvil and the flyer plate test shall be briefly introduced next.

7.4.1 Taylor Anvil-Test

The Classical Procedure - A Validation Test

Interested in characterizing the plastic behaviour of metal materials under impact loads Taylor [385] developed a post-impact evaluation for cylindrical specimen to serve as material test.

In its classical version, the Taylor anvil test means that a cylindrical rod of initial length L_0 and initial diameter D_0 is accelerated by an arbitrary device. Normal impact of the rod onto a rigid wall at an impact velocity v_p leads to a compressed length L_1 along with a plastic deformation zone covering a certain length L_{pl} measured along the axis of symmetry from the tip of the rod. Measured from the rear end of the rod, there is a zone of length L_{el} where only elastic deformations occurred. The dimensions of purely elastic and plastic zones, respectively, are usually determined by hardness measurement. Alternatively to the never perfectly rigid wall, a symmetric impact of two identical rods, one of which is at rest, can be performed.

From these dimensions in the initial and by the impact plastically deformed configurations, Taylor suggested a relation for the yield stress according to:

$$\sigma_Y^T = \frac{(L_0 - L_{pl})\,\rho\,\mathrm{v}_\mathrm{p}^2}{2\,(L_0 - L_1)\ln(L_0/L_{pl})} \qquad (7.2)$$

with the materials mass density ρ.

Part of the derivation that leads to equation (7.2) are the assumptions of constant propagation speed of the plastic front and uniform motion the rear rod end. Both of these assumptions must however be doubted as for example

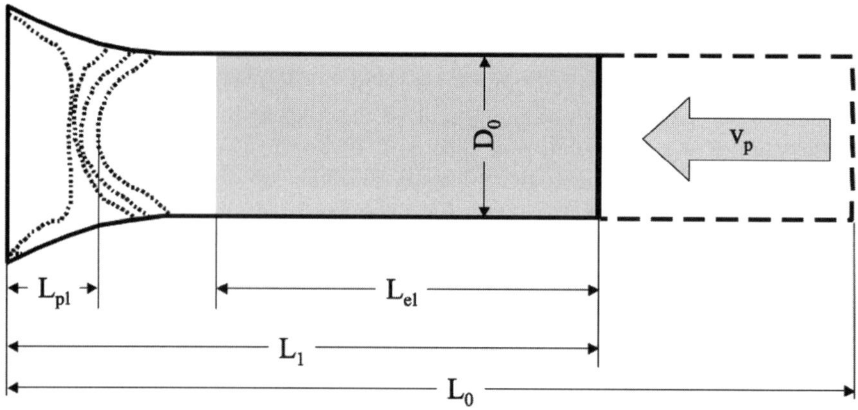

Fig. 7.11. Taylor anvil test: Set-up and dimensions for the classical derivation of yield stresses.

shown by Kuscher [239]. Moreover, as indicated in Figure 7.11, the plastic front is also mostly not a plane normal surface but rather of a curved or conical shape. Therefore the precise determination of L_{pl} and L_{el} would always be a problem.

To overcome these problems Wilkins and Guinan [426] improved the yield stress determination implementing the variable propagation speed of the plastic front. Their yield stress formulation is:

$$\sigma_Y^{WG} = \frac{\rho \, \mathrm{v}_\mathrm{p}^2}{2 \ln \frac{L_0 - L_{pl}}{L_1 - L_{pl}}} \tag{7.3}$$

which is still suffering from the difficult determination of the precise extension of the plastic deformation zone.

Taylor tests allow for large deformations and strain rates up to $\dot{\varepsilon} = 10^5\,[\mathrm{s}^{-1}]$. However, in addition to the uncertainties mentioned above, the actual local strain rate during the yield process cannot be determined precisely.

Thus, in the light of these circumstances the Taylor test in its classical set-up and with the yield stress determinations proposed by Taylor [386] or Wilkins and Guinan [426] must be seen as a validation rather than a material test. Still, the usefulness of the Taylor test is out of doubt. It clearly ranges beyond the characterization of plasticity related properties. A representative application was published by Chapman et al. [84] with a Taylor test based investigation of void nucleation under shock loading.

VISAR Instrumented Inverse Taylor Material Test

Rohr introduced a methodology that allows for a determination of stress, strain and current strain rate in a Taylor experiment that is modified with respect to its instrumentation. Its central component is the measurement of the free surface velocity at the rod's rear end by a *Velocity Interferometer for Any Reflector* (VISAR). The VISAR technique was originally developed by Barker and Hollenbach [35]. Introductions to the VISAR technology and remarks to its historical evolution are found in Arnold [17] and Barker [36], respectively.

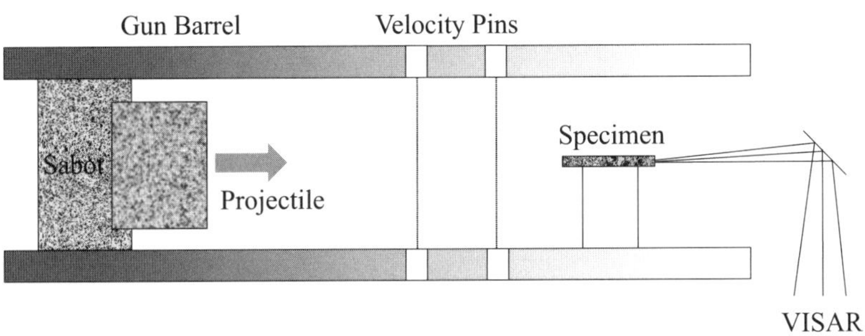

Fig. 7.12. Modified Taylor impact test as proposed by Rohr [331]. To allow for a VISAR instrumentation of the rear free surface of the rod, the "wall" is accelerated and impacted against the rod which is initially at rest.

With the set up illustrated in Figure 7.12, the wave propagations and reflections in the rod can be observed by VISAR in terms of velocity jumps. The free surface velocity measurement is enabled by an *inverse impact* where the rod is initially at rest and hit by the "rigid wall". A similar set-up was used by Rosenberg et al. [336] and later also by Rosenberg and Bless [335] to determine dynamic yield strengths in materials with manganin stress gauge instrumentation.

An example for the signals observed by Rohr is illustrated in Figure 7.13 (B). Typical for this set-up is the step-wise increasing free surface velocity due to repeated wave reflections. Part (A) of the same Figure explains the propagation of elastic waves due to the impact as well as reflections of the wave at the rear end. Each interaction with the free surface causes a reflection as rarefaction wave along with one step of acceleration.

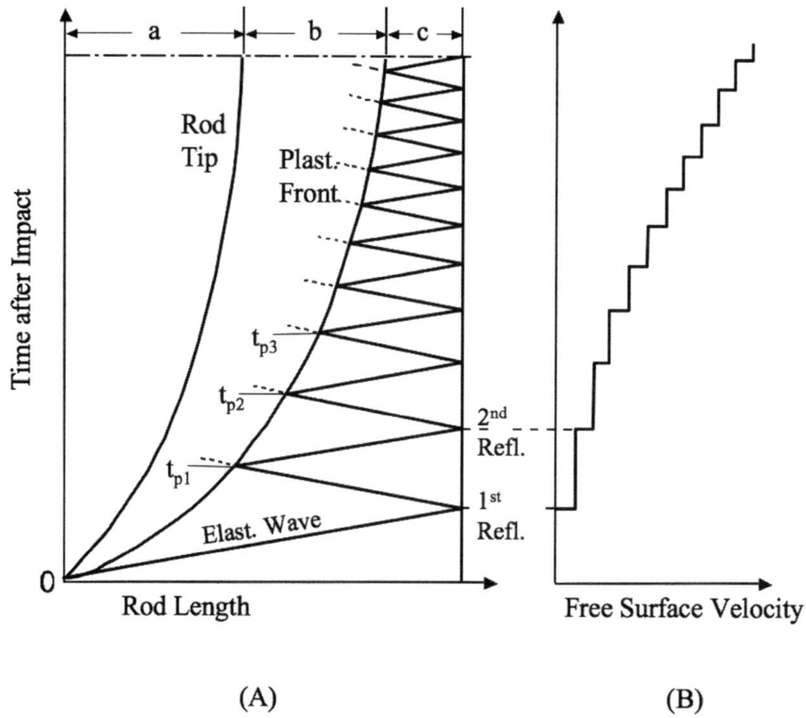

Fig. 7.13. Wave propagation and reverberation (A) as well as evolution of the free surface velocity (B) in an impacted Taylor rod.

Plastic deformation occurs along the propagation of a plastic wave front running behind the elastic wave. This plastic front in turn is an interface of changing material parameters (sound speed and density), and thus of acoustic impedance. Therefore, the elastic wave which is on its return path from the rear end is again reflected partly at the plastic front. The resulting run times of the elastic wave until its next arrival at the free surface allows for the detection of the exact position of the plastic front during the experiment.

Hence, a correlation between location of the plastic front, of the local strain rate at that position and of the yield stress is enabled. With the VISAR information on the velocity history, the yield stress is then evaluated as current stress at the location of the plastic front according to equation (4.28):

$$\sigma_Y = c_L \rho v_p = \frac{1}{2} c_L \, \rho \, v_{fs} \tag{7.4}$$

using the longitudinal sound speed c_L and the particle velocity v_p where the latter can be expressed by the free surface velocity as $v_p = 0.5\, v_{fs}$ assuming a total reflection at the free surface.

7.4.2 Flyer-Plate Experiments

Acceleration Techniques and Wave Effects

Characterization of material behavior under shock wave loading was originally focused on the determination of Hugoniot curves in order to derive equations of state for extreme conditions in terms of pressure and density. Impact tests with related high impact velocities to initiate waves of plane strain are an experimental challenge. Two flat plates impacting each other in a planar mode produce the needed plane waves. Early experiments were performed by shock wave accelerated plates. The needed well defined plane shock waves were originally initiated by explosives and specific *wave shapers*.

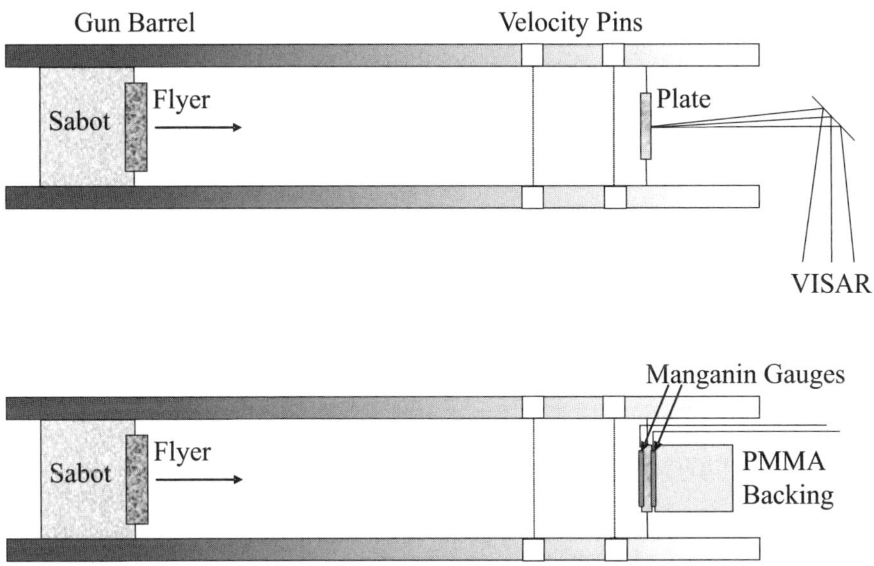

Fig. 7.14. Set-up of flyer-plate tests with sabot launched flyers impacting a plate that is instrumented by either VISAR or manganin stress gauges.

Alternative accelerators propel the plates by powder, gas or light-gas guns as well as by pulsed lasers (see for example Antoun et al. [10]) or nuclear devices as for example reported by Trunin [400]. Most advanced systems for plate impacts in terms of achievable impact velocities use magnetic accelerators as for example the Z-device at Sandia[3].

The impact induced plane shock waves are used in the so called *flyer-plate* or *planar-plate* impact test to load the investigated material at strain rates of 10^6 to 10^7 s^{-1}. Similar to the procedure in the inverse Taylor test, a sabot launched projectile is used to impact the specimen (see Figure 7.14). Both specimen and projectile, or plate and flyer, are planar discs that need to be impacted in a planar mode in order to trigger plane strain waves at the interface. These waves travel into flyer and plate and are reflected at the free surfaces.

Acceleration of the plate goes along with the shock wave propagation. The particle velocity v_1 associated with the shock can be observed at the free surface where it leads to a velocity jump of

$$v_{fs} = 2\, v_1 \tag{7.5}$$

This relation is an approximation. Actually, the free surface velocity v_{fs} equals the sum of the particle velocities due to shock wave v_S and reflected rarefaction wave v_R. Since the shock wave reflection on a free surface with almost infinite impedance difference leads to a rarefaction wave speed of $v_R \approx v_S$ and, thus, to the approximation (7.5).

Multiple reflections of the shock and rarefaction waves lead to the stepwise acceleration of the plate as indicated in the lower right diagram of Figure 7.15.

The particular usefulness of equation (7.5) is found in its application to relate measured free surface velocities to shock velocities and hence, to derive v_S-v_1 relation for equations of state. Figure 7.16 shows examples of v_S - v_1 relations derived from flyer-plate tests for two steel alloys and for copper, respectively.

Impedance Match Method

Knowing the material properties of one plate enables the derivation of Hugoniot data for the other one by the so called *impedance match method*, see e.g. Rinehart [328], Zel'dovich [433] or Rice et al. [320]. With the impedance match method, the construction of a Hugoniot curve is performed in the p-v_1 plane as follows:

---

[3] See for example Chhabildas and Knudson [90].

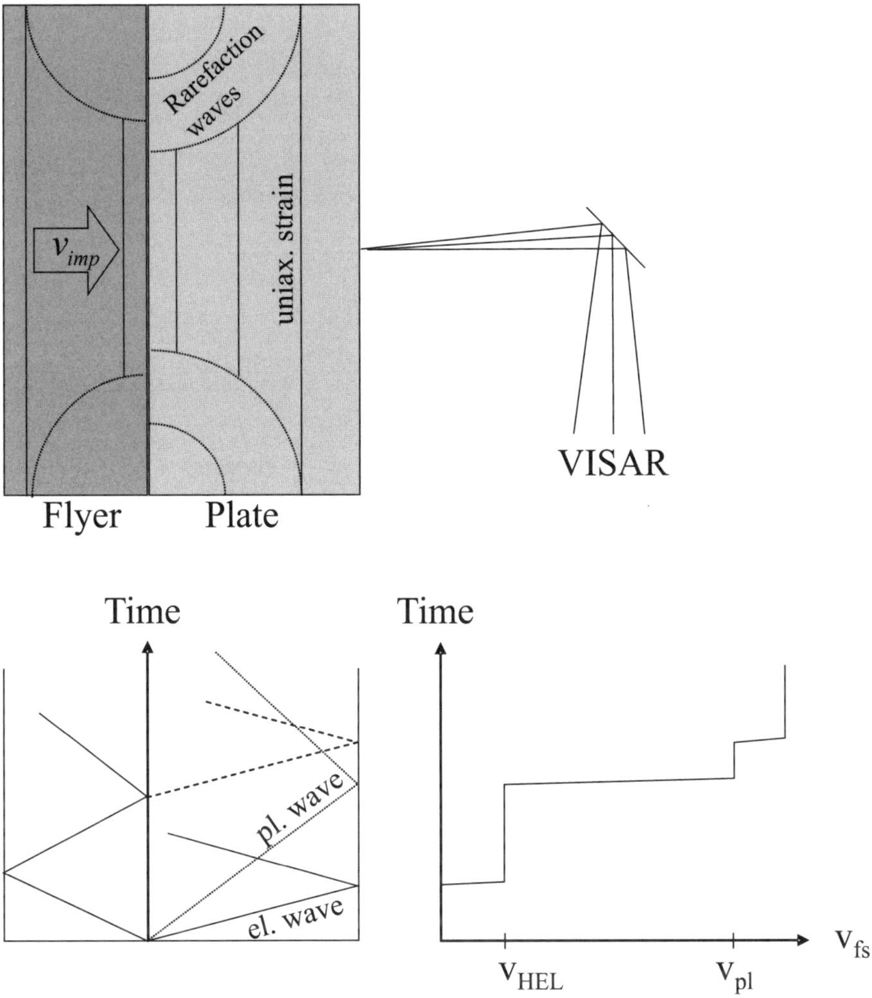

Fig. 7.15. Wave propagation in flyer and plate due to the impact at v_{imp}. Reflections at the rear free surface of the plate lead to the step-wise acceleration indicated the lower right diagram.

- Consider a material A for which the shock Hugoniot curve is known. In the p-v_1 plane this Hugoniot shall be represented by $p_A(v_1)$ as indicated in Figure 7.17.
- The reflected Hugoniots p_A^R represent material A being impacted at three particular impact velocities that would lead to particle velocities v_1^a, v_1^b and v_1^c.
- Consider next a planar plate experiment a with impact velocity v_{1^a}: material A is impacted by a plate made from material B with unknown Hugoniot

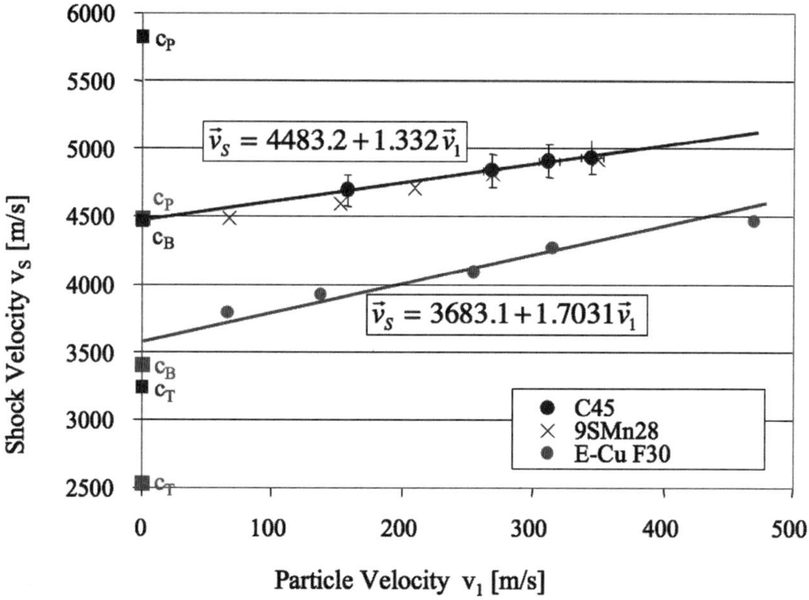

Fig. 7.16. Experimental results for two steel alloys and copper with resulting linear v_S - v_1 relations. (Reprint with permission from Nahme [290])

by known initial density ρ_0. The impact velocity leads to a shock state in both materials. In material B the pressure jump at the interface can be derived from equation (4.42) as

$$p_1^a = \rho_0 \, v_S \, v_1 \tag{7.6}$$

Pressure equilibrium at both surfaces means that the Hugoniots of both materials intersect at $p = p_1^a$. This pressure value can be derived graphically using the slope of the Rayleigh line in material B $\rho_0 \, v_S$ as indicated in Figure 7.17. Thus, the intersection point (1) is identified to be the first known point on the Hugoniot of material B.

- Further points of the Hugoniot can be derived with additional impact tests at varying impact velocities, e.g. v_{1^b} and v_{1^c} as illustrated in Figure 7.17.

Dynamic Yield Stresses and Spallation Pressures

Advanced instrumentation techniques enable more detailed investigations of material behavior in the strain rate regime beyond 10^5 s^{-1}. Hugoniot elastic

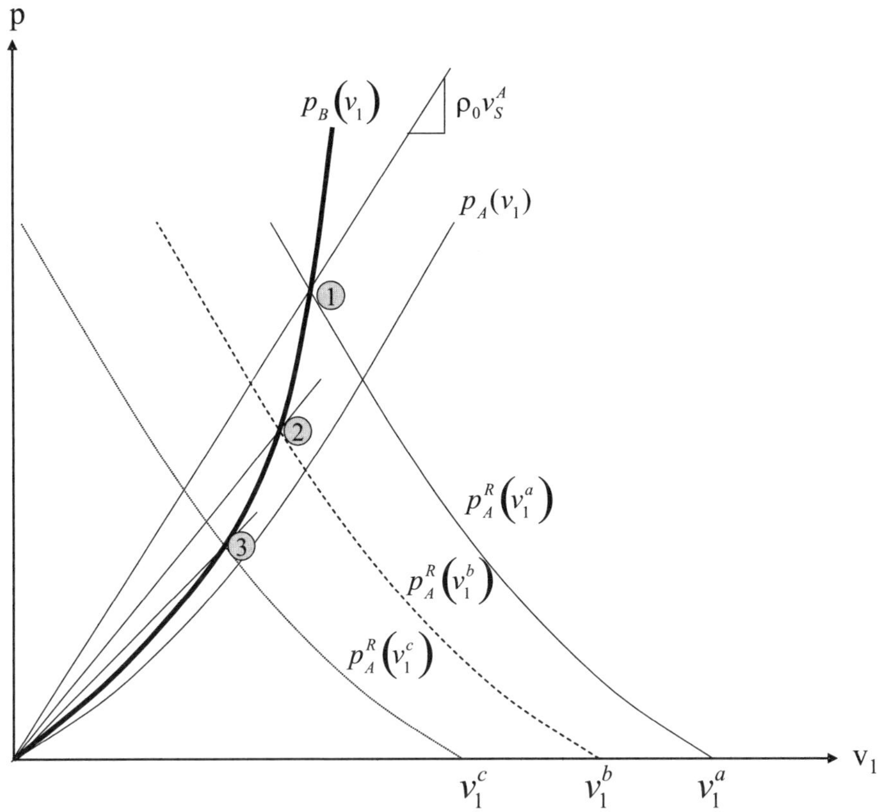

Fig. 7.17. Construction of a shock Hugoniot for material B from a known Hugoniot of material A and a number of plate impact tests.

limits (dynamic yield stresses) and spall failure criteria are important if structural response to complex loading conditions shall be investigated.

Two major types of instrumentation used for the flyer-plate test are the VISAR technique and manganin stress gauges. Whereas the VISAR option uses the free surface velocities at the rear side of the plate to derive stresses under the plane strain assumption, the manganin gauges[4] directly measure stresses on the front and rear sides of the plate. The latter instrumentation was used for flyer plate tests by Rosenberg et al. [337] as well as Rosenberg and Bless [335]. Early applications of stress gauges for uni-axial stress loaded long rods were published by Rosenberg et al. [336]. A dual instrumentation of flyer-plate tests with both VISAR and manganin stress gauges was reported

[4] Manganin gauges were first used by Bridgeman [66] for stress measurement.

by Bourne et al. [62].

Both methods are only applicable along with the describing theory as long as a plane strain state is present. At the same time as the plane waves travel back and forth through flyer and plate, rarefaction waves move inwards from the radial surfaces of the discs. Thus, the valid test duration is limited by the time the rarefaction waves need to interfere with the plane wave at the measurement locus. Thus, measurement of free surface velocities or stresses is performed in the center of the plate to gain the maximum measurement time.

A free surface velocity signal measured for 35NiCrMoV109 steel in a 918 $\mathrm{m\,s^{-1}}$ flyer plate test is illustrated in Figure 7.18. At 0.2 [μs] the first elastic wave arrives at the free surface leading to a velocity jump measured to be to v_{HEL}. Accordingly, the related particle velocity equals half the measured free surface velocity:

$$v_1 = \frac{1}{2}\,v_{HEL} \tag{7.7}$$

which, with equation (3.210) and the elastic longitudinal sound speed c_L, leads to the Hugoniot elastic limit:

$$\sigma_{HEL} = \frac{1}{2}\,\rho_0\,c_L\,v_{HEL} \tag{7.8}$$

The failure type spallation was mentioned already in section 6.2.5. In flyer plate tests reflected shock waves can lead to rarefaction waves that are strong enough to excess the limit of hydrostatic tensile failure. Figure 7.19 shows the cross-sections of two flyer plate specimen that were impacted to initial and extended spall failure, respectively. The reflected shock waves in the plate is a wave of uniaxial tensile strain. Superimposed by the rarefaction wave coming from the back of the flyer results in a strong tensile stress. The shock wave typical short rise times lead to a sharp plane of failure as can be seen in Figure 7.19.

Thus, spallation in the flyer plate test leads to the formation of a new free surface inside the plate. The remaining shock and rarefaction waves are reflected from that opening spall plane and lead to a characteristic signal in the measurement of the free surface velocity. As indicated in Figure 7.18, the process of building the spall plane leads to a deceleration of the free surface velocity from v_{max} to $v_{max} - \Delta\,v_{sp}$. When the first wave reflected from the spall plane arrives at the free surface, a new velocity jump occurs. With this re-rising signal of the free surface velocity the duration τ for the spall formation can be measured. The velocity difference $\Delta\,v_{fs}$ is called *velocity pullback*. Using this velocity difference the stress just before spalling and, thus, the stress criterion for the related failure mechanism.

A first approach to calculate the spall stress σ_{sp} would be:

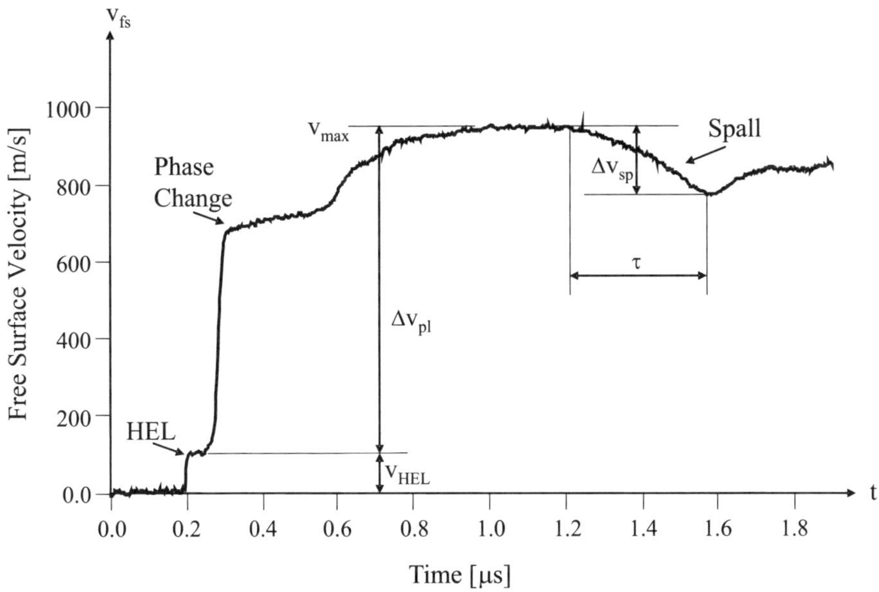

Fig. 7.18. Free surface velocity measured for 35NiCrMoV109 steel in a flyer plate test with a flyer velocity of 918 $\mathrm{m\,s^{-1}}$. (Reprint with permission from Rohr [331])

$$\sigma_{sp,0} = \frac{1}{2}\,\rho_0\,c_L\,\Delta v_{sp} \tag{7.9}$$

However, at the time when spallation occurs in the plate, an elastic-plastic pre-deformation has taken place already. To account for the related effects Stepanov [370] suggested to correct the prevailing sound speed by:

$$\sigma_{sp,c} = \rho_0\,\frac{c_L\,c_B}{c_L + c_B}\,\Delta v_{sp} \tag{7.10}$$

Moreover, an additional correction term is needed to account for the finite thickness of the spall plate. Romanchenko and Stepanov [334] suggested the correcting stress term

$$\Delta\sigma = \frac{1}{2}\,\frac{\partial\sigma}{\partial t}\,h_s\left(\frac{1}{c_B} - \frac{1}{c_L}\right) \tag{7.11}$$

to arrive at:

$$\sigma_{sp,C} = \rho_0\,\frac{c_L\,c_B}{c_L + c_B}\,\Delta v_{sp} + \Delta\sigma \tag{7.12}$$

Since the definition of the time derivative of the stress $\partial\sigma/\partial t$ poses additional complexity to the spall strength derivation, various approximations were derived[5]. A reasonable adjustment for finite spall plate thicknesses seems to be achieved by the following stress correction:

[5] See Antoun et al. [10] for a detailed discussion of the correction.

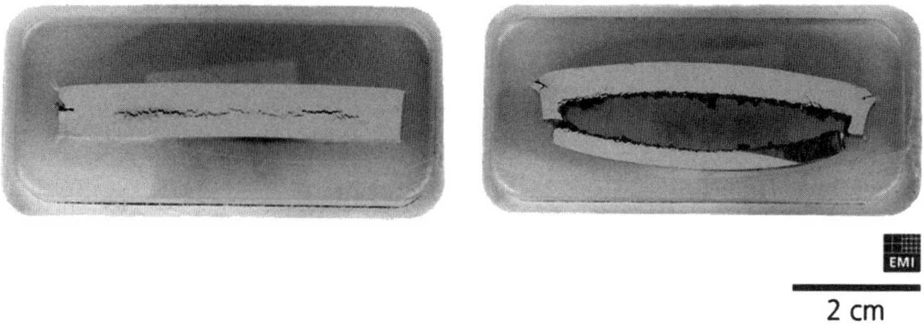

Fig. 7.19. Cross sections of two steel plates loaded at impact velocities of 384 [m s^{-1}] and 735 [m s^{-1}] leading to initial (left) and extended formation of spall planes (right), respectively. (Courtesy of Hartwig Nahme, Ernst-Mach-Institute).

$$\Delta\sigma = \frac{h_s}{\tau}\left(\frac{1}{c_B} - \frac{1}{c_L}\right)\sigma_{sp,c} \tag{7.13}$$

and thus the spallation strength:

$$\sigma_{sp,C} = \rho_0\,\frac{c_L\,c_B}{c_L + c_B}\,\Delta v_{sp}\,(1 + \Delta\sigma) \tag{7.14}$$

With the measured free surface velocities from five individual flyer plate tests illustrated in Figure 3.16 and using equation (7.14) Rohr [331] derived an average corrected spall strength for 35NiCrMoV109 steel of $\sigma_{sp,C} = 3.8$ [GPa].

The influence of different heat treatment on the high dynamic behaviour of steel becomes evident from the measured data for nitrided steel indicated in Figure 7.20. At first glance the difference in the Hugoniot elastic limit is observed. Evaluation of the spall strengths using equation (7.14) delivers values of 5.97[GPa] and 6.52 [GPa] for the 384 [m s^{-1}] and 735 [m s^{-1}] impact tests, respectively.

Flyer plate tests are performed in various configurations of flyer, specimens and backing plates. Specific choice of material impedances for the particular components lead to different wave propagation characteristics. High impedance backing of flyer and plate, for example, lead to a continuous reflection of compressive waves and thus to very high shock pressures. The *stress reverberation method* was suggested by Lysne et al. [258]. Applications can be found in Riedel et al. [324] as well as [326].

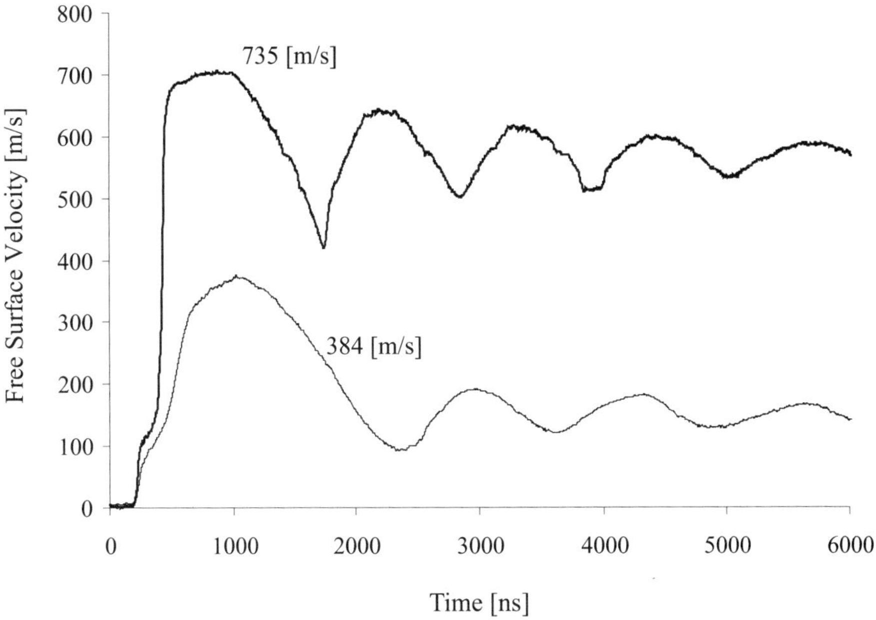

Fig. 7.20. Measured free surface for the two differently tempered steel plates illustrated in Figure 7.19. Loaded at impact velocities of 384 $[\mathrm{m\,s^{-1}}]$ and 735 $[\mathrm{m\,s^{-1}}]$ they show significantly different Hugoniot elastic limits and spall strengths. (Courtesy of Hartwig Nahme, Ernst-Mach-Institute).

7.4.3 Edge-On Impact Test

In the previous chapters the so called *edge-on impact test* was repeatedly mentioned. Although it is not a material test in the above defined sense for the derivation of stress-strain relations, it can be called material test for deriving crack propagation speeds in brittle materials. Its basic set-up as introduced by Straßburger and Senf [375] is indicated in Figure 7.21. A blunt projectile is accelerated and shot against the plane edge of a material sample. Adequate choice of impact velocity and planar impact enable the initiation of a failure wave in the brittle material.

Observation of the failure wave propagation by high speed cameras enables the identification of its propagation speed. Figure 7.22 illustrates crack propagation velocities in Al_2O_3 ceramic for various impact velocities.

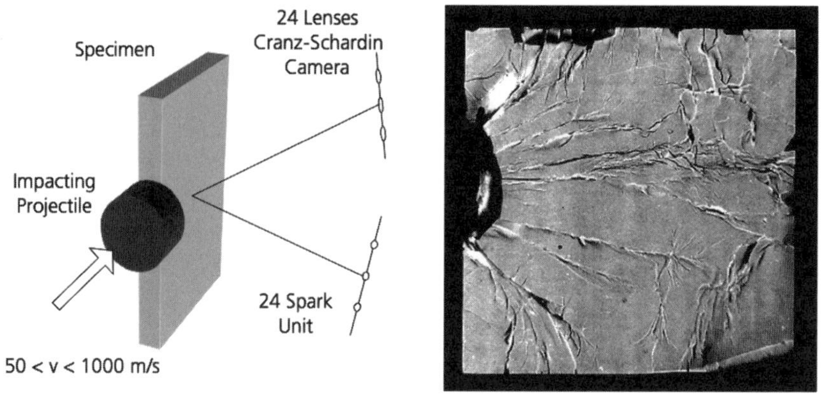

Fig. 7.21. Set-up and example picture of the edge-on-impact test for crack propagation in brittle materials as introduced by Straßburger and Senf [375]. (Reprint with permission from EMI Annual Report 1997)

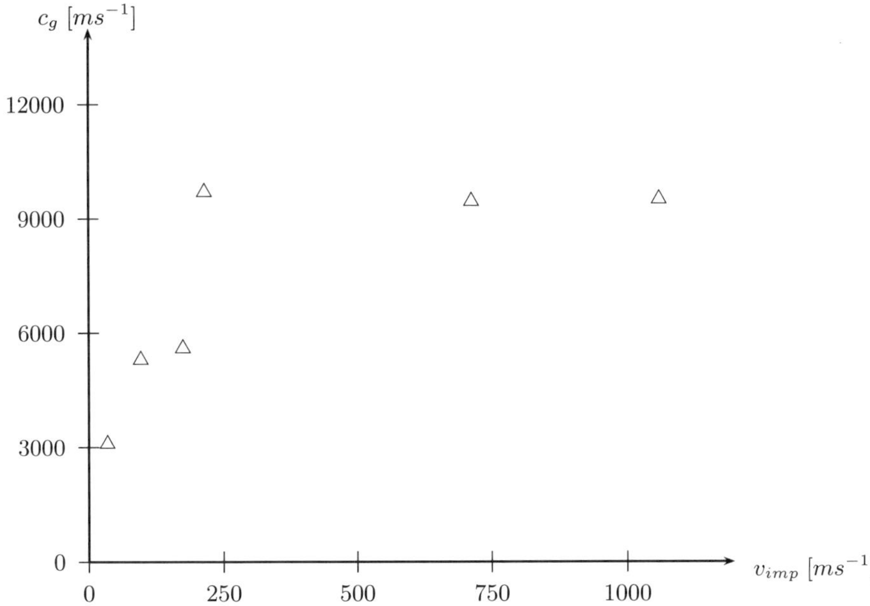

Fig. 7.22. Crack propagation velocities in Al_2O_3 ceramic for various impact velocities measured in edge-on impact tests by Straßburger and Senf [375]. (Reprint with permission from EMI Annual Report 1997)

References

1. Allix O, Hild F (eds.) (2002) Continuum Damage Mechanics of Materials and Structures. Elsevier, Amsterdam.
2. Allix O, Ladeveze P, Corigliano A (1995) Damage Analysis of Interlaminar Fracture Specimens. Composite Structures 31: 61–74
3. Alves M, Jones N (1999) Influence of Hydrostatic Stress on Failure of Axisymmetric Notched Specimens. Journal of Mechanics and Physics of Solids 47: 643–667
4. Anand L (1996) A Constitutive Model for Compressible Elastomeric Solids. Computational Mechanics 18: 339–355
5. Anand L, Ames N M (2006) On Modeling the Micro-Indentation Response of an Amorphous Polymer. International Journal of Plasticity 22: 1123–1170
6. Anand L, Gurtin M E (2003) A Theory of Amorphous Solids Undergoing Large Deformations, With Application to Polymeric Glasses. International Journal of Solids and Structures 40: 1465–1487
7. Anderson Jr. C E, Cox P A, Johnson G R, Maudlin P J (1994) A Constitutive Formulation for Anisotropic Materials Suitable for Wave Propagation Computer program-II Journal of Computational Mechanics 15: 201–223
8. Anderson Jr. C E, Mullin S A (1988) Hypervelocity Impact Phenomenology: Some Aspects of Debris Cloud Dynamics. In: Amman W J , Liu W K, Studer J A, Zimmermann T (eds.) Impact: Effects of Fast Transient Loadings. Balkema, Rotterdam.
9. Anderson J D (1995) Computational Fluid Dynamics. McGraw-Hill, New-York, London.
10. Antoun T, Seaman L, Curran D R, Kanel G, Razorenov S V, Utkin A V (2002) Spall Fracture. Springer, New York.
11. Arcan M, Hashin Z, Voloshin A (1978) A Method to Produce Uniform Plane-Stress States With Application to Fiber-Reinforced Materials. Experimental Mechanics 18:141–146
12. Archer J S (1963) Consistent mass matrix for distributed mass systems. Proceedings of the American Society of Civil Engineering, Journal of the Structural Division 89: 161–178.
13. Archer J S (1965) Consistent mass matrix formulation for structural analysis using finite element techniques. Journal of the American Institute of Aeronautics and Astronautics, 3: 1910–1918.

14. Argon A S (1973) A Theory for the Low-Temperature Plastic Deformation of Glassy Polymers. Philosophical Magazine 28: 839–865
15. Armstrong P J, Frederick C O (1966) A Mathematical Representation of the Multiaxial Bauschinger Effect. G.E.G.B. Report RD/B/N 741, Berceley Nuclear Laboratories.
16. Arnold D N (1981) Discretization by Finite Elements of a Model Parameter Dependent Problem. Numerische Mathematik 37: 405–421
17. Arnold W (1991) Dynamisches Werkstoffverhalten von Armco-Eisen bei Stoßwellenbelastung. Doctoral Thesis, Technical University Munich.
18. Arruda E M, Boyce M C (1993) A Three-Dimensional Constitutive Model for the Large Stretch Behavior of Rubber Elastic Materials. Journal of the Mechanics and Physics of Solids 41: 389–412
19. Arruda E M, Boyce M C (1993) Evolution of plastic anisotropy in amorphous polymers during finite straining. International Journal of Plasticity 9: 697–720
20. Atluri S N, Shen S (2002) The Meshless Local Petrov-Galerkin Method. Tech Science Press.
21. Atluri S N, Zhu T (1998) A New Meshless Local Petrov-Galerkin (MLPG) Approach in Computational Mechanics. Computational Mechanics 22: 117–127
22. Atluri S N, Shen S P (2002) The Meshless Local Petrov-Galerkin (MLPG) Method. Tech Science Pree, USA
23. Attaway S W, Heinstein M W, Swegle J W (1994) Coupling of Smooth Particle Hydrodynamics with the Finite Element Method. Nuclear Engineering and Design 150: 199–205
24. Attinger S, Koumoutsakos P (eds) (2004) Multiscale Modelling and Simulation - Lecture Notes in Computational Science and Engineering, 39. Springer Verlag, Berlin.
25. Auld B A (1973) Acoustic Fields and Waves in Solids. Volume 1. John Wiley & Sons Inc, New York.
26. Azzi V D, Tsai S W (1965) Anisotropic Strength of Composites. Journal of Experimental Mechanics 5: 283–288
27. Babuška I, Barnjee U, Osborn J E (2003) Meshless and Generalized Finite Element Methods: A Survey of Some Major Results. In: Griebel M, Schweitzer M A (eds.) Meshfree Methods for Partial Differential Equations. Springer, Berlin.
28. Babuška I, Melenk J (1997) The Partition of Unity Method. International Journal for Numerical Methods in Engineering 40: 727–758
29. Babuška I, Suri M (1992) On Locking and Robustness in the Finite Element Method. SIAM Journal on Numerical Analysis 29: 1261–1299
30. Babuška I, Suri M (1992) Locking Effects in the Finite Element Approximation of Elasticity Problems. Numerische Mathematik 62:439–463
31. Balsara D S (1995) Von Neumann Stability Analysis of Smoothed Particle Hydrodynamics - Suggestions for Optimal Algorithms. Journal of Computational Physics 121: 357–372
32. Bao Y (2003) Prediction of Ductile Crack Formation in Uncracked Bodies. PhD Thesis Massachusetts Institute of Technology, Department of Ocean Engineering, Impact and Crashworthiness Laboratory.
33. Bao Y, Wierzbicki T (2004) On Fracture Locus in the Equivalent Strain and Stress Triaxiality Space. International Journal of Mechanical Science 46: 81–98

34. Bardenheier R (1977) Versagensverhalten von PMMA bei ebener Beanspruchung - Diskussion hexagonaler Versagenskriterien. Zeitschrift für Werkstofftechnik 8: 379–388
35. Barker L M, Hollenbach R E (1965) Interferometer Technique for Measuring the Dynamic Mechanical Properties of Materials. Review of Scintific Instruments 36: 1617–1620
36. Barker L M (2000) The Development of the VISAR and its Use in Shock Compression Science. In: Furnish M D, Chhabildas L C, Hixson R S (eds.) Shock Compression of Condensed Matter - 1999. American Insitute of Physics.
37. Bathe K J (1996) Finite Element Procedures. Prentice Hall, Englewood Cliffs.
38. Bauer F, Graham R A (1995) Very High Pressure Behavior of Precisely-Poled PVDF, Ferroelectrics 171: 95–102
39. Bauer F (2000) PVDF Shock Sensors: Applications to Polar Materials and High Explosives. IEEE Transactions On Ultrasonics, Ferroelectrics, and Frequency Control 47: 1448–1454
40. Bažant Z P, Cedolin L (1991) Stability of Structures. Oxford University Press, New York and Oxford.
41. Bell J F (1973) The Experimental Foundation of Solid Mechanics. In: Handbuch der Physik, VIa/1. Springer, Berlin.
42. Belytschko T, Black T (1999) Elastic Crack Growth in Finite Elements with Minimal Remeshing. International Journal for Numerical Methods in Engineering 45: 601–620
43. Belytschko T, Guo Y, Liu W K, Xiao S P (2000) A Unified Stability Analysis of Meshless Particle Methods. International Journal for Numerical Methods in Engineering 48: 1359–1400
44. Belytschko T, Krongauz Y, Dolbow J, Gerlach C (1998) On the Completeness of Meshfree Particle Methods. International Journal for Numerical Methods in Engineering 43: 785–819
45. Belytschko T, Krongauz Y, Organ D, Fleming M, Krysl P (1996) Meshless methods: An Overview and Recent Developments. Computer Methods in Applied Mechanics and Engineering 139: 3–47
46. Belytschko T, Lin J I, Tsay C S (1984) Explicit Algorithms for the Nonlinear Dynamics of Shells. Computer Methods in Applied Mechanics and Engineering 42: 225–251
47. Belytschko T, Liu W K, Moran B (2001) Nonlinear Finite Elements for Continua and Structures. John Wiley & Sons Ltd, Chichester.
48. Belytschko T, Lu Y Y, Gu L (1994) Element-Free Galerkin Methods. International Journal on Numerical Methods in Engineering 37: 229–256
49. Belytschko T, Xiao S P (2002) Stability Analysis of Particle Methods With Corrected Derivatives. Computers and Mathematics with Applications 43: 329–350
50. Belytschko T, Yen H-J, Mullen R (1979) Mixed Methods for Time Integration. Computer Methods in Applied Mechanics and Engineering 17: 259–275
51. Ben-Dor G, Igra O, Elperin T (2001) Handbook of Shock Waves. Academic Press, San Diego.
52. Bennett L S, Tanaka K, Horie Y (1997) Developments in Constitutive Modelling of Shock-Induced Reactions in Powder Mixtures. In: Davison L, Horie Y, Shahimpoor (eds.) High-Pressure Shock Compression of Solids. Springer, New York.

53. Benson D J (1992) Computational Methods in Lagrangian and Eulerian Hydrocodes Computer Methods in Applied Mechanics and Engineering 99: 235–394
54. Benz W (1990) Smoothed Particle Hydrodynamics: A Review. In: Buchler J R (ed.) The Numerical Modelling of Nonlinear Stellar Pulsation. Kluwer Academic Publishers, Dordrecht, The Netherlands.
55. Benz W, Asphaug E (1994) Impact Simulation with Fracture: I. Methods and Tests. ICARUS 107: 98–116.
56. Benz W, Asphaug E (1995) Simulation of Brittle Solids Using Smooth Particle Hydrodynamics. Computer Physics Communications 87: 253–265
57. Berthelot J-M (1999) Composite Materials. Springer, New York.
58. Bethe H A (1942) On the Theory of Shock Waves for an Arbitrary Equation of State. In: Johnson J, Cheret R (eds) Classic Papers in Shock Compression Science, Springer, New York.
59. Binder K (1988) The Monte Carlo Method in Condensed Matter Physics. Springer, Berlin.
60. Birkhoff G, de Boor C, Swartz B, Wendroff B (1966) Rayleigh-Ritz Approximations by Piecewise Cubic Polynomials. SIAM Journal on Numerical Analysis 3: 188–203
61. Børvik T, Hopperstad O S, Berstad T (2003) On the Influence of Stress Triaxiality and Strain Rate on the Behaviour of a Structural Steel. Part II. Numerical Study. European Journal of Mechanics A/Solids 22: 15–32
62. Bourne N K, Rosenberg Z (1999) Manganin Gauge and VISAR Histories in Shock Stressed Polymethylmethacrylate. Proceedings Royal Society. Mathematical, Physical and Engineering Sciences 455: 1259–1266
63. Boyce M C(1987) Large Inelastic Deformation of Glassy Polymers. Ph.D. Thesis, Dept. of Mechanical Engineering, Massachusetts Institute of Technology, Cambridge, MA.
64. Boyce M C, Parks D, Argon A S (1988) Large Inelastic Deformations of Glassy Polymers. Part I: Rate Dependent Constitutive Model. Mechanics of Materials 7: 15–33
65. Brewer C J, Lagace P A (1988) Quadratic Stress Criterion for Initiation of Delamination. Journal of Composite Materials 22: 1141–1155
66. Bridgman P W (1940) The Measurement of Hydrostatic Pressure to 30000 kg/cm^2. Proceedings of the American Academy of Arts and Science 74: 1
67. Bridgman P W 1952 Studies in Large Plastic Flow and Fracture. Metallurgy and Metallurgical Engineering Series. McGraw Hill, New York.
68. Broek D (1986) Elementary Engineering Fracture Mechanics. Kluwer Academic Publications.
69. Brozzo P, Deluca B, Rendina R (1972) A New Method for the Prediction of Formability Limits in Metal Sheet, Sheet Metal Forming and Formability. Proceedings to the Seventh Biennial Conference of the International Deep Drawing Research Group, Amsterdam, The Netherlands.
70. Bulatov V V, Kaxiras E, Ghoniem N, Phillips R, Diaz De La Rubia T (eds) (1999) Proceedings to Symposium on Multiscale Modelling of Materials. Materials Research Society, Boston, MA.
71. Burk R C (1983) Standard Failure Criteria Needed for Advanced Composites. Journal of Composite Materials 21: 58–62
72. Bushman A V, Kanel' G I, Ni A L, Fortov V E (1993) Intense Dynamic Loading of Condensed Matter. Taylor & Francis, Washington DC.

73. Butcher B M, Karnes C H (1968) Sandia Labs. Research Report SC-RR-67-3040, Sandia Laboratory, Albuquerque, NM.
74. Caddell R M, Woodliff A R (1977) Macroscopic Yielding of Oriented Polymers. Journal of Materials Science 12: 2028–2036
75. Caddell R M, Raghava R S, Atkins A G (1974) Pressure Dependent Yield Criteria for Polymers. Materials Science and Engineering 13: 113–120
76. Campbell J, Vignjevic R, Libersky L (2000) A Contact Algorithm for Smoothed Particle Hydrodynamics. Computer Methods in Applied Mechanics and Engineering 184: 49–65
77. Campbell P M (1989) Some New Algorithms for Boundary Value Problems in Smoothed Particle Hydrodynamics. Technical Report DNA-TR-88-286, Mission Research Corporation.
78. Caramana E J, Shashkov M J, Whalen P P (1998) Formulations of Artificial Viscosity for Multi-dimensional Shock Wave Computations. Journal of Computational Physics 144: 70–97
79. Carter B J, Chen C-S, Ingraffea A R, Wawrzynek P A (1997) A Topology-Based System for Modeling 3-D Crack Growth in Solid and Shell Structures. Proceedings of the Ninth International Congress on Fracture, ICF9. Elsevier Science Publishers, Sydney, Australia.
80. Carrol M, Holt A C (1972) Suggested Modification of the P-α Model for Porous Materials. Journal of Applied Physics 43: 759–761
81. Chang F-K, Chang K-Y (1987) Post-Failure Analysis of Bolted Composite Joints in Tension or Shear-Out Mode Failure. Journal of Composite Materials 21: 809–833
82. Chang F-K, Chang K-Y (1987) A Progressive Damage Model for Laminated Composites Containing Stress Concentrations. Journal of Composite Materials 21: 834–855
83. Chang F-K, Scott R A, Springer G S (1984) Failure Strength of Nonlinearly Elastic Composite Containing a Pin Loaded Hole. Journal of Composite Materials 18: 464–477
84. Chapman D J, Radford D D, Reynolds M, Church P D (2005) Shock Induced Void Nucleation During Taylor Impact. International Journal of Fracture 134: 41–57
85. Chatelin F, Lemordant M J (1975) La méthode de Rayleigh-Ritz appliquée à des opérateurs différentiels elliptiques – ordres de convergence des éléments propres. Numerische Mathematik 23: 215–222
86. Chen W F, Han D J, (1988) Plasticity for Structural Engineers. Springer, New York.
87. Chessa J, Belytschko T (2004) Arbitrary Discontinuities in Space-Time Finite Elements by Level Sets and X-FEM. International Journal for Numerical Methods in Engineering 61: 2595–2614
88. Chhabildas L C, Asay J R (1992) Dynamic Yield Strength and Spall Strength Measurement Under Quasi-Isentropic Loading. In: Meyers M A, Murr L E, Staudhammer K P (eds.) Shock-Wave and High-Strain-Rate Phenomena in Materials. Marcel Dekker, Inc., New York, NY.
89. Chhabildas L C, Barker L M (1988) Shock Waves in Condensed Matter - 1987. Elsevier Science Publishers B. V., North Holland.
90. Chhabildas L C, Knudson M D (2005) Techniques to Launch Projectile Plates to Very High Velocities. In: Chhabildas L C, Davison L, Horie Y (eds) High Pressure Shock Compression of Solids VIII. Springer Verlag.

91. Chisolm E D, Crockett S D, Wallace D C (2003) Test of Theoretical Equation of State for Elemental Solids and Liquids. Physical Review B 68, 104103.
92. Ciarlet P G, Schultz M H, Varga R S (1968) Numerical Methods of High-Order Accuracy for Nonlinear Boundary Value Problems III. Numerische Mathematik 12: 120–133
93. Clough R W (1960) The Finite Element Method in Plane Stress Analysis. Proceedings to the Second Conference on Electronic Computation, A.S.C.E. Structural Division, Pittsburgh, PA.
94. Cockcroft M G, Latham D J (1968) Ductility and the Workability of Metals. Journal of the Institute of Metals 96: 33–39
95. Coleman B and Gurtin M (1967) Thermodynamics of Internal State Variables. Journal of Chemical Physics 47: 597–613
96. Cook R D, Malkus D S, Plesha M E (1989) Concepts and Application of Finite Element Methods. 3rd edition, Wiley, New York.
97. Corley J (2004) Characterization and Modeling of the Dynamic Mechanical Properties of a Particulate Composite Material. Doctoral Thesis, Fraunhofer Institute for High Speed Dynamics, Ernst-Mach-Institute, Issue 7 of the scientific series Thoma K, Hiermaier S (eds), $\dot{\varepsilon}$ - Forschungsergebnisse aus der Kurzzeitdynamik. ISBN: 3-8167-6343-3.
98. Courant R, Friedrichs K O, Lewy H (1928) Über die partiellen Differenzengleichungen der mathematischen Physik. Mathematische Annalen 100: 32–74. More recently also in: IBM Journal of Research Developments (1967) 11: 215–234
99. Curran D R, Seaman L (1985) Computational Models for Nucleation, Growth and Coalescence of Adiabatic Shear Bands. Journal de Physique, Colloque C5, Supplement on No. 8, Tome 46.
100. Curran D R, Seaman L, Shockey D A (1987) Dynamic Failure of Solids. Physics Reports 147: 253–388
101. Daniel W J T (1998) A Study of the Stability of Subcycling Algorithms in Structural Dynamics. Computational Methods in Applied Mechanics and Engineering 156: 1–13
102. Daux C, Moës N, Dolbow J, Sukumar N, Belytschko T (2000) International Journal for Numerical Methods in Engineering 48:1741–1760
103. Davis W (1993) Equation of State for Detonation Products. Proceedings to the Tenth International Detonation Symposium: 369–376
104. Davis W (1998) Equation of State for Detonation Products. Proceedings to the Eleventh International Detonation Symposium: 303–308
105. Davison L, Horie Y, Shahinpoor M (1997) High-Pressure Shock Compression of Solids IV - Response of Highly Porous Solids to Shock Loading. Springer, New York.
106. Dellinger L, Vasicek D, Sondergeld C (1998) Kelvin notation for stabilizing elastic constant inversion. Revue de l'institute Francaise de Petrol 53: 709–719.
107. Dick E (1996) Introduction to Finite Element Techniques in Computational Fluid Dynamics. In: Wendt J F (ed.) Computational Fluid Dynamics. Springer, Berlin.
108. Dilts G A (1999) Moving Least Squares Particle Hydrodynamics I: Consistency and Stability. International Journal for Numerical Methods in Engineering 44: 1115–1155

109. Dilts G A (2000) Moving Least Squares Prickle Hydrodynamics II: Conservation and Boundaries. International Journal for Numerical Methods in Engineering 48: 1503–1524
110. Dilts G A (2001) Some Recent Developments for Moving Least Squares Particle Methods. Proceedings First MIT Conference on Computational Fluid and Solid Mechanics, Massachusetts Institute of Technology, Cambridge, MA
111. Donea J, Huerta A (2003) Finite Element Methods for Flow Problems. John Wiley & Sons Ltd., Chichester, England.
112. Doppler C A (1848) Über den Einfluß der Bewegung des Fortpflanzungsmittels auf die Erscheinung der Äther-, Luft- und Wasserwellen. Abhandlung der Königlich Böhmischen Gesellschaft der Wissenschaften in Prag 5 [V]: 293–306
113. Doyle T C, Ericksen J L (1956) Nonlinear Elasticity. Advances in Applied Mechanics 4: 53–115
114. Dow J O (1999) Finite Element Method and Error Analysis Procedures. Academic Press, San Diego.
115. Doyoyo M, Wierzbicki T (2003) Experimental Studies on the Yield Behavior of Ductile and Brittle Aluminum Foams. International Journal of Plasticity 19: 1194–1214
116. Drumheller D S (1998) Introduction to Wave Propagation in Nonlinear Fluids and Solids. Cambridge University Press.
117. Dugdale J S, MacDonald D K C (1953) The Thermal Expansion of Solids. Physical Review 89: 832–834
118. Duhem P (1909) Sur la Propagationdes Ondes de Choc au Sein des Fluides. Zeitschrift für Physikalische Chemie 69: 169–186
119. Dyka C T, Ingel R P (1995) An Approach for Tensile Instability in Smoothed Particle Hydrodynamics. Computers and Structures 57: 573–580
120. Dyka C T, Randles P W, Ingel R P (1997) Stress Points for Tension Instability in Smoothed Particle Hydrodynamics. International Journal for Numerical Methods in Engineering 40: 2325–2341
121. Engelmann B E, Whirley R G (1991) A new Explicit Shell Element Formulation for Impact Analysis. In: Kulak L F, Schwer L E (eds.): Computational Aspects of Contact, Impact and Penetration. Elmepress International, Lausanne, Switzerland.
122. Evans M W, Harlow F H (1957) The Particle-in-Cell Method for Hydrodynamic Calculations. Los Alamos Scientific Laboratory Report LA-2139
123. Fahrenthold E P, Koo J C (1997) Energy based particle hydrodynamics for hypervelocity impact simulation. International Journal of Impact Engineering 20: 253–264
124. Fahrenthold E P, Koo J C (1999) Discrete Hamilton's Equations for Viscous Compressible Fluid Dynamics. Computer Methods in Applied Mechanics and Engineering 178: 1–22
125. Feucht M, Sun D-Z, Erhart T, Frank T (2006) Recent Development and Application of the Gurson Model. Proceedings to the Fifth LS-DYNA Anwenderforum, Ulm, Germany.
126. Flanagan D, Belytschko T (1981) A uniform Strain Hexahedral and Quadrilateral with Orthogonal Hourglass Control. International Journal of Numerical Methods in Engineering 17: 679–706
127. Flory P J (1961) Thermodynamic Relations for High Elastic Materials. Transactions of the Faraday Society 57: 829–838

128. Flory P J, Rehner Jr J (1943) Statistical Mechanics of Cross-Linked Polymer Networks. 1. Rubberlike Elasticity. Journal of Chemical Physics 11: 512–520.
129. Follansbee P S (1985) The Hopkinson Bar. American Society of Metals.
130. Fowles R, Williams R F (1970) Plane Stress Wave Propagation in Solids. Journal of Applied Physics 41: 360–363
131. Fu Y B, Ogden R W (ed) (2001) Nonlinear Elasticity. Cambridge University Press.
132. Galerkin B G (1915) Series Solutions of Some Problems of Elastic Equilibrium of Rods and Plates. Vest. Inzh. Tech. 19: 897–908 (Russian)
133. Gearing B P (2002) Constitutive Equations and Failure Criteria for Amorphous Polymeric Solids. Ph.D. Thesis, Massachusetts Institute of Technology, Cambridge, MA.
134. Gese H, Werner H, Hooputra H, Dell H, Health A (2004) CrachFEM - A Comprehensive Failure Model for Metallic Structures in Sheet Metal Forming and Crash Simulation. Proceedings EuroPAM 2004, Paris, France.
135. Gingold R A, Monaghan J J (1977) Smoothed Particle Hydrodynamics: Theory and Application to Non-Spherical Stars. Monthly Notices Royal Astronomical Society 181: 375–389
136. Godunov S K (1969) A Difference Scheme for Numerical Computation of Discontinuous Solution of Hydrodynamic Equations. Russian Original: Math. Sbornik 47: 271–306. Translated to English: US Public Research Service, JPRS 7226.
137. Gol'denblat I I, Kopnov V A (1966) Strength of Glass Reinforced Plastics. Journal of Polymer Mechanics 1: 54–59
138. Gourma M (2003) Towards better understanding of the Smoothed Particle Hydrodynamic Method. PhD thesis, Cranfield University, UK.
139. Grady D E (1993) Impact Compression Properties of Concrete. Proceedings to Sixth International Symposium on the Interaction of the Effects of Nonnuclear Munitions with Structures, Panama City, FL.
140. Grady D E (1995) Shock and Release Data for SAC-5 Concrete to 25 GPa. Sandia National Laboratories Memorandum - TMDG0595.
141. Gray III G T (2000) Classic Split Hopkinson Pressure Bar Technique. ASM - Mechanical Testing 8: 1–36
142. Greeff C W, Trinkle D R, Albers R C (2001) Shock-Induced $\alpha - \omega$ Transition in Titanium. Journal of Applied Physics 90: 2221–2226
143. Greeff C W, Graf M J (2004) Lattice Dynamics and the High Pressure Equation of State of Au. Physical Review B 69, 054107.
144. Gregson V R (1971) A Shock Wave Study of Fondu-Fyre WA-1 and Concrete. General Motors Materials and Structures Laboratory Report MSL-70-30.
145. Greve L, Pickett A K (2006) Delamination Testing and Modelling for Composite Crash Simulation. Composites Science and Technology 66: 816–826
146. Griffith A A (1921) The phenomena of Rupture and Flow in Solids. Philosophical Transactions of the Royal Society of London, Series A, 221: 163–198
147. Gross D, Seelig T (2006) Fracture Mechanics. Springer.
148. Grüneisen E (1912) Theorie des festen Zustandes einatomiger Elemente. Annalen der Physik 39: 257–306
149. Grüneisen E (1926) Zustand des festen Körpers. In: Greiger H, Scheel K (eds) Handbuch der Physik. Vol. 10: 1–59. Springer, Berlin

150. G'Sell C, Hiver J M (1999) Principles, Developments and Limitations of Video Control Material Testing Systems. International Workshop on Video-Controlled Materials Testing and In-Situ Micromechanical Characterization, Nancy, France.
151. G'Sell C, Hiver J M, Dahoun A, Souahi A (1992) Video-Controlled Tensile Testing of Polymers and Metals Beyond the Necking Point. Journal of Materials Science 27: 5031–5039
152. G'Sell C, Jonas J J (1979) Determination of the Plastic Behavior of Solid Polymers at Constant True Strain Rate. Journal of Materials Science 14: 583–591
153. Gumbsch P (ed) (2006) Proceedings on the Third International Conference on Multiscale Materials Modelling. Fraunhofer IRB Verlag. ISBN 978-3-8167-7206-4
154. Guenther C, Hicks D L, Swegle J W (1994) Conservative Smoothing Versus Artificial Viscosity. Technical Report SAND94-1853, Sandia National Laboratories, Albuquerque, NM.
155. Gurtin M E (2003) On a Framework for Small Deformation Visco-Plasticity: Free Energy, Microforces, Plastic Strain Gradients. International Journal of Plasticity 19: 47–90
156. Guth E, Mark H (1934) Zur innermolekularen Statistik, insbesondere bei Kettenmolekülen I. Monatshefte für Chemie 65: 93–121
157. Haigh B P (1920) Elastic Limit of a ductile metal. Engineering 109, 158–160
158. Hallquist J O (ed.) (2006) LS-DYNA Theory Manual. Livermore Software Technology Corporation, Livermore, CA.
159. Hancock J W, Mackenzie A C (1976) On the Mechanism of Ductile Failure in High-Strength Steels Subjected to Multi-Axial Stress States. Journal of the Mechanics and Physics of Solids 24: 147–169
160. Hansen N R, Schreyer H L (1994) A Thermodynamically Consistent Framework for Theories of Elastoplasticity Coupled with Damage. International Journal of Solids and Structures 3: 359–389
161. Hao S, Brocks W (1997) The Gurson-Tvergaard-Needleman-Model for Rate and Temperature-Dependent Materials With Isotropic and Kinematic Hardening. Computational Mechanics 20: 34–40
162. Harlow F H (1956) The Particle-in-Cell Method for Two Dimensional Hydrodynamic Problems. Technical Report No. LAMS-2082, Los Alamos Scientific Laboratory.
163. Harlow F H (1963) The Particle-In-Cell Method for Numerical Solution of Problems in Fluid Dynamics. Proceedings Symposium Applied Mathematics 15: 269–288
164. Hashin Z (1980) Failure Criteria for Unidirectional Fiber Composites. Journal of Applied Mechanics 47: 329–334
165. Haupt P (2000) Continuum Mechanics and Theory of Materials. Springer, Berlin Heidelberg
166. Haupt P, Lion A (1995) Experimental Identification and Mathematical Modelling of Viscoplastic Material Behaviour. Journal of Continuum Mechanics and Thermodynamics 7: 73–96
167. Haupt P, Lion A, Backhaus E (2000) On the Dynamic Behaviour of Polymers During Finite Strains: Constitutive Modelling and Identification of Parameters. International Journal of Solids and Structures 37: 3633–3646

168. Haupt P, Tsakmakis Ch (1989) On the Application of Dual Variables in Continuum Mechanics. Journal of Continuum Mechanics and Thermodynamics 1: 165–196
169. Hayes D B (1975) Wave propagation in a condensed medium with N transforming phases: application to solid-I–solid-II-liquid bismuth. Journal of Applied Physics 46: 3438–3443
170. Hayhurst C J, Clegg R A (1997) Cylindrical Symmetric SPH Simulations of Hypervelocity Impacts on Thin Plates. International Journal of Impact Engineering 20: 337–348
171. Haward R N, Thackray G (1968) The use of a Mathematical Model to Describe Isothermal Stress-Strain Curves in Glassy Thermoplastics. Proceedings of the Royal Society of London, A, 302: 453–472
172. Henderson L F (2001) General Laws for Propagation of Shock Waves Through Matter. In: Ben-Dor G, Igra O, Elperin T (eds.) Handbook of Shock Waves. Academic Press, San Diego.
173. Henderson L F, Menikoff R (1998) Triple-Shock Entropy Theorem and its Consequences. Journal of Fluid Mechanics 366: 179–210
174. Hernquist L, Katz N (1989) Tree-SPH - A Unification of SPH with the Hierarchical Tree Method. The Astrophysical Journal Supplement Series, 70: 419–446
175. Herrmann W (1969) Constitutive Equation for the Dynamic Compaction of Ductile Porous Materials. Journal of Applied Physics 40: 2490–2499
176. Hiermaier S (1996) Numerische Simulation von Impaktvorgängen mit einer netzfreien Lagrangemethode (Smooth Particle Hydrodynamics). Doctoral Thesis, Universität der Bundeswehr München, Heft 8 aus der Reihe "Mitteilungen des Instituts für Mechanik und Statik".
177. Hiermaier S, Könke D, Thoma K (1997) Numerical Simulation of Protective Design. In: Meskouris K, Wittek U (eds.), Aspects in Modern Computational Structural Analysis. Balkema Press, Rotterdam.
178. Hiermaier S, Könke D, Stilp A J, Thoma K (1997) Computational Simulation of the Hypervelocity Impact of Al-Spheres on Thin Plates of Different Materials. International Journal of Impact Engineering 20: 363–374
179. Hiermaier S, Peter J, Sauer M, Thoma K (2001) Coupled FE-Particle Codes Applied to Material Characterization and Crash Simulation. Proceedings to the European Conference on Computational Mechanics, Krakow, Poland.
180. Hiermaier S, Riedel W (1997) Numerical Simulation of Failure in Brittle Materials using Smooth Particle Hydrodynamics. Proceedings to the International Workshop on New Models and Numerical Codes for Shock Wave Processes in Condensed Media, St Catherine's College, Oxford, UK.
181. Hiermaier S, Riedel W, Hayhurst C J, Clegg R A, Wentzel C M (1999) Advanced Material Models for Hypervelocity Impact Simulations. Final report to European Space Agency project No. 12400/97/NL/PA(SC). EMI report E 43/99, Ernst-Mach-Institute, Freiburg, Germany.
182. Hiermaier S, Schäfer F (1999) Hypervelocity Impact Fragment Clouds in High Pressure Gas - Numerical and Experimental Investigations. International Journal on Impact Engineering 23: 391–400
183. Hill R (1948) A Theory of the Yielding and Plastic Flow of Anisotropic Metals. Royal Society of London, Proceedings Series A, Vol. 193.
184. Hill R (1950) The Mathematical Theory of Plasticity. Oxford University Press, New York.

185. Hill R (1970) Constitutive Inequalities for Isotropic Elastic Solids Under Finite Strain. Proceedings of the Royal Society of London A314: 457–472
186. Hill R (1990) Constitutive Modelling of Orthotropic Plasticity in Sheet Metals. J. Mech. Phys. Solids 38, Nr. 3: 405–417
187. Hirsch C (1995) Numerical Computation of Internal and External Flows, Vol 1: Fundamentals of Numerical Discretization. Wiley, Chichester, New-York.
188. Hirsch C (1995) Numerical Computation of Internal and External Flows, Vol 2: Computational Methods for Inviscid and Viscous Flows. John Wiley & Sons Ltd, Chichester, New-York.
189. Hoffman O (1967) The Brittle Strength of Orthotropic Materials. Journal of Composite Materials 1: 200–206
190. Holian K S (1984) T-4 Handbook of Material Properties Data Base, Vol Ic: Equation of State. Los Alamos National Laboratory Report LA-10160-MS, Los Alamos, NM.
191. Holian K S, Burkett M W (1987) Sensitivity of Hypervelocity Impact Simulations to Equations of State. International Journal of Impact Engineering 5: 333–341
192. Holian K S, Holian B L (1989) Hydrodynamic Simulations of Hypervelocity Impact. International Journal of Impact Engineering 8: 115–132
193. Hörmann M (2002) Nichtlineare Versagensanalyse von Faserverbundwerkstoffen. Doctoral Thesis, Universität Stuttgart, Institut für Baustatik, ISBN 3-00-010896-3
194. Holzapfel G A (2000) Nonlinear Solid Mechanics. John Wiley & Sons Ltd, Chichester.
195. Hooputra H, Gese H, Dell H, Werner H (2004) A Comprehensive Failure Model for Crashworthiness Simulation of Aluminum Extrusions. International Journal of Crashworthiness 9: 449–463
196. Hopkinson B (1914) A Method of Measuring the Pressure Produced in the Detonation of High Explosives or by the Impact of Bullets. Philosophical Transactions of the Royal Society of London A 213: 437
197. Hopperstad O S, Børvik T, Langseth M, Labibes K, Albertini C (2003) On the Influence of Stress Triaxiality and Strain Rate on the Behaviour of a Structural Steel. Part I. Experiments. European Journal of Mechanics A/Solids 22: 1–13
198. Horie Y (ed) (2007) Shock Wave Science and Technology Reference Library, Vol. 2, Solids I. Springer, Berlin, Heidelberg.
199. Huber M T (1904) Czasopismo Techniczne, Lemberg, Austria, 22: 181.
200. Huberth F, Gerster T, Guth S, Hiermaier S (2006) Kunststoffe in der Crashsimulation: Eine Betrachtung aus experimenteller Sicht unter Berücksichtigung numerischer Problemstellungen. Proceedings Werkstoffprüfung 2006, Stahlinstitut VDEh, Bad Neuenahr, Germany.
201. Hughes T J R (1987) Finite Element Method - Linear Static and Dynamic Finite Element Analysis. Prentice-Hall, Englewood Cliffs.
202. Hughes T J R, Liu W K (1981) Nonlinear Finite Element Analysis of Shells. Part I: Three dimensional Shells. Computational Methods in Applied Mechanic and Engineering 26: 331–362
203. Hugoniot P H (1887) On the Propagation of Motion in Bodies and in Perfect Gases in Particular. Journal de l'Ecole Polytechnique 57: 3–97
204. Hugoniot P H (1889) Sur la propagation du mouvement dans les corps et plus specialement dans les gaz parfaits. Journal de l'Ecole Polytechnique 57 e(3) and 58 e(1).

205. Irons B M (1966) Engineering Application of Numerical Integration in Stiffness Method. Journal of the American Institute of Aeronautics and Astronautics 14: 2035–2037
206. Irwin G R (1949) Fracture Dynamics. In: Fracturing of Metals, ASM: 147–166
207. James H M, Guth E (1943) Theory of elastic properties of rubber. Journal of Chemical Physics 11: 455–481
208. Jansen J (2007) Ein Werkstoffmodell für eine Aluminium Druckgusslegierung unter statischen und dynamischen Beanspruchungen. Doctoral Thesis, Fraunhofer Institute for High Speed Dynamics, Ernst-Mach-Institute, Issue 13 of the series Thoma K, Hiermaier S (eds), $\dot{\varepsilon}$ - Forschungsergebnisse aus der Kurzzeitdynamik. ISBN: 978-3-8167-7382-5.
209. Jirásek M, Belytschko T (2002) Computational resolution of Strong Discontinuities. Proceedings to the Fifth World Congress on Computational Mechanics, WCCM-V, Vienna, Austria.
210. Johnson A F (2001) Modelling Fabric Reinforced Composites Under Impact Loads. Composites Part A 32A: 1197–1206
211. Johnson A F, Pickett A K and Rozycki P (2001) Computational Methods for Predicting Impact Damage in Composite Structures. Composites Science and Technology 61: 2183–2192
212. Johnson G R (1994) Linking of Lagrangean Particle Methods to Standard Finite Element Methods for High Velocity Impact Computations. Nuclear Engineering and Design 150: 265–274
213. Johnson G R, Beissel S R (1996) Normalized Smoothing Functions for SPH Impact Computations. International Journal of Numerical Methods in Engineering 39: 2725–2741
214. Johnson G R, Cook W H (1983) A Constitutive Model and Data for Metals Subjected to Large Strains, High Strain Rates and High Temperatures. Proceedings to the Seventh International Symposium on Ballistics, Den Haag.
215. Johnson G R, Cook W H (1985) Fracture Characteristic of Three Metals Subjected to Various Strains, Strain Rates, Temperatures and Pressures. Engineering Fracture Mechanics 21: 31–48
216. Johnson G R, Holmquist T J (1992) A Computational Model for Brittle Materials Subjected to Large Strains, High Strain Rates and High Pressures. In: Meyers M A, Murr L E, Staudhammer K P (eds.): Shock-Wave and High-Strain-Rate Phenomena in Materials. Marcel-Dekker, New York.
217. Johnson G R, Holmquist T J (1994) An Improved Computational Constitutive Model for Brittle Materials. In: Schmidt S C, Shaner J W, Samara G A, Ross M (eds.): High Pressure Science and Technology. American Institute of Physics, New York.
218. Johnson G R, Stryk R A, Beissel S R (1996) SPH for High Velocity Impact Computations. Computer Methods in Applied Mechanics and Engineering 139: 347–374
219. Jones D F, Treloar L R G (1975) The Properties of Rubber in Pure Homogeneous Strain. Journal of Physics D: Applied Physics, 8: 1285–1304.
220. Junginger M (2004) Charakterisierung und Modellierung unverstärkter thermoplastischer Kunststoffe zur numerischen Simulation von Crashvorgängen. Doctoral Thesis, Fraunhofer Institute for High Speed Dynamics, Ernst-Mach-Institute, Issue 3 of the series Thoma K, Hiermaier S (eds), $\dot{\varepsilon}$ - Forschungsergebnisse aus der Kurzzeitdynamik. ISBN: 3-8167-6339-1.

221. Kachanov (1958) Time of Rupture Process Under Creep Conditions. Izv. Akad. Nauk. SSSR 8: 26–31
222. Kawai H (1969) The Piezoelectricity of Polyvinylidene Fluoride. Journal of Applied Physics 8: 975–976
223. Khan A S, Huang S (1995) Continuum Theory of Plasticity. John Wiley & Sons Inc., New York.
224. Kan J van, Segal A (1995) Numerik Partieller Differentialgleichungen. B. G. Teubner Verlag, Stuttgart.
225. Kipp M E, Chhabildas L C, Reinhart W D (1997) Elastic Shock Response and Spall Strength of Concrete. AIP Conference Proceedings: Tenth American Physics Society Topical Conference on Shock Compression and Condensed Matter, Amhurst, MA: 557–560
226. Kirchner H O, Kubin L P, Pontikis V (eds) (1996) Computer Simulation in Materials Science. Kluwer Academic Publisher, Dordrecht, Boston, London.
227. Klisimski M, Mostrom A (1998) On Stability of Multi-Time-Step Integration Procedures. Journal of Engineering Mechanics 124: 783–793
228. Klomfass A, Warken D, Private Communication on Spin Stabilized Projectiles.
229. Kolling S, Haufe A, Feucht M, Du Bois P A (2005) SAMP-1: A Semi-Analytical Model for the Simulation of Polymers. Proceedings to the LS-DYNA Anwenderforum, Bamberg, Germany.
230. Kolsky H (1949) An Investigation of the Mechanical Properties of Materials at Very High Rates of Loading. Proceedings of the Physical Society, Section B, 62: 676–700
231. Krajcinovic D (1989) Damage Mechanics. Mechanics of Materials 8: 117–197
232. Krehl P (2006) Shock Waves, Explosions and Impact Phenomena, A Chronological and Biographical Reference. Springer, Heidelberg & Berlin
233. Krongauz Y, Belytschko T (1997) Consistent Pseudo Derivatives in Meshless Methods. Computer Methods in Applied Mechanics and Engineering 146: 371–386
234. Kuhn W (1934) Über die Gestalt fadenförmiger Moleküle in Lösungen. Kolloidzeitschrift 68: 2–15
235. Kuhn W (1936) Beziehungen zwischen Molekülgrösse, statistischer Molekülgestalt und elastischen Eigenschaften hochpolymerer Stoffe. Kolloidzeitschrift 76: 258–271
236. Kuhn W, Grün F (1942) Beziehungen zwischen elastischen Konstanten und Dehnungsdoppelbrechungen hochelastischer Stoffe. Kolloidzeitschrift 101: 248–271.
237. Kuhnert J (2003) An Upwind Finite Pointset Method (FPM) for Compressible Euler and Navier-Stokes Equations. In: Griebel M, Schweitzer M A (eds.) Meshfree Methods for Partial Differential Equations. Springer, Berlin.
238. Kuhnert J (1999) General Smoothed Particle Hydrodynamics. Doctoral Thesis, University Kaiserslautern, Germany.
239. Kuscher G (1985) Nichtlineare Ausbreitung elasto-plastischer Wellen in Stäben. PhD Thesis RWTH-Aachen, Germany.
240. Ladeveze P (1994) Inelastic Strains and Damage. In: Talreja R (ed) Damage Mechanics of Composite Materials. Composite Materials Series 9: 117–136. Elsevier, Amsterdam.
241. Ladeveze P, Le Dantec E (1992) Damage Modelling of the Elementary Ply for Laminated Composites. Composites Science and Technology 43: 257–267

242. Lancaster P, Salkauskas K (1981) Surfaces Generated by Moving Least Squares Methods. Mathematics of Computation 37: 141–158
243. Lanczos C (1970) The Variational Principles of Mechanics. Dover Publications, Inc., New York.
244. Landau L D, Lifshitz E M (1980) Statistical Physics. Pergamon Press, Oxford.
245. Landshoff R (1955) A Numerical Method for Treating Fluid Flow in the Presence of Shocks. Los Alamos Scientific Laboratory Report LA-1930.
246. Lemaitre J (1971) Evaluation of Dissipation and Damage in Metal Submitted to Dynamic Loading. Proceedings First International Conference on the Mechanical Behaviour of Materials, ICM-1, Kyoto, Japan.
247. Lemaitre J (1992) A Course on Damage Mechanics. Springer, Berlin.
248. Lepinoux J, Maziere D, Pontikis V, Saada G (2000) Multiscale Phenomena in Plasticity: From Experiments to Phenomenology, Modelling and Materials Engineering. Kluwer Academic Publishers.
249. LeRoy G, Embury J D, Edward G, Ashby M F (1981) A Model of Ductile Fracture Based on the Nucleation and Growth of Voids. Acta Metallurgica 29: 1509–1522
250. Libersky L D, Randles P W (1999) The Stress Point Technique as a Meshless Method. Proceedings 5th US National Conference on Computational Mechanics, Boulder, CO
251. Libersky L D, Petschek A G (1990) Smooth Particle Hydrodynamics with Strength of Materials. Advances in the Free Lagrange Method, Lecture Notes in Physics 395: 248–257
252. Libersky L D, Randles P W (2004) A Dual Particle Computational Method for Continua. Proc. XXI International Congress of Theoretical and Applied Mechanics. Warsaw, Poland, August 15-21.
253. Liu G R, Liu M B (2005) Smoothed Particle Hydrodynamics - A Meshfree Particle Method. World Scientific Publishing Co. Pte. Ltd., New Jersey.
254. Lu G, Kaxiras E (2005) Overview of Multiscale Simulations of Materials. In: Rieth M, Schommers W (eds) Handbook of Theoretical and Computational Nanotechnology. American Scientific Publishers.
255. Lucy L B (1977) A Numerical Approach to the Testing of the Fission Hypothethis. The Astronomical Journal 82 (12): 1013–1024
256. Ludwik P (1909) Elemente der technologischen Mechanik. Springer, Berlin
257. Lukyanov A A, Vignjevic R, Panov V, Bourne N (2004) Modelling of Ductile Failure in Metals under High Velocity Impact Loading. Proceedings to The First International Conference on Computational Mechanics (CM'04). Belgrade, Serbia and Montenegro, November 15-17.
258. Lysne P C, Boade R R, Percival C M, Jones O E (1969) Determination of Release Adiabats and Recentered Hugoniot Curves by Shock Reverberation Techniques. Journal of Applied Physics 40: 3786
259. Mach E, Salcher P (1887) Photographische Fixierung der durch Projectile in der Luft eigeleiteten Vorgänge. Sitzungsbericht, Akad. Wiss. Wien 95 (IIa): 764–780
260. Macvean D B (1968) Die Elementararbeit in einem Kontinuum und die Zuordnung von Spannungs- und Verzerrungstensoren. ZAMP 19: 157–185
261. Maers D R, Pae K D, Sauer J A (1969) Effects of Hydrostatic Pressure on the Mechanical Behavior of Polyethylene and Polypropylene. Journal of Applied Physics 40: 4229–4237

262. Maier M (1990) Experimentelle Untersuchung und numerische Simulation de Crashverhaltens von Faserverbundwerkstoffen. Doctoral Thesis, Universität Kaiserslautern.
263. Malvern L E (1969) Introduction to the Mechanics of a Continuous Medium. Prentice Hall, Englewood Cliffs, New Jersey.
264. Mandell D A, Wingate C A, Swegle L A (1996) Simulation of a Ceramic Impact Experiment Using the SPHINX Smooth Particle Hydrodynamics Code. Proceedings Sixteenth International Symposium on Ballistics, San Francisco, CA.
265. Marczyk J, Deshpande B (2006) Measuring and Tracking Complexity in Science. Proceedings Sixth International Conference on Complex Systems, Boston, MA.
266. Marklund P-O, Nilsson L (2002) Simulation of Airbag Inflation Processes Using a Coupled Fluid Structure Approach. Computational Mechanics 29: 289–297
267. Marsh S P (ed) (1980) LASL Shock Hugoniot Data. University of California Press, Berkeley and Los Angeles, CA.
268. Martha L, Wawrzynek P A, Ingraffea A R (1993) Arbitrary Crack Propagation Using Solid Modeling. Engineering with Computers 9: 63–82
269. Meenken T (2007) Charakterisierung niederimpedanter Werkstoffe unter dynamischen Lasten. Doctoral Thesis, Fraunhofer Institute for High Speed Dynamics, Ernst-Mach-Institute, Issue 14 of the scientific series Thoma K, Hiermaier S (eds), $\dot{\varepsilon}$ - Forschungsergebnisse aus der Kurzzeitdynamik.
270. Mehra V, Chaturvedi S (2006) High Velocity Impact of Metal Sphere on Thin Metallic Plates: A Comparative Smooth Particle Hydrodynamics Study. Journal of Computational Physics 212: 318–337
271. Melosh H J (2000) A new and improved equation of state for impact computations. Lunar and Planetary Science 31: # 1903
272. Mertins U (1991) Zur Konvergenz des Rayleigh-Ritz Verfahrens bei Eigenwertaufgaben. Numerische Mathematik 59: 667–682
273. Menikoff R (2007) Empirical Equations of State for Solids. In: Horie Y (ed) Shock Wave Science and Technology Reference Library, Vol. 2, Solids I. Springer, Berlin, Heidelberg.
274. Menikoff R, Kober E (2000) Equation of State and Hugoniot Locus for Porous Materials: P-α Model Revisited. In: Furnish M D, Chhabildas L C, Hixson R S (eds.) Shock Compression of Condensed Matter - 1999. American Insitute of Physics.
275. Menikoff R, Plohr B J (1989) The Riemann Problem for Fluid Flow of Real Material. Reviews of Modern Physics 61: 75–130
276. Meyers M A (1994) Dynamic Behavior of Materials. John Wiley - Interscience Publication, New York.
277. Meyers M A, Armstrong R W, Kirchner H O K (eds) (1999) Mechanics and Materials. Fundamentals and Linkages. John Wiley & Sons, Inc., New York.
278. Michalke W, Heym A, Pyttel T (2004) Investigation of the Finite Point Method for OoP Simulations. Proceedings EUROPAM, Paris
279. Mie G (1912) Grundlagen einer Theorie der Materie. Annalen der Physik 2: 1–40
280. Mises R v (1913) Mechanik der festen Körper im plastisch deformablen Zustand. Nachr. Königl. Ges. Wiss. Göttingen, Math.-phys. Kl., 582–592

281. Mitchell A R (2001) Finite Difference and Related Methods for Differential Equations. John Wiley and Sons, Chichester, UK.
282. Mohr D, Doyoyo M (2002) Analysis of the Arcan Apparatus in the Clamped Configuration. Journal of Composite Materials 36: 2583–2594
283. Monaghan J J (1982) Kernel Estimates as a Basis for General Particle Methods in Hydrodynamics. International Journal of Computational Physics 46: 429–453
284. Monaghan J J (1988) An Introduction to SPH. Computer Physics Communication 48: 89–96
285. Monaghan J J (2005) Smoothed Particle Hydrodynamics. Reports on Progress in Physics 68: 1703–1759
286. Monaghan J J, Gingold R A (1983) Shock Simulation by the Particle Method SPH. Journal of Computational Physics 52: 374–389
287. Monaghan J J, Lattanzio J C (1985) A Refined Particle Method for Astrophysical Problems. Astronomy and Astrophysics 149: 135–143
288. Moës N, Dolbow J, Belytschko T (1999) A Finite Element Method for Crack Growth Without Remeshing. International Journal for Numerical Methods in Engineering 46: 131–150
289. Nagayama K, Mori Y (1994) Simple Method of Calculating Grüneisen Parameter Based on the Shock Hugoniot Data for Solids. J. Phys. Soc. Jap. 63: 4070–4077
290. Nahme H (1991) Equation of State Measurements of 9SMn28 and C45 Steel. Ernst-Mach-Institute report E11/91, EMI, Freiburg
291. Narayanaswami R, Adelman H (1977) Evaluation of the Tensor Polynomial and Hoffman Strength Theories for Composite Materials. Journal of Composite Materials 11: 366.
292. Neal M O, Belytschko T (1989) Explicit-Explicit Subcycling With Non-Integer Time Step Ratios for Structural Dynamic Systems. Computers and Structures 31: 871–880
293. Needleman A, Tvergaard V (1984) Analysis of Ductile Rupture in Notched Bars. Journal of the Mechanics and Physics of Solids 35: 151–183
294. Neumann J von (1942) Theory of Shock Waves. Progress Report to the National Defense Research Committee, Div. 8, U.S. Dept. Comm. Off. Tech. Serv., PB 32719. Also in: Taub A H (1963) John von Neumann Collected Works. Pergamon Press, Oxford, England.
295. Neumann J von, Richtmyer R D (1950) A Method for the Numerical Calculation of Hydrodynamic Shocks. Journal of Applied Physics 21: 232–237
296. Neuwald P, Reichenbach H (1993) A New Colour Schlieren Setup. EMI report E 27/93, Ernst-Mach-Institute, Freiburg, Germany.
297. Noh W F (1964) Fundamental Methods in Hydrodynamics. In: Adler B et al. (eds) Methods in Computational Physics Vol. 3: 117. Academic Press, New York.
298. Ockewitz A, Sun D-Z (2006) Damage Modelling of Automobile Components of Aluminum Materials under Crash Loading. Proceedings to the Fifth LS-DYNA Anwenderforum, Ulm, Germany.
299. Ogden R W (1972) Large deformation isotropic elasticity - On the correlation of theory and experiment for incompressible rubberlike solids. Proceedings of the Royal Society of London, A326: 565–584.

300. Ogden R W (1982) Elastic Deformations of Rubber-like Solids. In: Hopkins H G, Swell M J (eds) Mechanics of Solids. The Rodney Hill 60th Anniversary Volume. Pergamon Press, Oxford.
301. Ogden R W (1997) Non-Linear Elastic Deformations. Dover Publications Inc., New York.
302. Orowan E (1950) Fatigue and Fracture of Metals. MIT Press, Cambridge.
303. Oertel H(1966) Stoßrohre. Springer, Wien
304. Parks D M, Argon A S, Bagepalli P (1985) Large Elastic-Plastic Deformation of Glassy Polymers, Part I: Constitutive Modeling. MIT, Program in Polymer Science and Technology Report.
305. Peter J (2004) Experimentelle und numerische Untersuchungen zum Crashverhalten von Strukturbauteilen aus kohlefaserverstärkten Kunststoffen. Doctoral Thesis, Fraunhofer Institute for High Speed Dynamics, Ernst-Mach-Institute, Issue 8 of the scientific series Thoma K, Hiermaier S (eds), $\dot{\varepsilon}$ - Forschungsergebnisse aus der Kurzzeitdynamik. ISBN: 3-8167-6748-6.
306. Prager W (1955) The Theory of Plasticity: A Survey of Recent Achievements. Proceedings of the Institution for Mechanical Engineers 169: 41–57
307. Prandtl L (1906) Zur Theorie des Verdichtungsstoßes. Zeitschrift f. d. ges. Turbinenwesen 3: 241–245
308. Rabinowitz S, Ward I M, Parry J S C (1970) The Effect of Hydrostatic Pressure on the Shear Yield Behavior of Polymers. Journal of Material Science 5: 29–39
309. Rabotnov Y N (1963) On the Equation of State for Creep. In: Koiter, W T (ed) Progress in Applied Mechanics (Prager Anniversary Volume):307–315. MacMillan, New York
310. Raghava R, Cadell R M, Yeh G S Y (1973) The Macroscopic Yield Behavior of Polymers. Journal of Material Science 8: 225–232
311. Randles P W, Libersky L D (2000) Normalized SPH With Stress Points. International Journal for Numerical Methods in Engineering 48: 1445–1462
312. Randles P W, Libersky L D (1996) Smoothed Particle Hydrodynamics: Some Recent Improvements and Applications. Computational Methods in Applied Mechanics and Engineering 139: 375–408
313. Randles P W, Libersky L D, Petschek A G (1999) On Neighbours, Derivatives and Viscosity in Particle Codes. Proceedings of the European Conference of Computational Mechanics (ECCM), Munich, Germany.
314. Randles P W, Petschek A G, Libersky L D, Dyka C T (2002) Stability of DPD and SPH. In: Griebel M, Schweitzer M (eds) Lecture Notes in Computational Science and Engineering, vol. 26: 339. Springer, Berlin.
315. Rankine W J (1870) On the Thermodynamic Theory of Waves of Finite Longitudinal Disturbances. Philosophical Transactions 160 II: 277–288
316. Rabczuk T, Belytschko T (2006) Application of Particle Methods to Static Fracture of Reinforced Concrete Structures. International Journal of Fracture 137: 19–49
317. Rabczuk T, Belytschko T, Xiao S P (2004) Stable Particle Methods Based on Lagrangean Kernels. Computer Methods in Applied Mechanics and Engineering 193: 1035–1063
318. Rayleigh J W (1912) Aerial Plane Waves of Finite Amplitude. Scientific Papers 5: 573–616.
319. Retting W (1991) Mechanik der Kunststoffe. Hanser Verlag, München, Wien, New York.

320. Rice M H, McQueen R G, Walsh J M (1958) Compression of Solids by Strong Shock Waves. In: Seitz F, Turnbull D (eds) Solid State Physics, Volume 6, Academic Press Inc., Publishers, New York and London.
321. Rice D L, Neuwald P (1999) A Study of Blast Wall Effectiveness Using Small Scale Experiments and Hydrocodes Simulation. Proceedings Ninth International Symposium on the Interaction of the Effects of Munitions with Structures, Berlin-Straussberg, Germany.
322. Rice J R, Tracey D M (1969) On the Ductile Enlargement of Voids in Triaxial Stress Fields. Journal of Mechanics and Physics of Solids 17: 201–217
323. Riedel W (2000) Beton unter dynamischen Lasten. Meso- und makromechanische Modelle und ihre Parameter. Doctoral Thesis, Fraunhofer Institute for High Speed Dynamics, Ernst-Mach-Institute, Issue 5 of the scientific series Thoma K, Hiermaier S (eds), $\dot{\varepsilon}$ - Forschungsergebnisse aus der Kurzzeitdynamik. ISBN: 3-8167-6340-5.
324. Riedel W, Harwick W, White D M, Clegg R A (2003) Advanced Material Damage Models for Numerical Simulation Codes. Final Report to the European Space Agency Project No. 12400/97/NL/PA(SC). EMI Report I 75/03, Ernst-Mach-Insitute, Freiburg, Germany.
325. Riedel W, Kawai N (2007) Numerical Simulation of Mortar Strength Measurements at Highest Strain Rates. Proceedings to the 2007 International Symposium on Interaction of the Effects of Munitions with Structures, ISIEMS-07, Orlando, FL.
326. Riedel W, Wicklein M, Thoma K (2007) Shock Properties of Conventional and High Strength Concrete: Experimental and Mesomechanical Analysis. International Journal of Impact Engineering, In Press, Available online 1 March 2007.
327. Riemann B (1860) Über die Fortpflanzungen ebener Luftwellen von endlicher Schwingungsweite. In: Weber H (ed) (1953) Collected Works of Bernhard Riemann. Dover Publications, New York.
328. Rinehart J S (1975) Stress Transients in Solids. Hyper Dynamics, Santa Fe, NM.
329. Ritz W (1909) Über eine neue Methode zur Lösung gewisser Variationsprobleme der mathematischen Physik. Journal für die reine und angewandte Mathematik 135: 1–61
330. Rivlin R S (1948) Large Elastic Deformations of Isotropic Materials, I. Fundamental Concepts. Philosophical Transactions of the Royal Society of London, A 240: 459–490.
331. Rohr I (2004) Integrale Charakterisierung und Modellierung von duktilem Stahl unter dynamischen Lasten. Doctoral Thesis, Fraunhofer Institute for High Speed Dynamics, Ernst-Mach-Institute, Issue 2 of the scientific series Thoma K, Hiermaier S (eds), $\dot{\varepsilon}$ - Forschungsergebnisse aus der Kurzzeitdynamik. ISBN: 3-8167-6397-9.
332. Rohr I (2000) Ermittlung von statischen und dynamischen Modellparametern für HZB996(L) am Beispiel des Johnson-Cook Modells. EMI report E-64/00, Ernst-Mach-Insitute, Freiburg, Germany.
333. Rohr I, Nahme H, Thoma K (2001) A modified Taylor Impact Test in Combination with Numerical Simulation - A new Approach for Obtaining Material Properties Under High Dynamic Loads. Proceedings $3^{r}d$ International Workshop on New Models and Hydrocodes for Shock Wave Processes, Paris.

334. Romanchenko V I, Stepanov G V (1980) The Dependence of Critical Stresses Upon the Time Parameters of Load at Spalling in Copper, Aluminum and Steel. Journal of Applied Mechanics and Technical Physics 21: 141–147
335. Rosenberg Z, Bless S J (1986) Determination of Dynamic Yield Strengths with Embedded Manganin Gages in Plate Impact and Long-Rod Experiments. Experimental Mechanics 26: 279–282
336. Rosenberg Z, Mayseless M, Partom Y (1984) The Use of Manganin Stress Transducers in Impulsively Loaded Long Rod Experiments. Journal of Applied Mechanics 51: 202–204
337. Rosenberg Z, Yaziv D, Partom Y (1980) Calibration of Foil-Like Manganin Gauges in Planar Impact Experiments. Journal of Applied Physics 51: 3702–3705
338. Ryan S (2007) Spacecraft Disturbances from Hypervelocity Impact - Test Results. EMI report I-10/07 to European Space Agency Contract 18583/04/NL/CH, Ernst-Mach-Institute, Freiburg, Germany.
339. Ryan S, Schäfer F, Riedel W (2006) Numerical Simulation of Hypervelocity Impact on CFRP/Al HC SP Spacecraft Structures Causing Penetration and Fragment Ejection. International Journal of Impact Engineering 33: 703–712
340. Ryan S, Schäfer F, Guyot M, Hiermaier S, Lambert M (2007) Characterizing the Transient Response of CFRP/AL HC Spacecraft Structures Induced by Space Debris Impact at Hypervelocity. Proceedings to the 2007 Hypervelocity Impact Symposium, Williamsburg, VA
341. Sauer M (2000) Adaptive Kopplung des netzfreien SPH-Verfahrens mit finiten Elementen zur Berechnung von Impaktvorgängen. Doctoral Thesis, Berichte aus dem Konstruktiven Ingenieurbau, 00/5, University of the German Armed Forces, Munich.
342. Sauer J A, Pae K D, Bhateja S K (1973) Influence of Pressure on Yield and Fracture in Polymers. Journal of Macromolecular Science - Physics B8: 631–654
343. Sawas O, Brar N S, Brockman R A (1998) High Strain Rate Characterization of Low-Density Low-Strength Materials. Shock Compression of Condensed Matter 429: 855–858
344. Schang O, Billon N, Muracciole J M, Fernagut F (1996) Mechanical Behavior of a Ductile Polyamid During Impact. Polymer Engineering and Science 36: 541–550
345. Scheffer U, Hiermaier S (1999) Improving a SPH Code by Alternative Interpolation Schemes. Proceedings 7. Fachtagung Baustatik-Baupraxis, Aachen, Germany
346. Schleicher F (1926) Der Spannungszustand an der Fließgrenze. Zeitschrift für angewandte Mathematik und Mechanik 6: 199–216
347. Schlimmer M (1984) Zeitabhängiges mechanisches Werkstoffverhalten. Springer, Berlin
348. Schonberg W P (1993) Characterizing the Material in a Debris Cloud Created in a Hypervelocity Impact. Proceedings First European Conference on Space Debris, Darmstadt, Germany.
349. Seaman L, Curran D R, Shockey D A (1983) Calculation of Failure by Shear Bands. Proceedings of the 29th Sagamore Conference, Lake Placid, NY
350. Seki W, Fukahori Y, Iseda Y, Matsunaga T (1987) A Large-Deformation Finite Element Analysis for Multilayer Elastomeric Bearings. Rubber Chemistry and Technology 60: 856–869.

351. SESAME (1983) Report on the Los Alamos Equation of State Library. Report LANL-83-4, Los Alamos National Laboratories, Los Alamos, NM.
352. Setchell R E, Taylor P A (1988) A Refined Equation of State for Unreacted Hexanitrostilbene. Journal of Energetic Materials 6: 157–199
353. Seth B R (1964) Generalized Strain Measure with Application to Physical Problems. In Reiner M, Abir D (eds) Second Order Effects in Elasticity, Plasticity and Fluid Dynamics. Pergamon Press, Oxford.
354. Shaw S (2002) Direct Simulation of Detonation Products Equation of State by a Composite Monte Carlo Method. Proceedings to the Twelfth International Detonation Symposium
355. Sheffield S A, Hayes D B, Mitchell D E (1977) An equation of state and chemical kinetics for hexanitrostilbene (HNS) explosive. 6th Symposium (International) on Detonation: 748–754
356. Shepard D (1968) A Two-Dimensional Interpolation Function For Irregularly Spaced Points. Proceedings of the 23rd National Conference of the Association for Computing Machinery, Brabdon Syst. Press: 517–524
357. Shockey D A, Seaman L, Curran D R (1979) Microfracture Models and Their Application to Nonlinear Dynamic Fracture Problems. Nonlinear Fracture Mechanics, ASME, AMD 35: 79
358. Siebel E, Pomp A (1927) Die Ermittlung der Formänderungsfestigkeit von Metallen duch den Stauchversuch. Mitteilungen aus dem Kaiser-Wilhelm Institut für Eisenforschung 9: 157
359. Slater J C (1939) Introduction to Chemical Physics. McGraw-Hill, New York
360. Sogabe Y, Yokoyama T, Yokoyama T (1995) A Split Hopkinson Bar Method for Testing Materials with Low Characteristic Impedance. ASME-Dynamic Fracture, Failure, and Deformation 300: 137–143
361. Spencer A J M (1984) Continuum Theory of the Mechanics of Fibre-Reinforced Composites. Springer Verlag, Wien, New York.
362. Sperling L H (2006) Introduction to Physical Polymer Science. John Whiley & Sons, Inc., Hoboken, New Jersey.
363. Spitzig W A, Richmond O (1979) Effect of Hydrostatic Pressure on the Deformation Behavior of Polyethylene and Polycarbonate in Tension and in Compression. Polymer Engineering and Science 19: 1129–1139
364. Stassi-D'Alia F (1969) Limiting Conditions of Yielding for Anisotropic Materials. Meccanica 4: 349–364
365. Steinberg D J, Cochran S G, Guinan M W (1980) A Constitutive Model for Metals Applicable at High Strain Rate. Journal of Applied Physics 51: 1498–1504
366. Steinberg D J, Lund C M (1989) A Constitutive Model for Strain Rates from 10^{-4} to 10^{6} s^{-1}. Journal of Applied Physics 65: 1528–1533
367. Steinhauser M O, Grass K, Thoma K, Blumen A (2006) Impact Dynamics and Failure of Brittle Solid States by Means of Nonequilibrium Molecular Dynamics Simulations. Europhysics Letters 72: 62–69
368. Steinhauser M, Thoma K (2004) MMM-Tools: Multiskalen Modellierung und Simulation. Congress Proceedings Intelligente Leichtbausysteme, Autovision, Wolfsburg, Germany.
369. Stellingwerf R F, Wingate C A (1993) Impact Modelling with Smoothed Particle Hydrodynamics. International Journal of Impact Engineering 14: 707–718
370. Stepanov G V (1976) Spall Fracture of Metals by Elastic-Plastic Loading Waves (In Russian Language). Problemy Prochnosty 8: 66–70

371. Sternstein S S, Myers F A (1973) Yielding of Glassy Polymers in the Second Quadrant of Principal Stress Space. Journal of Macromolecular Science - Physics B8: 539–571
372. Sternstein S S, Ongchin L (1969) Yield Criteria for Plastic Deformation of Glassy High Polymers in General Stress Fields. Polymer Preprints 10: 1117–1124
373. Steward D S, Yoo S, Davis W C (2002) Equation of State for Modelling the Detonation Reaction Zone. Proceedings to the Twelfth International Detonation Symposium.
374. Stillinger F H, Rahmann A (1974) Improved Simulation of Liquid Water by Molecular Dynamics. Journal of Chemical Physics 60: 1545–1557
375. Straßburger E, Senf H (1994) Experimental Investigation of Wave and Fracture Phenomena in Impacted Ceramics. Ernst-Mach-Institute Report 3/94, EMI, Freiburg, Germany.
376. Strikwerda J C (2004) Finite Difference Schemes and Partial Differential Equations. Cambridge University Press.
377. Strobl G (1996) The Physics of Polymers. Springer, Berlin.
378. Stroud A H, Secrest D (1966) Gaussian Quadrature Formulas. Prentice Hall Inc., NJ.
379. Swegle J W (1980) Constitutive Equation for Porous Materials with Strength. Journal of Applied Physics 51: 2574–2580
380. Swegle J W, Grady D E (1985) Shock Viscosity and the Prediction of Shock Wave Rise Times. Journal of applied physics 58: 693.
381. Swegle J W, Attaway S W, Heinstein M W, Mello F J, Hicks D L (1994) An Analysis of Smooth Particle Hydrodynamics. Sandia Report SAND93-2513, Sandia National Laboratories, Albuquerque, NM.
382. Swegle J W, Hicks D L, Attaway S W (1995) Smoothed Particle Hydrodynamics Stability Analysis. Journal of Computational Physics 116: 123.
383. Taig I C (1961) Structural Analysis by the Matrix Displacement Method. English Electric Aviation Report Nr. S017.
384. Tamma K K, D'Costa J F (1992) A New Explicit Variable Time-Integration Self-Starting Methodology for Computational Structural Dynamics. International Journal for Numerical Methods in Engineering 33: 1165–1180
385. Taylor G I (1946) Testing of Materials at High Rates of Loading. Journal of the Institute of Civil Engineers 26: 486–518
386. Taylor G I (1948) The Use of Flat Ended Projectiles for Determining Yield Stress 1: Theoretical Considerations. Proceedings of the Royal Society of London A 194: 289–299
387. Thoma K, Riedel W, Hiermaier S (1999) Mesomechanical Modeling of Concrete Shock Response. Proceedings European Conference on Computational Mechanics, Munich, Germany.
388. Thoma K, Vinckier D, Maier M, Altstädt V (1991) Entwicklung eines mikromechanischen Modells auf Finite-Element-Basis zur Untersuchung des Versagens von Faserverbundwerkstoffen bei transienten Belastungen. Final report to the EU project CARMAT-2000.
389. Thompson P A (1971) A Fundamental Derivative in Gas Dynamics. Physics of Fluids 14: 1843–1849
390. Tillotson J H (1962) Metallic Equations of State for Hypervelocity Impact. General Atomic Division of General Dynamics Report GA-3216, San Diego, CA.

391. Timmel M, Kaliske M, Kolling S (2004) Modellierung gummiartiger Materialien bei dynamischer Beanspruchung. Proceedings, LS-DYNA Anwenderforum 2004, Bamberg, Germany.
392. Toro E F (1997) Riemann Solvers and Numerical Methods for Fluid Dynamics. Springer Verlag.
393. Treloar L R G (1944) Stress-Strain Data for Vulcanized Rubber under Various Types of Deformation. Transactions of the Faraday Society, 39: 241–246.
394. Treloar L R G (1975) The Physics of Rubber Elasticity. Oxford University Press, Oxford.
395. Tresca H 1864 Mémoires sur l'écoulement des corps solides soumis à des fortes pressions. C. R. Acad. Sci. Paris, 59: 754–758
396. Troost A, Schlimmer M (1977) Fließverhalten plastisch kompressibler Werkstoffe. Rheologica Acta 16: 340–351
397. Truesdell C (1966) The Elements of Continuum Mechanics. Springer, New York
398. Truesdell C and Noll W (1965) The Non-Linear Field Theories of Mechanics. In: Flgge S (ed) Encyclopedia of Physics, Vol. III/3. Springer, Berlin
399. Truesdell C and Toupin R A (1960) The Non-Linear Field Theories of Mechanics. In: Flgge S (ed) Encyclopedia of Physics, Vol. III/1. Springer, Berlin
400. Trunin R F (1998) Shock Compression of Condensed Matter. Cambridge University Press
401. Tsai S W, Wu E M (1971) A General Theory of Strength for Anisotropic Materials. Journal of Composite Materials 5: 58–80
402. Turner M J, Clough R W, Martin W C, Topp L J (1956) Stiffness and Deflection Analysis of Complex Structures. Journal of Aeronautical Science 23: 805–824
403. Tvergaard V (1981) Influence of Voids on Shear Band Instabilities under Plane Strain Conditions. International Journal of Fracture 17: 389–407
404. Tvergaard V (1982) On Localization in Ductile Materials Containing Spherical Voids. International Journal of Fracture 18: 237–252
405. Valanis K C (1968) The Viscoelastic Potential and its Thermodynamic Foundations. Journal of Mathematical Physics 48: 262
406. Valanis K C (1971) Irreversible Thermodynamics of Continuous Media. Springer, Wien.
407. Valanis K C (1968) Unified Theory of Thermomechanical Behavior of Viscoelastic Materials. In: Lindholm U S ()ed) Mechanical Behavior of Materials under Dynamic Loads. Springer, Berlin.
408. Valanis K C, Landel R F (1967) The Strain Energy Function of a Hyperelastic Material in Terms of the Extension Ratios. Journal of Applied Physics 38: 2997–3002
409. Vignjevic R (2004) Review of Development of the Smooth Particle Hydrodynamics (SPH) Method. In Hobbs S (ed.): Proceedings Sixth Conference on Dynamics and Control of Systems and Structures in Space, Rimaggiore, Italy.
410. Vignjevic R, Campbell J, Bourne N, Djordjevic N (2007) Modelling Shock Waves in Composite Materials. Proceedings 2007 American Physical Society Topical Conference on Shock Compression of Condensed Matter.
411. Vignjevic R, Campbell J, Libersky L (2000) A Treatment of Zero Energy Modes in the Smoothed Particle Hydrodynamics Method. Computer Methods in Applied Mechanics and Engineering 184: 67–85

412. Vignjevic R, Millet J C F, Bourne N K, Meziere Y, Lukyanov A (2006) The Behaviour of Carbon-Fibre Epoxy Composite Under Shock Loading. Shock Compression of Condensed Matter - 2005, American Insitute of Physics, Melville, NY.
413. Vinson J R, Sierakowski R L (1986) The Behavior of Structures Composed of Composite Materials. Martinus Nijhoff Publishers, Dordrecht.
414. Wall F T (1942) Statistical Thermodynamics of Rubber II. Journal of Chemical Physics 10: 485-488
415. Walsh R T (1972) Finite Difference Methods. In: Chou P C, Hopkins A K (eds.) Dynamic Response of Materials to Intense Impulsive Loading. Air Force Materials Laboratory, USA.
416. Weirick L J (1992) Shock Characterization of Epoxy-42 Volume Percent Glass Microballoons. In: Meyers M A, Murr L E, Staudhammer K P (eds.) Shock Wave and High Strain Rate Phenomena in Materials. Marcel Dekker, Inc., New York.
417. Wendt J F (ed.) (1996) Computational Fluid Dynamics. Springer, Berlin.
418. Westergaard H M (1920) On the Resistance of Ductile Materials to Combined Stresses in Two or Three Directions Perpendicular to one Another. Journal Franklin Institute, 189, 627–640.
419. Weyl H (1949) Shock Waves in Arbitrary Fluids. Communications on Pure and Applied Mathematics 2: 103–122.
420. Wicklein M (2006) Zellulares Aluminium: Entwicklung eines makromechanischen Materialmodells mittels mesomechanischer Simulation. Doctoral Thesis, Fraunhofer Institute for High Speed Dynamics, Ernst-Mach-Institute, Issue 9 of the scientific series Thoma K, Hiermaier S (eds), $\dot{\varepsilon}$ - Forschungsergebnisse aus der Kurzzeitdynamik. ISBN: 3-8167-7018-5.
421. Wierzbicki T, Bao Y, Lee Y-W, Bai Y (2005) Calibration and Evaluation of Seven Fracture Models. International Journal of Mechanical Science 47: 719–743
422. Wierzbicki T, Xue L (2005) On the Effect of the Third Invariant of the Stress Deviator on Ductile Fracture. Impact and Crashworthiness Laboratory, Lab Report 136, Massachusetts Institute of Technology, Cambridge, MA
423. Wilkins M L (1964) Fundamental Methods in Hydrodynamics. In: Adler B et al. (eds) Methods in Computational Physics Vol. 3: 211. Academic Press, New York.
424. Wilkins M L (1992) Computer Simulation of Dynamic Phenomena. Springer, Berlin.
425. Wilkins M L (1980) Use of Artificial Viscosity in Mutlidimensional Shock Wave Problems. Journal of Computational Physics 36: 281–303
426. Wilkins M L, Guinan M W (1973) Impact of Cylinders on a Rigid Boundary. Journal of Applied Physics 44: 1200–1206
427. Wilkins M L, Streit R D, Reaugh J E (1980) Cumulative Strain Damage Model of Ductile Fracture: Simulation and Prediction of Engineering Fracture Tests. Technical Report UCRL-53058, Lawrence Livermore National Laboratory, Livermore, CA.
428. Wingate C A, Stellingwerf R F (1995) Los Alamos SPHINX Manual Version 7.6. Los Alamos National Laboratory Report LAUR 93-2476, Los Alamos, NM.
429. Wu H-C (2005) Continuum Mechanics and Plasticity. Chapman and Hall/CRC, London.

430. Wu P D, van der Giessen E (1993) On Improved Network Models for Rubber Elasticity and their Application to Orientation Hardening of Glassy Polymers. Journal of the Mechanics and Physics of Solids 41: 427–456
431. Yeoh O H (1990) Characterization of Elastic Properties of Carbon-Black-Filled Rubber Vulcanizates. Rubber Chemistry and Technology 63: 792–805.
432. Zel'dovich Ya B (1946) On the possibility of rarefaction shock waves. Zh. Eksp. Teor. Fiz. 4: 363–364.
433. Zel'dovich Ya B, Raizer Yu P (1967) Physics of Shock Waves and High-Temperature Hydrodynamic Phenomena. Academic Press, New York.
434. Zerilli F J, Armstrong R W (1987) Dislocation-Mechanics-Based Constitutive Relations for Materials Dynamics Calculations. Journal of Applied Physics 61: 1816–1825
435. Zienkiewicz O C, Taylor R L, Zhu J Z (2005) The Finite Element Method: Its Basis and Fundamentals. Butterworth-Heinemann, Amsterdam.
436. Zukas J, Nicholas T, Swift H, Greszczuk L, Curran D (1992) Impact Dynamics. Krieger Publishing, Malabar, FL.

Index

Printed in the United States of America